THEORY OF
PROBABILITY

Written for Statistics [Hons/Major] & Mathematics [Hons/Major] students of different Indian Universities. Also useful to Postgraduate students of Statistics and Mathematics.

THEORY OF PROBABILITY

Parimal Mukhopadhyay
Indian Statistical Institute [ISI]
Kolkata, India

New Central Book Agency (P) Ltd
LONDON
HYDERABAD ERNAKULAM BHUBANESWAR
NEW DELHI KOLKATA PUNE GUWAHATI

NCBA

REGD OFFICE
8/1 Chintamoni Das Lane, Kolkata 700 009, India
email: ncbapvtltd@eth.net

OVERSEAS
NCBA (UK) Ltd, 149 Park Avenue North, Northampton, NN3 2HY, UK
email: ncbauk@yahoo.com

EXPORT
NCBA Exports Pvt. Ltd, 212 Shahpur Jat, New Delhi 110 049
email: ncbaexports@ncbapvtltd.com

BRANCHES

2/27 Ansari Road, 1st Floor
Daryaganj, **New Delhi** 110 002
email: ncbadel@ncbapvtltd.com

House No. 3-1-315, 1st Floor
Nimboliadda, Kachiguda, **Hyderabad** 500 027
email: ncbahydb@ncbapvtltd.com

Shop Nos. 3 & 4, Vinayak Towers
681/B Budhwar Peth, Appa Balwant Chowk
Pune 411 002
email: ncbapune@ncbapvtltd.com

GSS Shopping Complex, 1st Floor
Opposite College Post Office, Convent Road
Ernakulam 682 035
0484-4021054/53

Shop No. 15, Bharati Towers, Ground Floor
Forest Park
Bhubaneswar 751 001
email: ncbabub@ncbapvtltd.com

Radhamayee Complex (Second Floor)
Opp. SBI, Panbazar, Jaswanta Road
Guwahati 781 001
email: ncbaguwahati@ncbapvtltd.com

THEORY OF PROBABILITY • Parimal Mukhopadhyay

First Published: September 1991
Reprinted: January 2007
Reprinted: January 2012
Reprinted: July 2017

PUBLISHER
New Central Book Agency (P) Ltd
8/1 Chintamoni Das Lane, Kolkata 700 009

TYPESETTER
Books & Allied (P) Ltd
8/1 Chintamoni Das Lane, Kolkata 700 009

PRINTER
New Central Book Agency (P) Ltd
Web-Offset Division, Dhulagarh, Sankrail, Howrah

ISBN: 81 7381 157 1

Price ₹ 325.00

Dedicated

to the memory of my father

and

to my mother

PREFACE

The theory of probability has found a major role in the understanding and explanation of various phenomena in different natural, physical and social sciences e.g. physics, engineering, genetics, geology, economics, psychology, business management, operations research.

This book is intended mainly for the students of Statistics (Honours) courses and graduate and post-graduate students in Mathematics who learn statistics in their curricula. However, this book should also be useful to the students of other disciplines who have to learn the theory of probability as a part of their major courses.

The basic requirement in reading the book is knowledge in mathematics up to pre-graduation level and an elementary course in calculus. The ideas have been illustrated at each step and supplemented by a number of exercises.

The chapters 1 through 3 deal with the frequency theorists' view of probability, classical theory of probability and axiomatic theory of probability. Chapter 4 considers expectation, variance, etc. of a discrete random variable. Chapter 5 examines the notion of a random variable and its probability distribution more closely. Chapter 6 considers some characteristics of a probability distribution, e.g. its location parameters, dispersion, moments. Chapter 7 considers different generating functions. Chapters 8 and 9 examine several one-dimensional discrete and continuous distributions. Chapters 10 and 11 consider properties of n-dimensional distributions and distributions of functions of several random variables. Convergence of sequences of random variables have been looked into in the last chapter.

The readers interested only in solving problems with finite number of possible cases would find chapters 2 and 4 useful.

In preparing this book I have unhesitatingly taken the help of a number of publications a list of which has been enclosed as reference.

My sincere thanks are due to my colleagues in the ISI for their support during the preparation of the manuscript.

The book would not have been written at all but for the initiative of Shri Arunabha Sen of New Central Book Agency who took great care in bringing out the same in a very short time.

I must acknowledge my indebtedness to my wife and my sons, Pabak and Pralay, who bestowed invaluable encouragement in preparing the book.

June 5, 1991
Indian Statistical Institute
Calcutta, INDIA

Parimal Mukhopadhyay

CONTENTS

1 BASIC CONCEPTS : STATISTICAL REGULARITY

1.1. Introduction

If a coin is tossed nobody knows for sure whether a head or a tail will turn up, though, it is certain either of a head or a tail will occur. Similarly if a die is thrown either of the faces 1, 2, ⋯, 6 will turn up though it is not known beforehand which one will actually occur. Experiments of this type where the outcome cannot be predicted exactly are called 'random' experiments as opposed to 'deterministic' experiments where the results of the experiments can be foretold beforehand. Examples of deterministic experiments are throwing a stone upwards where it is known that the stone will definitely fall to the ground (due to the force of gravitation), throwing a missile with a definite velocity and at a given angle with the ground when the position of the missile at any point of time can be exactly foretold. The problems of deterministic experiments are dealt with in subjects like physics, statics, dynamics etc. Random experiments are subjects of study of the theory of probability. The theory of probability analyses the results of such experiments, states which one is more 'probable' than the other, which one is impossible under the conditions of the given experiment, relations between different kinds of outcomes, so on and so forth.

'Probability' or 'Chance' is a word we often encounter in our day-to-day life. We often say that it is very probable that it will rain tonight meaning thereby we very much expect to have downpour this night. This expectation, of course, comes from our general knowledge about the conditions of weather in this month of the season. One may say that chances of England and West Indies winning the World Cup in Cricket, before the start of the game, are in the ratio 50 : 50 meaning thereby that one expects to see either of the teams coming out winner with equal expectation. Again, this expectation is based on one's present knowledge and belief about the relative performance of the teams. One may also say it is more likely to have a good yield of

Aman paddy in district X than in district Y, meaning one expects better yield from X than from Y. In short, the word 'probability' or 'chance' here measures one's degree of belief about the occurrence of certain events or truth of certain statements. This degree of belief cannot be verified by making repeated experiments under identical conditions, as detailed below. The 'probability' of this type may be called 'subjective' probability.

If a symmetric coin (no asymmetry in physical properties like distribution of weight, materials between the two sides being allowed) is tossed, we say that head and tail are equally likely to occur. In this case our belief that both faces have the same chance of occurrence can be objectively verified. If we repeat this experiment a large number of times (say 100, 200, 500 times) under identical conditions we will notice on an average 50% of the results are 'heads' and 50% tails. Similarly in throwing a six-faced 'balanced' die we may say that each face is equally likely to turn up. Repetition of this experiment a large number of times under identical conditions will show on an average $\frac{1}{6}$ proportion of experiments resulting in each of the faces 1, ..., 6. This type of probability may be called 'objective probability.' The objective probability is very much a property of an experiment as 'volume', 'mass', 'temperature' are properties of a substance. Throughout this treatise we shall be concerned with this objective measurement of probability.

The principal impetus for the development of theroy of probability originated from an interest in games of chance, — which involve an element of randomness. The rudiments of a mathematical theory to describe these elements of randomness, probably, took shape in the sixteenth century as is evidenced from a short note written in the early seventeenth century by the famous astronomer and mathematician Galiles Galilei. A French nobleman Chevalier de Méré asked Blaise Pascal (1623–62), a great mathematician to explain some apparent contradictions between his reasonings and experience in gambling. Pascal solved these problems and some other problems brought to him by de Méré. Pierre de Fermat (1601–65), another mathematician, subsequently became interested in the subject and in the private correspondance between Pascal and Fermat in 1654 was laid down the foundation of theory of probability. Subsequently Christianus Huygens (1629–95), a great Dutch Scientist published in 1657 the first book on probability, entitled De Ratiociniis in Aleae Ludo. Another pioneering contribution is the work, Ars Conjectandi, by James Bernoulli (1654–1705) [published posthumously in 1713, but written sometime around 1690]. Mathematicians like A. de Moivre (1667–1754), Laplace (1749–1827) enriched the science of probability through their important pioneering contributions.

The development of the subject took an important turn with the works of Tshebysheff (1821–1894), A. Markoff (1856–1922), and A. Liapounoff (1858–1918). The fundamental definition of probability which was accepted in the earlier period remained the classical definition until in the recent times mathematicians like S. Bernstein, A. Khintchine and A. Kolmogoroff developed axiomatic formulations providing important extensions and generalisations to the earlier results.

1.2. Statistical Regularity

Although, as noted before, the result of a single random experiment can never be correctly predicted, if the random experiment is carried out a large number of times under identical conditions it will be seen that the relative frequency of an event stablises to a certain value. Relative frequency (r.f.) of an outcome 0 of an experiment is the number of times 0 occurs, $f_N(0)$, divided by the total number of times, N, the experiment is carried out. Thus *r.f.* of 0, $r_N(0)$ (say) $= f_N(0)/N$. Clearly, $0 \leq r_N(0) \leq 1$.

It is often seen that when the experiment is repeated N times under identical conditions, as $N \to \infty$, $r_N(0) \to$ a certain value p (say) where $0 \leq p \leq 1$. That it is so, can be seen from the following experiments.

❑ **EXPERIMENT 1.** *A coin was tossed several times and the number of times it fell heads was noted. The following table shows the number of heads (H) obtained in sets of N experiments.*

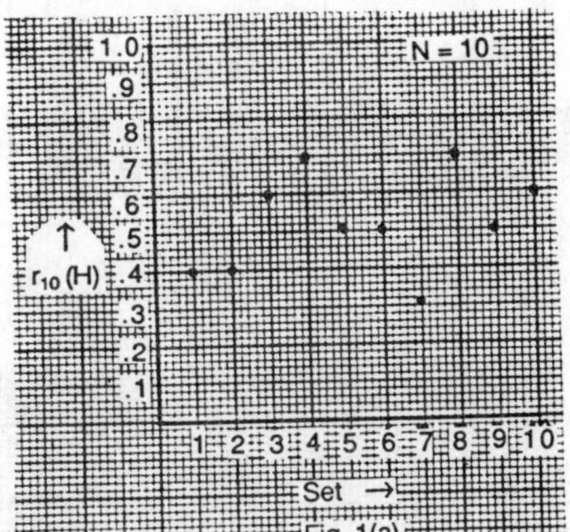

Fig. 1(a)

Table 1.1

Set	$N =$ 10	50	100
1	4	29	47
2	4	22	52
3	6	24	54
4	7	27	49
5	5	31	53
6	5	26	51
7	3	25	48
8	7	28	52
9	5	21	47
10	6	23	55
Total	52	256	508

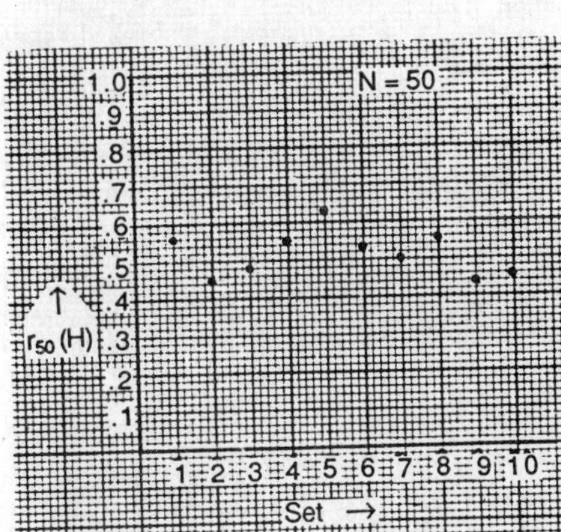

Fig. 1(b)

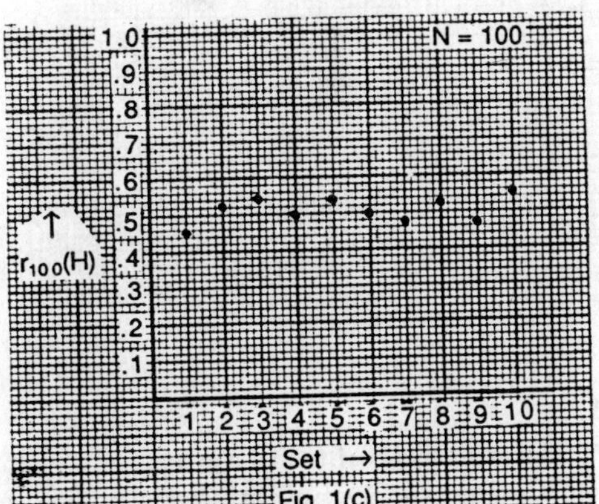

Fig. 1(c)

Figures 1(a)–1(c) show the relative frequencies of H for various sets for different values of N. For $N = 10, r(H)$ varied from .3 to .7. For $N = 50$, extreme values of $r(H)$ become closer being .42 and .62. For $N = 100$, relative frequencies varied between (.47 and .55). Thus for increasing N the $r_N(H)$-values were observed to be gradually more stable. The average values of $r(H)$ were .520, .512, .508 for $N = 10$, 50 amd 100 respectively. Thus one may conclude that as N increases relative frequency of H will be expected to be very close to .5.

❐ **EXPERIMENT 2.** *The following table shows the number of digits $x \leq 3$ occuring in sets of sequences of N random numbers.*

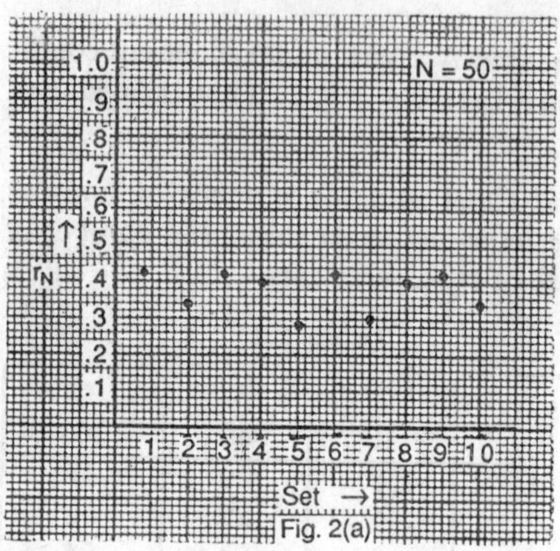

Table 1.2

Set	$N =$ 50	200	250
1	21	86	106
2	17	82	98
3	21	83	105
4	20	71	92
5	14	75	95
6	21	94	110
7	15	75	95
8	20	87	103
9	21	86	108
10	18	78	101
Total	188	817	1013

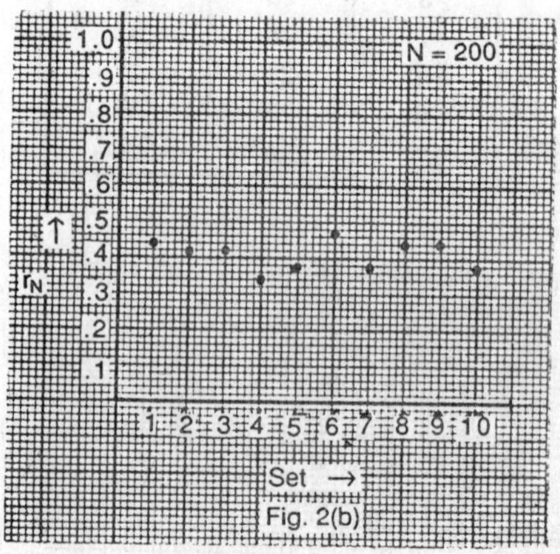

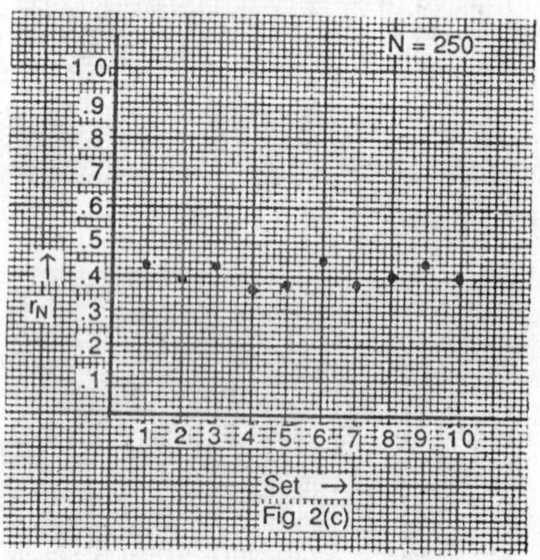

Extremes of values of r_N for $N = 50$, 200 and 250 are (.28, .42), (.355, .470), (.368, 440), the ranges (maximum value – minimum value) being .14, .115, 072 respectively. Also the averages are .376, .408, .405 respectively. Thus it may be inferred that as N increases relative frequency would be very close to .40. Figures 2(a)–(c) show the r_N values for various sets for different N.

❐ **EXPERIMENT 3.** *The following data have been taken from Kerrich* (1946). A coin was spun 10,000 times. Table 1.3 gives n, the number of spins made; m, the number of heads in n spins and* $\frac{m}{n}$ *the proportion of heads in n tosses.*

We find that $\frac{m}{n}$ fluctuates in value and at first the fluctuations are large. For the values of n noted, the fluctuation is between .400 and .567 upto $n = 100$, between .473 and .526 between $n = 150$ and $n = 1000$; between .496 and .510 between $n = 1200$ and 3000; between .503 and .511 between $n = 3200$ and $n = 4800$, between .499 and .507 between $n = 5000$ and $n = 10000$. The decrease in fluctuations in the values of $\frac{m}{n}$ was very rapid in the region of increasing values of n and the ratio was found to be increasingly closer to .5 with increasing value of n (Fig. 3).*

Table 1.3

n	m	m/n
10	4	.400
20	10	.500
30	17	.567
40	21	.524
50	25	.500
75	34	.453
100	44	.440
150	71	.473
200	98	.490
250	125	.500
300	146	.487
400	199	.498
500	255	.510
600	312	.520
700	368	.526
800	413	.516

cont.

cont.

n	m	m/n
900	458	.509
1000	502	.502
1200	596	.496
1400	704	.503
1600	810	.506
1800	918	.510
2000	1013	.507
2200	1122	.510
2400	1220	.508
2600	1316	.506
2800	1409	.503
3000	1510	.504
3200	1609	.503
3400	1722	.506
3600	1826	.507
3800	1926	.507

cont.

cont.

n	m	m/n
4000	2029	.507
4200	2135	.508
4400	2242	.510
4600	2346	.510
4800	2447	.510
5000	2533	.507
5500	2772	.504
6000	3009	.502
6500	3245	.499
7000	3516	.502
7500	3780	.504
8000	4034	.504
8500	4284	.504
9000	4538	.504
9500	4803	.505
10000	5067	.507

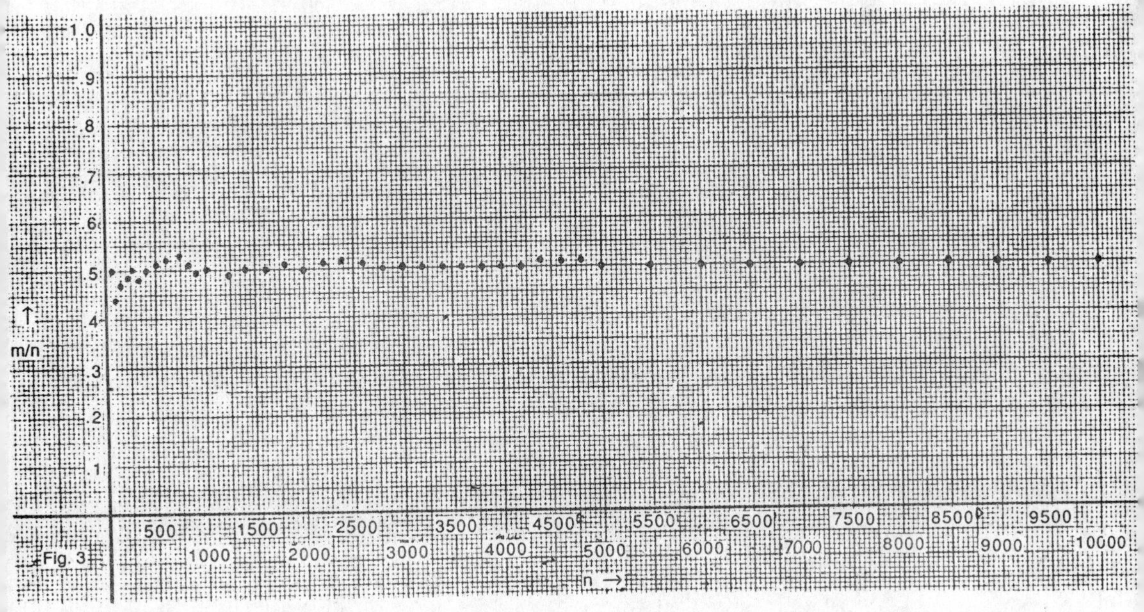

Fig. 3

---

* Kerrich, J. E. : An experimental introduction to the theory of probability. Kimar Munksgaard, Copenhagen (1946).

In all the above cases and in similar cases it is seen that the relative frequency r_N stabilises to a specific value p at a fairly definite rate as the number of trials N increases. For a series of given experiments one may thus make the relative frequency as nearly constant as one pleases by making N sufficiently large. Again for fairly large N one may reasonably expect that the outcome 0 will occur in Np number of trials. Thus if a fair die is thrown 2000 times one may expect that each of the faces 1, 2, ..., 6 will turn 333 times. The data collected from such random experiments, conducted under identical conditions show, therefore, the following stability property of frequencies : for sequences of sufficient length relative frequency of each outcome is practically constant. This constant value may be called the long-term relative frequency of the outcome. The only condition required for the stability property is that the experiments must be conducted under identical conditions. The value of p to which the relative frequency stabilises may depend on the condition of the experiment. It should be noted that this statistical regularity or stability property of relative frequency is an empirical deduction based on millions of observations. This fact is the empirical basis for the concept of probability.

1.3. Exercise

1. *Consider 10 experiments each consisting of throwing a fair die 100* $[=n]$ *times and calculate the relative frequency of face value* $i[=r_{in}]$ *for each experiment and overall relative frequency of* i, $i = 1, \cdots, 6$ *for the 10 experiments. Also find the range of relative frequency of* i *over these 10 experiments. Repeat the whole exercise for values of* n = *120, 140, ... Observe that* r_{in} *stabilises to* $\frac{1}{6}$ *as* n *increases.*

2 CLASSICAL THEORY AND ITS LIMITATIONS

2.1. Introduction

We shall first consider the classical concept of probability. For this we examine a few definitions.

(a) *Event* : An event is the result of an experiment. Thus in throwing a die, the results — face 5 turns up, an even number turns up are both events. Events may be of two types :

(a.1) *Elementary Event (or Case)* : It is an event which cannot be broken further into smaller events. It is an 'atom' of an event. Thus, in throwing a die, each of the events 1, $\cdots$, 6 is an elementery event. Symbolically, an elementary event will often be represented by small case letters a, b, etc.

(a.2) *Random (contingent) Event* : A random (contingent) event is obtained through the combination of several elementary events. Thus, in throwing a die, the event that an odd number turns up is a combination of three elementary events : 1 or 3 or 5 turns up. This event occurs if any of these three cases takes place. Symbolically, a compound event will often be represented by large case letters A, B, etc.

We shall often use simply the word event, whether it is elementary or compound would be clear from the context.

(b) *Mutually Exclusive Events* : Events are said to be mutually exclusive (or incompatible) with respect to an experiment if the occurrence of one of them precludes the occurrance of all the others every time the experiment is performed; in other words the two events cannot materialise simultaneously. Thus in throwing a die, the events, '2 occurs' and '5 occurs' are mutually exclusive.

(c) *Exhaustive Events* : A set of events in relation to a random experiment is said to be exhaustive if one of them must necessarily materialise everytime the experiment is performed. Thus in throwing a die cases 1, $\cdots$, 6 form an exhaustive set of cases.

(d) *Equally likely Cases* : Cases are said to be equally likely when we have no reason to believe that one is more likely to occur than the other. Thus in drawing a card at random from a full pack of well-shuffled cards, one may believe that each of the cards is equally likely to appear and in this case, each of the 52 cases is equally likely. This concept, therefore, presupposes the simplest hypothesis regarding the possibility of occurrences of different cases viz. each case is equally likely to materialise.

(e) *Favourable Cases* : *A* case '*a*' is said to be favourable to an event A if whenever '*a*' occurs, '*A*' occurs. Thus in throwing a die, each of the cases 1, 3, 5 is favourable to the event, 'an odd face turns up'.

❑ **EXAMPLE 2.1.** *A head and a tail being denoted by H and T respectively, the cases HH, HT, TH, TT are mutually exclusive, exhaustive and equally likely cases for the random experiment of throwing a fair coin twice.*

❑ **EXAMPLE 2.2.** *In throwing two balanced dice, mutually exclusive, exhaustive and equally likely cases are 36 in number, namely, (1,1), (1,2), $\cdots$, (1.6), (2,1), $\cdots$, (2,6), $\cdots$, (6,1), $\cdots$, (6,6), where in (x, y), x, y denote the result of first and second throw respectively.*

2.2. Representation in Set Theory

We consider a few definitions from the theory of sets.

(i) *Set* : A set is a collection of some elements which are its members. Thus we may have a set of college students, a set of some books, a set of some digits, a set of houses etc. A set can be denoted either by enumerating all its elements given in braces or by indicating (also in braces) a rule which associates an element with a set. Thus the set of all odd numbers not exceeding 15 can be written as $\{1, 3, 5, 7, 9, 11, 13, 15\}$ or $\{2x+1;\ x = 1, 2, \cdots, 7\}$. A set is often pictured as an oval shaped space with points denoting its elements [Fig. 3a]. The diagram representing a set is called a Venn diagram.

Depending on how many elements it has, a set may be finite or infinite. The set $M = \{1, 2, \cdots, 50\}$ is finite and contains 50 elements. The set of all natural numbers $N_1 = \{1, 2, \cdots, n, \cdots\}$ is infinite. The set of all even numbers $N_2 = \{2, 4, \cdots, 2n, \cdots\}$ is also infinite. An infinite set is said to be countable if all its elements can be enumerated. Both the sets N_1, N_2 above are countable. The set C of all points within or on a circle of radius $r > 0$, $C = \{(x, y)\ :\ x^2 + y^2 \leq r^2\}$ is infinite and uncountable. Its elements can not be enumerated.

(ii) *Universal Set* : It is the largest possible set of interest under a given condition. All other sets are parts (subsets) of the universal set. Thus if we are interested in some students of a college, a set containing all the students of this college may constitute an universal set. The set of all first year commerce students, set of all second year chemistry-honours students are all subsets of the above universal set. An universal set is often represented as $\mathcal{U}$ (Fig. 3b).

(iii) *Null Set* : It is a set containing no element. It is also called an empty set and is often represented as ϕ. It is a member of all other sets (Fig. 3b).

(iv) *Disjoint Sets* : Two sets A, B are disjoint if they have no common element. Similarly sets $A_1, \cdots, A_m$ are mutually disjoint if no two of them have any common element (Fig. 3c).

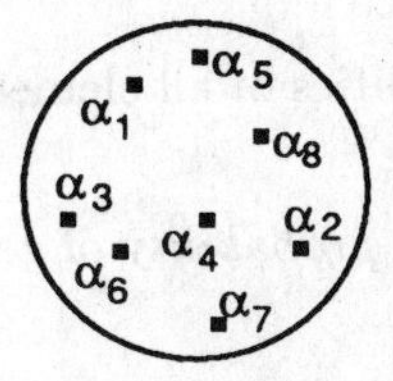

Fig. 3(a)

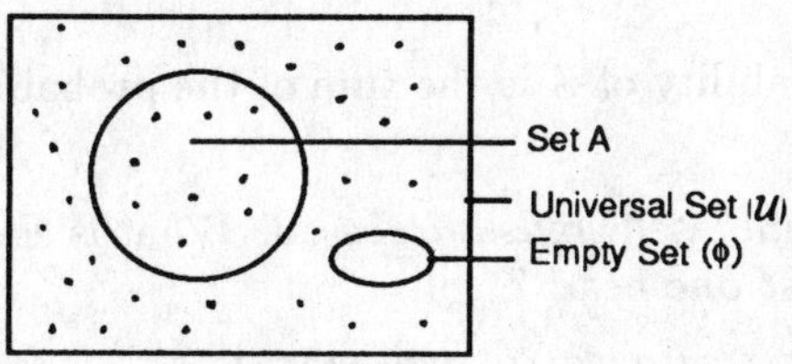

Fig. 3(b)

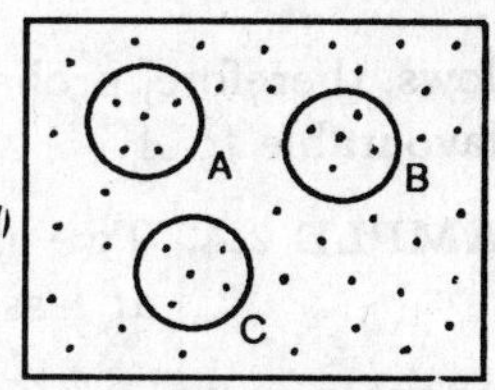

Fig. 3(c)

In the probability theory an elementary event or a case resulting from a random experiment is represented as a point in a set. A compound event A is represented as a set containing all the points which stand for the elementary events of which A is made up (i.e. which are favourable to A). Thus in throwing a die, each of the cases $1, \cdots, 6$ is a point in the set. The compound event, an odd face turns up is represented by the set A(Fig. 4a). Two events A and B are mutually exclusive if the sets representing them are disjoint. In Fig. 3(c), events represented by the sets A, B, C are mutually exclusive.

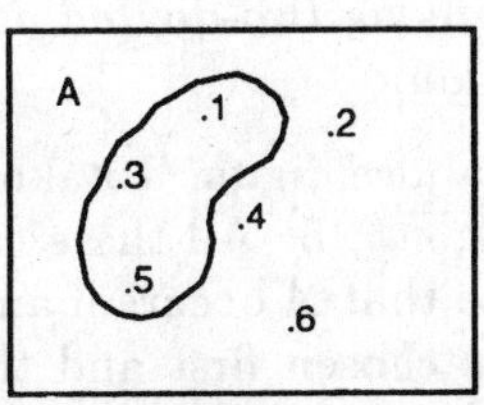

Fig. 4(a)

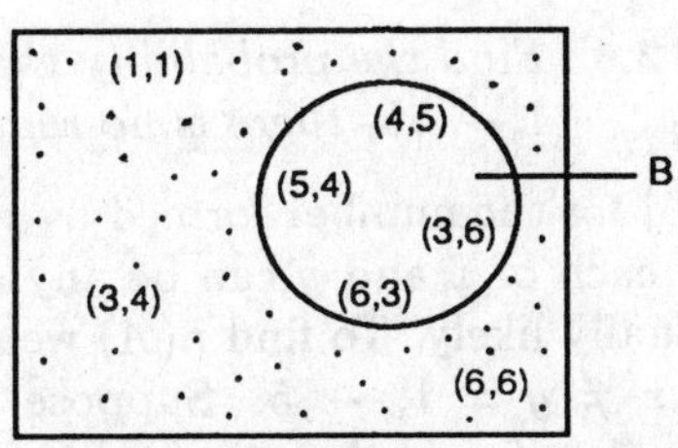

Fig. 4(b)

The set containing all the possible points representing the elementary events of a random experiment i.e. the universal set is called the 'sample space'. It is often represented as $\mathcal{S}$. Thus in tossing a coin once, $\mathcal{S} = \{H, T\}$. In tossing a coin twice, $\mathcal{S} = \{HH, HT, TH, TT\}$. In throwing two dice, $\mathcal{S} = \{(1,1), (1,2), \cdots, (6,6)\}$. Fig. 4(b) represents the sample space for the throw of two dice, set B denoting the event that the total of the two faces is 9.

❑ **EXAMPLE 2.3.** *A coin is tossed until a head appears. Here the sample space consists of elementary events H, TH, TTH, $TTTH$, $\cdots$. These points are countable and infinite in number. The sample space consists of countably infinite number of cases.*

2.3. The Classical Definition of Probability

We assume that the total number of elementary events N in the sample space of the experiment is finite. We also assume that all the elementary events are equally likely to occur i.e. it is assumed that the circumstances of the experiment are such that no elementary event is more likely to occur than the other. Under these assumptions we have the following classical or a priori definition of probability.

If an experiment can result in N mutually exclusive, exhaustive and equally likely cases of which $n(A)$ are favourable to an event A then the probability of occurrence of A is $P(A) = \frac{n(A)}{N}$. (2.1)

It follows, therefore, probability of A is the sum of the probabilities of all elementary events favourable to A.

❑ **EXAMPLE 2.4.** *Two unbiased coins are tossed. What is the probability of getting at least one head ?*

The possible cases are HH, HT, TH, TT. All these cases are mutually exclusive and equally likely. Thus $N = 4$. The event A (at least one head) occurs if either of HH, HT, TH occurs.

Thus $n(A) = 3$. Hence $P(A) = \frac{3}{4}$.

❑ **EXAMPLE 2.5.** *Two unbiased dice are thrown. What is the probability that the sum of the top faces is 9.*

The possible cases are (1,1), (1,2), $\cdots$, (6,6), being 36 in number. All these cases are mutually exclusive and equally likely. The event A (sum of the face values is 9) occurs if any of (3,6), (4,5), (5,4), (6,3) occurs. Thus $n(A) = 4$. Hence $P(A) = \frac{4}{9}$.

❑ **EXAMPLE 2.6.** *Find the probability that among two-digited numbers formed by* $1, \cdots, 5$, *there is no repetition.*

Let (x, y) stand for the number formed by the above digits. Total number of possible cases is 25 since each of x and y can be any of $1, \cdots, 5$. All these cases are mutually exclusive and equally likely. To find $n(A)$ we note that A occurs if any of the following occurs : (x, y), $x \neq y = 1, \cdots, 5$. Suppose x is chosen first and then y. x may be anything between 1 to 5 and thus x may be chosen in 5 ways. After x has been chosen, y may be chosen in any of the remaining 4 ways. Since with each choice of x there are 4 ways of chosing y, the total numbers of cases, (x, y), $x \neq y = 1, \cdots, 5$ is $5 \cdot 4 = 20$. Thus $n(A) = 20$ and hence $P(A) = \frac{4}{5}$.

The following properties immediately follow from equation (2.1).

(i) Since $0 \leq n(A) \leq N$, probability $P(A)$ of an event A satisfies $0 \leq P(A) \leq 1$. If $P(A) = 0$ then $n(A) = 0$ i.e., there is no favourable case for A and hence A is impossible under the given circumstances. If $P(A) = 1$, $n(A) = N$ i.e., all the cases are favourable to A and hence A is sure to occur. A is now a certain event.

(ii) The number of cases which are not favourable for A is $N - n(A)$. Hence probability of non-occurrence of A is $1 - P(A)$.

(iii) If occurrence of an event A implies occurrence of an event B, then the cases favourable for A from a subset of the cases favourable for B, hence $n(A) < n(B)$. Thus $P(A) < P(B)$.

The classical definition of probability, however, suffers from some serious disadvantages. The statistical or empirical definition of probability takes care of these drawbacks to some extent. These have been explained in details in section 2.13. However, most of the problems on finding probabilities of simple events can be calculated using a priori definition of probability. As such we shall consider the development of classical theory in further details.

Since calculation of the value of probability often involves knowledge about the arrangement of different objects, selection of some objects out of a number of objects, etc. we shall first consider some elements of combinatorial analysis. Their applications in solving problems of probability will be illustrated.

2.4. Elements of Combinatorial Analysis

The results 2.1, 2.2 below are obvious. The proof of the results below can be obtained from any elementary book of algebra.

Result 2.1. If there are two groups G_1 and G_2, G_1 consisting of m elements $a_1, \cdots, a_m$ i.e., if $G_1 = (a_1, \cdots, a_m)$ and similarly $G_2=(b_1, \cdots, b_n)$ consisting of n elements, number of pairs (a_i, b_j) formed by taking an element 'a' from G_1 and an element 'b' from G_2 is mn.

❑ **EXAMPLE 2.7.** *A college has 27 students in the first year honours class and 20 students in the second year honours class. Number of pairs that can be formed by taking one student from each class is* $27 \times 20 = 540$.

Result 2.2. This is an extention of result 2.1. If there are k groups $G_1, G_2, \cdots, G_k$ containing $m_1, m_2, \cdots, m_k$ elements respectively, i.e. if

$$\begin{aligned} G_1 &= \{x_1, x_2, \cdots, x_{m_1}\} \\ G_2 &= \{y_1, y_2, \cdots, y_{m_2}\} \\ &\cdots \cdots \cdots \cdots \\ G_k &= \{t_1, t_2, \cdots, t_{m_k}\} \end{aligned}$$

then the number of multiplets $(x_i, y_j, \cdots, t_r)$ formed by taking one element from each group is $m_1 m_2 \cdots m_k$.

❑ **EXAMPLE 2.8.** *Number of 3 digited numbers that can be formed out of 7 digits* $1, 2, \cdots, 7$ *is* 7^3. *Here each of the groups* G_1, G_2, G_3 *contains 7 elements.*

Result 2.3. The number of ways in which r things can be chosen out of n different things and arranged among themselves is

$$n(n-1) \cdots (n-r+1) = {}^nP_r \qquad (r \le n),$$

the permutation of n things taken r at a time.

If $r = n$, the number of ways in which n different things can be arranged among themselves is

$$n(n-1)\cdots 1 = {}^{n}P_n = n!$$

We shall often denote ${}^{n}P_r$ as $(n)_r$.

❑ **EXAMPLE 2.9.** *An elevator contains 5 passengers and stops at 10 floors. What is the probability that no two passengers get down at the same floor?*

Number of possible cases is 10^5 since each of the 5 passengers has 10 choices to get down. All these cases are mutually exclusive and equally likely. Number of favourable cases for the event under consideration is $(10)_5$, since the first passenger can choose his floor in 10 ways, the second in 9 ways and so on. Hence probability is $(10)_5/10^5$.

❑ **EXAMPLE 2.10.** *If n distinct balls are placed in n cells what is the probability that each cell will be occupied?*

Total numbers of possible arrangements is n^n because each ball can go to any cell and thus each ball has n choices. All these cases are mutually exclusive and equally likely.

The number of ways each cell may be kept occupied is the number of ways in which n balls may be arranged among themselves and this number is $n!$ Hence probability is $n!/n^n$.

❑ **EXAMPLE 2.11.** *The numbers 1, 2, $\cdots$, n are arranged in a random order. Find the probability that digits (i) 1 and 2 (ii) 1, 2, 3 appear as neighbours in the order named.*

Total number of mutually exclusive and equally likely arrangements is $n!$

(i) Assume the digits 1 and 2 are tied together. Then the total number of digits taking 1 and 2 as a single digit is $(n-1)$ and these can be arranged among themselves in $(n-1)!$ ways. Hence probability is $(n-1)!/n! = 1/n$.

(ii) Similarly probability is $1/\{n(n-1)\}$.

Result 2.4. The number of ways in which a set of r objects can be formed out of n different objects is

$$\frac{n!}{r!(n-r)!} = \binom{n}{r}$$

This is the number of combinations of r things taken out of n different things. Two sets are different if and only if each has at least one uncommon element. Here arrangement of elements within a set is not considered. Thus (a, b, c) and (a, b, d) are two different sets; but (a, b, c) and (b, a, c) are the same set. If follows that

$$\text{(i)}\ \binom{n}{r} + \binom{n}{r-1} = \binom{n+1}{r}, \qquad \text{(ii)}\ \sum_{i=0}^{n}\binom{n}{i} = 2^n,$$

$$\text{(iii)}\ \sum_{i=0}^{n}(-1)^i\binom{n}{i} = 0, \qquad \text{(iv)}\ \sum_{i=0}^{n} i\binom{n}{i} = n\,2^{n-1}$$

[(iv) since, $i\binom{n}{i} = n\binom{n-1}{i-1}$. Alternatively, differentiating both sides of $\binom{n}{0} + \binom{n}{1}t + \cdots + \binom{n}{n}t^n = (1+t)^n$ with respect to t and setting $t = 1$, the result follows.]

(v) $\binom{n}{0}^2 + \binom{n}{1}^2 + \cdots + \binom{n}{n}^2 = \binom{2n}{n}$.

❑ **EXAMPLE 2.12.** *What is the probability that the 2 cards drawn at random from an ordinary deck will be all spades.*

Total number of possible cases is the total number of ways in which a set of 2 cards can be formed out of 52 cards. This is $\binom{52}{2}$. All these cases are mutually exclusive and equally likely.

Number of favourable cases is the number of ways in which 2 spades can be chosen out of 13 spades. This number is $\binom{13}{2}$.

Hence probability is $\binom{13}{2}\bigg/\binom{52}{2}$.

❑ **EXAMPLE 2.13.** *What is the probability that a bridge hand* will be a complete suit ?*

Total* number of mutually exclusive and equally likely cases is $\binom{52}{13}$. Since there are only 4 suits number of faourable cases is 4. Hence probability is $4\bigg/\binom{52}{13}$.

❑ **EXAMPLE 2.14.** *The face cards are removed from a full pack. Out of the remaining cards, 4 are drawn. (i) What is the probability that they belong to different suits ? (ii) What is the probability that the 4 cards belong to different suits and different denominations ?*

Total number of mutually exclusive and equally likely cases is $\binom{40}{4}$.

(i) If the cards are (a, b, c, d) where 'a' is a spade, 'b' a hearts, 'c' a club, 'd' a diamond, then each of a, b, c, d has 10 choices. Thus total number of choices for (a, b, c, d) is 10^4. Hence probability is $10^4\bigg/\binom{40}{4}$.

(ii) Here total number of favourable cases is $(10)_4$. Hence probability is $(10)_4\bigg/\binom{40}{4}$ $= .05515$.

* In a bridge game 13 cards are distributed to each of four players.

Result 2.5. Let a population consist of two types of elements of sizes N_1 and N_2 respectively. The number of ways in which n_1 elements of the first type and n_2 elements of the second type may be chosen is

$$\binom{N_1}{n_1}\binom{N_2}{n_2}.$$

Similarly if the population consists of k types of elements of sizes $N_1, N_2, \cdots, N_k$ respectively, the number of ways in which n_1 elements of the first type, $\cdots$, n_k elements of the kth type may be chosen is

$$\binom{N_1}{n_1}\binom{N_2}{n_2}\cdots\binom{N_k}{n_k}.$$

❑ **EXAMPLE 2.15.** *What is the probability that of 6 cards taken from a full pack 3 will be black and 3 will be red ?*

Number of possible mutually exclusive and equally likely cases is $\binom{52}{6}$. Number of ways in which 3 black and 3 red cards may be chosen is $\binom{26}{3}\binom{26}{3}$. Hence probability $= \binom{26}{3}\binom{26}{3}\Big/\binom{52}{6}$.

❑ **EXAMPLE 2.16.** *Find the probability that a hand at bridge will consist of 3 spades, 5 hearts, 2 diamonds and 3 clubs.*

There are 13 cards of each type in the pack. Hence the probability is $\binom{13}{3}\binom{13}{5}\binom{13}{2}\binom{13}{3}\Big/\binom{52}{13}$.

Result 2.6. The number of ways in which N elements can be grouped into k specific groups $G_1, G_2, \cdots, G_k$ of sizes $N_1, N_2, \cdots, N_k$ respectively $(N_1+N_2+\cdots+N_k = N)$ is

$$\frac{N!}{N_1!\,N_2!\,\cdots\,N_k!}.$$

If the groups are of equal sizes, $N_i = M$, $i = 1, \cdots, k$, the number of unordered groups into which N elements can be grouped is $\dfrac{N!}{k!(M!)^k}$.

❑ **EXAMPLE 2.17.** *What is the probability that in a throw of 12 dice each face occurs twice ?*

Number of all possible, equally likely and mutually exclusive cases is 6^{12} since each die may result in one of the 6 faces.

Number of favourable cases is the number of ways in which 12 dice may be arranged in 6 groups of size 2 each, group 1 consisting of two dice each showing 1, group 2 of two dice each showing 2 etc. and this number is $\frac{12!}{(2!)^6}$.

Hence the probability is $12!/\{2^6 \cdot 6^{12}\} = .00344$.

❑ **EXAMPLE 2.18.** *Given thirty people, find the probability that among the twelve months there are six containing two birthdays and six containing three.*

The number of ways 30 people can be grouped into twelve groups according to the month of birth is

$$\frac{30!}{(2!)^6\,(3!)^6}$$

Hence the probability is $\{30!/(2!)^6\,(3!)^6\}/12^{30} \simeq .00035$.

2.5. Models in Statistical Mechanics

So far we have assumed when n balls are distributed in m cells each of the m^n sample points has the probability $1/m^n$. The balls here are assumed to be each distinct.

When the balls are indistinguishable, an outcome of an experiment is described by a sequence of m members, $(k_1, \cdots, k_m)$, $k_1 + \cdots + k_m = n$, giving the number of balls in each cell. The probability of such an outcome, using assumption of equal likelihood of each of m^n sample points (as in the case of distinguishable balls as above) is

$$P(k_1, \cdots, k_m) = \frac{n!}{k_1! \cdots k_m!} \cdot \frac{1}{m^n}$$

This is called the Maxwell-Boltzman statistic. (If the balls were distinct there were $\frac{n!}{k_1! \cdots k_m!}$ distinct arrangements for the event $(k_1, \cdots, k_m)$ all of which have now become indistinguishable).

Under Bose-Einstein statistic only distinguishable arrangements of n indistinguishable balls in m cells are considered and each is given equal probability. There are $\binom{m+n-1}{n}$ such distinguishable arrangements and probability of each distinct arrangement is $\left[\binom{m+n-1}{n}\right]^{-1}$

Proof. Suppose m cells are represented by a sequence of $(m + 1)$ bars (|), each bar representing an wall of a cell. Within these cells the balls (crosses) are placed. For example, | × | × ×||| shows a configuration of 3 balls in 4 cells with $k_1 = 1, k_2 = 2, k_3 = k_4 = 0$. Any arrangement of n balls in m cells is thus represented by a sequence of $(m + 1)$ bars and n crosses provided the sequence must start with a bar and end with a bar. Hence the number of distinguishable arrangements equal the number of choices of n positions for crosses out of $(m+n-1)$ available positions. This is $\binom{m+n-1}{n}$.

The Fermi-Dirac statistic is based on the following hypothesis : (a) it is impossible for two or more balls to be in the same cell; (b) all distinguishable arrangements satisfying (a) have equal probability. With $F-D$ statistic there are thus $\binom{m}{n}$ arrangements each having probability $1\Big/\binom{m}{n}$, $n \le m$.

The model of distributing balls over cells is an appropriate one for various problems in statistical mechanics. The n balls are physical particles and the m cells represent m equal parts into which a spacial region is divided. $B-E$ statistic hold good for photons, nuclei and atoms containing an even number of elementary particles. $F-D$ statistic apply to electrons, neutrons and protons. However, different situations fit different models and it is impossible to select or justify probability models by a priori arrangements.

2.6. Exercises

1. *Considering each of the following groups of events state whether they are groups of : (i) elementary events, (ii) exhaustive events, (iii) mutually exclusive events, (iv) equally likely events.*

(a) Two coins are tossed; the events are : A_1 = {two heads}, A_2 = {two tails}, A_3 = {one head, one tail}. [No, Yes, Yes, No]

(b) A die is tossed. The events are : B_1 = {1 or 2}, B_2 = {2 or 3}, B_3 = {3 or 4}, B_4 = {4 or 5}, B_5 = {5 or 6}. [No, Yes, No, Yes]

(c) A shot is fired at a target; the events are : C_1 = {a hit}, C_2 = {a miss} [No, Yes, Yes, No]

2. *Three unbiased coins are tossed. Write down the sample space. Find the probability of the following events; A = {the first coin comes up head}, B = {exactly two heads have occurred}, C = {not more than two heads have occurred}.*

$$\left[S = \{HHH,\ HHT,\ HTH,\ HTT,\ THH,\ THT,\ TTH,\ TTT\}\right.$$

$$\left.P(A) = \frac{1}{2},\quad P(B) = \frac{3}{8},\quad P(C) = \frac{7}{8}\right]$$

3. *A club has five members A, B, C, D. It is required to select a Chairman and a Secretary. Assuming that one member cannot occupy both positions, write the sample space associated with the selection. Assuming all selections are equally likely, what is the probability that the member A is an office bearer.*

[$S = \{(x,y);\ x \neq y$, both x, y belong to $(A, B, C, D)\}$; here x, y represent Chairman and Secretary respectively. Probability $= \frac{1}{2}$]

4. *The morse code consists of a sequence of dots and dashes with repetitions permitted. (a) How many letters can be coded by using exactly n symbols ? (b) What is the number of letters that can be coded using n or fewer symbols ?*

[(a) 2^n because each place can be filled up in 2 ways; (b) letters can be formed by using 1 symbol, 2 symbols, $\cdots$, n symbols. Number of possible letters are $2 + 2^2 + \cdots + 2^n = 2(2^n - 1)$]

5. *Find the probability that among three random digits there occur 0, 1, 2 repetitions.* [.72, .27, .001]

6. *There is a 3-volume dictionary among 30 books arranged in a shelf in a random order. Find the probability of these volumes standing in increasing order from left to right (the volumes are not necessarily side by side).* [1/6]

7. *Five cards are marked 1, $\cdots$, 5 and randomly laid out in a row. Find probabilities of: (i) card 1 appears in the first position, (ii) card 1 is immediately followed by card 2, (iii) card 1 is not in the first position and card 2 not in the second position, (iv) there are exactly 3 matches (a match means card i is in position i).*

$$\left[\text{(i) } \frac{4!}{5!} = \frac{1}{5},\quad \text{(ii) } \frac{1}{5},\quad \text{(iii) } \frac{13}{20},\quad \text{(iv) } \frac{1}{12}\right]$$

8. *The last three digits of a telephone number begining 584 have been erased. Assuming that all combinations of last three digits are equally likely, find the probability of the following events : $A = \{$distinct digits different from 5, 8, 4 have been erased$\}$, $B = \{$identical digits have been erased$\}$, $C = \{$two of the missing digits coincide$\}$.*

$$\left[P(A) = (7)_3 / 10^3 = .21, P(B) = .01, P(C) = .27\right]$$

9. *What is the chance that a 4-digited number on the licence plate of a car chosen at random in a city : (i) consists of different digits, (ii) includes only two identical digits, (iii) includes two pairs of identical digits, (iv) contains only three identical digits, (v) consists of identical digits.*

$$\left[\text{(i) } (10)_4 a,\quad \text{(ii) } \binom{4}{2}(10)_3 a,\quad \text{(iii) } 90\binom{4}{2} a,\quad \text{(iv) } 360a,\quad \text{(v) } 10a \text{ where } a = 10^{-4}\right].$$

10. *In a city seven accidents occur per week. What is the probability there will be at least one day in a week without any accident ?* $[1 - (7!/7^7)]$

11. *In a random number table what is the probability that five consecutive digits are all different ?* $[(10)_5/10^5]$

12. *What is the probability that the birthdays of r randomly selected persons will be all different ?*
$$\left[(365)_r/365^r;\ \text{for } r = 30,\ p = .294\right]$$

13. *Six dice are tossed. What is the probability every possible number will appear?* [$6!/6^6$]

14. *If n men, among whom are A and B, stand in a row : (a) what is the probability there will be exactly r men between A and B ? (b) If they stand in a ring instead of in a row, show that the probability is independent of r and is $1/(n-1)$. [In the circular position consider only the arc leading from A to B in the positive direction].*

(a) Suppose A and B are kept fixed. r persons can be chosen out of $(n-2)$ persons and arranged among themselves in $(n-2)_r$ ways. The remaining $(n-r-2)$ persons can be arranged among themselves in $(n-r-2)!$ ways. Again if A occupies a position on the left of B, then A may choose any of the positions, first, second, $\cdots$ $(n-r-1)$th from the left. Similar is the case if B is on the left of A. Hence probability $= \dfrac{1}{n!}\{2(n-r-1)!(n-2)_r\}$ $= \dfrac{2(n-r-1)}{n(n-1)}$.

(b) If they form a ring the total number of possible arrangements is $(n-1)!$ which is obtained by keepng the position of any person fixed and arranging the remaining persons in all possible ways. Hence probability $= \dfrac{\{(n-r-2)!(n-2)_r\}}{(n-1)!} = \dfrac{1}{(n-1)}$, because A can now occupy any position allowing only r persons between A and B and we are considering only the arc leading from A to B in the positive direction.

15. *If n balls are placed at a random order into n cells find the probability exactly one cell remains empty.*

[The empty cell can be chosen in n ways; the two balls to be kept in the same cell in $\binom{n}{2}$ ways; the n balls, including the two balls put together, can be arranged in $(n-1)$ cells in $(n-1)!$ ways. Hence the probability $= n\binom{n}{2}(n-1)!/n^n = \binom{n}{2}\cdot n!/n^n$]

16. *If n balls are tossed into m boxes so that each ball is equally likely to fall in any box, show that the probability that a specified box will contain r balls is*

$$p_r = \binom{n}{r}\frac{(m-1)^{n-r}}{m^n}$$

Show that the most probable number of balls in a cell is r which satisfies

$$\frac{(m-1)^2}{(n-r)(n-r+1)} < \frac{m-1}{r(n-r)} > \frac{1}{r(r+1)} \qquad \text{(i)}$$

Also show that

$$\frac{1}{r!}\left(\frac{n}{m}\right)^r\left(1-\frac{1}{m}\right)^{n-r}\left(1-\frac{r}{n}\right)^{(r-1)/2} \le p_r \le \frac{1}{r!}\left(\frac{n}{m}\right)^r\left(1-\frac{1}{m}\right)^{n-r}\left(1-\frac{r}{2n}\right)^{r-1} \quad \text{(ii)}$$

For $m = n$ with $m \to \infty, p_r \to e^{-1}/r!$.

[r balls can be chosen out of n balls in $\binom{n}{r}$ ways. The remaining $(n-r)$ balls can go into $(m-1)$ boxes in $(m-1)^{n-r}$ ways. Hence the expression for p_r. The most probable number of ball is r which satisfies $p_{r-1} < p_r > p_{r+1}$. Hence (i) on simplification. Now, $p_r = \left(1-\frac{1}{m}\right)^{n-r}\frac{Q}{r!}\left(\frac{n}{m}\right)^r$ when $Q = \left(1-\frac{1}{n}\right)\left(1-\frac{2}{n}\right)\cdots\left(1-\frac{r-1}{n}\right)$.

For $1 \le k \le r-1$, $\left(1-\frac{r}{n}\right) < \left(1-\frac{k}{n}\right)\left(1-\frac{r-k}{n}\right) < \left(1-\frac{r}{2n}\right)^2$.

Hence $\left(1-\frac{r}{n}\right)^{(r-1)/2} < Q < \left(1-\frac{r}{2n}\right)^{r-1}$. Hence (ii).

For $m = n$, with $n \to \infty$, $Q \to 1$, $\left(1-\frac{1}{m}\right)^{m-r} \to e^{-1}$ when $p_r \to e^{-1}/r!$]

17. *What is the probability that among k random digits 0 appears (i) exactly three times, (ii) at most three times.* [(i) $\binom{k}{3}9^{k-3}\Big/10^k$, (ii) $\sum_{r=0}^{3}\binom{k}{r}9^{k-r}\Big/10^k$]

18. *From a population of n elements a sample of size r is taken. Find the probability that none of the N prescribed elements will be included in the sample assuming the sample to be: (a) without replacement, (b) with replacement.*

[(a) $\binom{n-N}{r}\Big/\binom{n}{r}$, (b) $(n-N)^r/n^r$]

19. *In a parking lot there are 12 places arranged in a row. A man observed that there were eight cars parked and the four empty spaces were adjacent to each other. Find the probability of this event.*

[Probability $= 9\Big/\binom{12}{8} = .0182$, because eight cars can choose their positions in $\binom{12}{8}$ ways of which 9 are favourable to the given event as the four consecutive vacant positions may start from the first, second, $\cdots$, ninth position in the parking lot].

20. *Each of the 12 districts has 2 representatives. Find the probability that in a committee of 12 representatives chosen at random : (a) a given district is represented, (b) all districts are represented.*

[(a) probability that a given district is not represented $= \binom{22}{12} \Big/ \binom{24}{12} = q$ (say); required probability $= 1-q$, (b) Probability $= 2^{12} \Big/ \binom{24}{12}$, because each district may be represented in two ways]

21. *What is the probability that : (a) the birthdays of twelve people will fall in twelve calender months, (b) the birthdays of six people will fall in exactly two calender months.*

[(a) $12!/12^{12}$, (b) $\binom{12}{2}(2^6-2)/12^6$, because two months can be chosen in $\binom{12}{2}$ ways, the 6 persons can have their birthdays in these two months so that both months are occupied in (2^6-2) ways and the remaining 6 persons can have their birthdays in any of the 12 months in 12^6 ways]

22. *A closet contains n pairs of shoes. If 2r shoes are chosen at random ($2r < n$), what is the probability that there will be : (a) no complete set, (b) exactly one complete set (c) exactly two complete pairs among them ?*

[(a) Each pair must be represented only by one shoe. Hence probability $= \binom{n}{2r} 2^{2r} P$ where $P = \left[\binom{2n}{2r}\right]^{-1}$, because from each pair one has two choices. (b) $n\binom{n-1}{2r-2} 2^{2r-2}P$, becuase there are n choices for the complete set. (c) $\binom{n}{2}\binom{n-2}{2r-4} 2^{2r-4}P$.]

23. *What is the probability that two throws with three dice each will show the same configuration if (i) the dice are distinguishable, (ii) they are not.*

[(i) The first throw of 3 dice may occur in 6^3 ways; the second throw must show the same configuration. Hence probability $= \dfrac{6^3}{6^6} = \dfrac{1}{216}$.

(ii) The first three throws may show results of the following types : (a) all face values are different, (b) two face values are same, (c) all face values are same.

(a) Total number of such cases = $(6)_3.3! = 720$, because the first three throws may occur in $(6)_3$ ways and thereafter the same face values may occur in the second three throws in 3! ways.

(b) Total number of possible cases is $6.5.\binom{3}{2}^2 = 270$, becuase two different face values can be chosen in 6.5 ways and two different throws to show the same face value out of the first three throws in $\binom{3}{2}$ ways and similarly for the second set of three throws.

(c) The total number of possible cases is 6. Hence probability $= \frac{996}{6^6}$.]

24. *(a) From a pack of 52 cards, two are chosen at random. Find the chance that one is a king and the other queen.*

(b) What is the probability that a nine-card hand will contain cards of the same suit ?

$$\left[\text{(a) } 4^2\Big/\binom{52}{2} = \frac{8}{663},\ \text{(b) } 4\binom{13}{9}\Big/\binom{52}{9}\right]$$

25. *What is the probability that in a game of poker, which requires drawing of 5 cards from a full pack, all the cards drawn will be : (a) of the same colour, (b) of the same suit ?*

$$\left[\text{(a) } 2\binom{26}{5}\Big/\binom{52}{5} = \frac{253}{4998},\ \text{(b) } 4\binom{13}{5}\Big/\binom{52}{5} = \frac{33}{16660}\right]$$

26. *What is the probability that a five-card hand will contain exactly two aces ? at least two aces ?*

$$\left[\binom{48}{3}\binom{4}{2}\Big/\binom{52}{5};\ \left\{\binom{48}{3}\binom{4}{2}+\binom{48}{2}\binom{4}{3}+48\right\}\Big/\binom{52}{5}\right]$$

27. *Ten cards are taken from a full pack. What is the probability of finding among them (a) at least one ace ? (b) at least two aces ?*

$$\left[\text{(a) } 1-\binom{48}{10}\Big/\binom{52}{10} = \frac{349}{595},\ \text{(b) } 1-\left[\binom{48}{10}+4\binom{48}{9}\right]\Big/\binom{52}{10} = \frac{1257}{7735}\right]$$

28. *Five cards are taken from a full pack. Find the probability : (a) they are of different denominations, (b) that 2 are of the same denomination and 3 scattered, (c) that one pair is of one denomination and the another pair of a different denomination and one odd, (d) that 3 are of the same denomination and 2 scattered, (e) that 2 are of one denomination and 3 of another (f) that 4 are of one denomination and 1 of another.*

$\Big[$(a) $\binom{13}{5}4^5\Big/\binom{52}{5} = .50708$ (because 5 denominations (d) can be chosen in $\binom{13}{5}$ ways and each d is represented in 4 ways.)

(b) $13\binom{4}{2}\binom{12}{3}4^3\Big/\binom{52}{5} = .42257$ (because the d with 2 cards can be chosen in 13 ways, these 2 cards in $\binom{4}{2}$ ways, remaining 3 d's in $\binom{12}{3}$ ways, these 3 cards in 4^3 ways)

(c) $\binom{13}{2}\binom{4}{2}^2 11.4\Big/\binom{52}{5} = .04754$ (because the two d's with 2 cards each in $\binom{13}{2}$ ways, two pairs of cards, each pair with the same d in $\binom{4}{2}^2$ ways, the remaining d in 11 ways and the corresponding card in 4 ways.

(d) $13\binom{4}{3}\binom{12}{2}4.4\Big/\binom{52}{5} = .02113$

(e) $13\binom{4}{2}\cdot 12\binom{4}{3}\Big/\binom{52}{5} = .00144$

(f) $13{\cdot}12.4\Big/\binom{52}{5} = .00024.$

29. ***Two boxes each have r balls labelled 1, 2, $\cdots$, r. A random sample of size $n(\le r)$ is drawn without replacement from each box. Find the probability that the sample contains exactly k balls having the same number in common.***

$\Big[\binom{r}{n}\binom{n}{k}\binom{r-n}{n-k}\Big/\binom{r}{n}^2 = \binom{n}{k}\binom{r-n}{n-k}\Big/\binom{r}{n}$ because the n balls can be chosen from the first box in $\binom{r}{n}$ ways; from the second box, k balls have to be chosen out of these n balls and the $(n-k)$ balls out of the remaining $(r-n)$ balls.$\Big]$

30. ***From the population of 5 symbols a, b, c, d, e a sample of size 25 is taken. Find the probability that the sample will contain 5 symbols of each kind.***

$$\left[\frac{25!}{(5!)^5 5^{25}} = .00209\right]$$

31. ***Suppose that each of n sticks is broken into one long and one short part. The 2n parts are arranged into n pairs from which new sticks are formed. Find the probability that : (a) the pairs will be joined in the original order, (b) all the long parts are paired with the short parts.***

[Number of possible arrangements $= \dfrac{2n!}{n!2^n} = N$ (say). (a) Probability $= \dfrac{1}{N}$ (b)

Keeping the n long parts fixed in n different groups, the n short parts can be allotted to n different groups in $n!$ ways. Probability $= \dfrac{n!}{N} = 2^n \Big/ \dbinom{2n}{n}$.]

32. *In a town of $(n+1)$ inhabitants, a person tells a rumour to a second person, who in turn repeats it to a third person, etc. At each step the receipient of the rumour is chosen at random from the n people available. Find the probability that the rumour will be told r times without : (a) returning to the originator, (b) being repeated to any person.*

[(a) At each step a receipient may choose his disciple in $(n-1)$ ways out of a total number of n persons, because any person except the originator may receive the remour. Hence probability is $\left(\dfrac{n-1}{n}\right)^{r-1}$ because after the originator has told the rumour it must be repeated $(r-1)$ times. (b) The rumour will not be repeated at any step. Probability $= \dfrac{(n-1)_{r-1}}{n^{r-1}} = \dfrac{(n)_r}{n^r}$]

33. *Compute the probability $p_0(r), p_1(r), \cdots, p_4(r)$ that among r bridge cards drawn at random there are 0, 1, $\cdots$, 4 aces respectively. Verify that $p_0(r) = p_4(52-r)$.*

[$p_0(r) = \dbinom{48}{r} \Big/ \dbinom{52}{r} = (52-r)_4 \big/ (52)_4$,

$p_1(r) = 4\dbinom{48}{r-1} \Big/ \dbinom{52}{r} = 4r(52-r)_3 \big/ (52)_4$,

$p_2(r) = 6r(r-1)(52-r)_2 \big/ (52)_4$, $p_3(r) = 4r(r-1)(r-2)(52-r) \big/ (52)_4$,

$p_4(r) = (r)_4/(52)_4$]

34. *Calculate the probabilities of the following events in a game of bridge : (i) a specified player, say S, has at least two aces, irrespective of other players' hands, (ii) some one of the four has at least two aces, irrespective of the other players' hands, (iii) S and no other player has at least two aces, (iv) exactly one player of the four has at least two aces ?*

[(i) $p_1 = \left\{\dbinom{4}{2}\dbinom{48}{11} + \dbinom{4}{3}\dbinom{48}{10} + \dbinom{48}{9}\right\} \Big/ \dbinom{52}{13} \simeq 0.257$

(ii) $p_2 = 1 -$ Prob {each player has an ace} $= 1 - \left\{\dfrac{4!48!}{(12!)^4} \Big/ \dfrac{52!}{(13!)^4}\right\} \simeq 0.895$

$$\text{(iii)}\quad p_3 = p_1 - \frac{3\binom{4}{2}\binom{48}{11}\binom{37}{11}\binom{26}{13}}{\binom{52}{13}\binom{39}{13}\binom{26}{13}} \simeq 0.190$$

(because we must subtract from p_1 the probability of the event (S has two aces, some other player has two aces)

(iv) $p_4 = 4p_3 \simeq 0.759$ (because p_3 gives the probability for a specified player)]

35. ***Find the probability that in a bridge game a specified player has : (i) only one ace, (ii) at least one ace, (iii) the ace of hearts, (iv) only the ace of hearts. Suppose the player asserts that he has one ace. Find the probability that he has : (v) another ace, (vi) only another ace.***

[(i) $4\binom{48}{12}\Big/\binom{52}{13} \simeq 0.43885$ (ii) $1-\left\{\binom{48}{13}\Big/\binom{52}{13}\right\} \simeq 0.69618$

(iii) $\binom{51}{12}\Big/\binom{52}{13} \simeq 0.25$ (iv) $\binom{48}{12}\Big/\binom{52}{13} \simeq 0.10971$

(v) $$\text{Probability} = \frac{\text{No. of cases the player has more than one ace}}{\text{Total no. of cases he has at least one ace } (= N \text{ (say)})}$$

$$= 1 - \frac{\text{No. of cases he has one ace}}{N}$$

$$= 1 - \frac{4\binom{48}{12}}{\binom{52}{13} - \binom{48}{13}} \simeq 0.36964.$$

(vi) $$\text{Probability} = \frac{\text{No. of cases he has two aces}}{N}$$

$$= 6\binom{48}{11}\Big/\left\{\binom{52}{13} - \binom{48}{13}\right\} \simeq 0.30666$$

36. ***Cards are taken one by one from a full pack. What is the probability that 10 cards will precede the first ace. If cards are continued to be dealt until a second ace appears find the probability that exactly 15 cards are dealt in all before the second ace appears.***

$$\left[4 \cdot (48)_{10}\big/(52)_{11};\quad \frac{4\binom{48}{14}}{\binom{52}{15}} \cdot \frac{3}{37}\quad \text{because the first 15 cards must contain an ace.}\right]$$

37. *There are N tickets numbered $1, \cdots, N$ of which n are taken at random and arranged in an increasing order of their magnitudes : $x_1 < x_2 < \cdots < x_n$. What is the probability that $x_m = M$?*

$$\left[\binom{M-1}{m-1}\binom{N-M}{n-m}\Big/\binom{N}{n},\right.$$

because there are $(m-1)$ tickets with numbers between 1 and $M-1$ (both inclusive) and $(n-m)$ tickets with numbers between $M+1$ and N (both inclusive)]

38. *A pack of 36 playing cards has been well-shuffled. Find the probabilities of the following events : A : {4 aces are drawn in succession}, B : {the places occupied by the aces form an arithmatic progression with a common difference 7}.*

[$P(A) = 33!\,4!/36!$ (assuming the 4 aces are tied together there are 33 cards)

$P(B) = 32!\,4!\,15/36!$ (because the position of aces may be (1, 8, 15, 22), $\cdots$, (15, 22, 29, 36)]

39. *The digits 1, $\cdots$, 7 are written in a random order such that 7! arrangements are all equally likely to give 7 digited numbers. Find the probability that the number is divisible by : (i) 2, (ii) 4, (iii) 8.*

[(i) $\dfrac{3 \times 6!}{7!}$ because the last digit must be an even number.

(ii) $10 \times 5!/7!$ because the last two digits must be divisible by 4.

(iii) $22 \times 4!/7!$ because the last three digits must be divisible by 8.]

40. *A number 'a' is chosen at random between 0 and 6 and a number 'b' is chosen at random between 0 and 9. Find the probability that the equation $x^2 - ax + b = 0$ has two real roots.*

[We require $a^2 \geq 4b$; given $0 \leq a \leq 6, 0 \leq b \leq 9$, there are 29 combinations for which $a^2 \geq 4b$. Hence probability $= 29/70$.]

41. *Find the probability p_N that a natural number chosen at random from the set $1, \cdots, N$ is divisible by a fixed natural number k. Find $\lim\limits_{N\to\infty} p_N$.*

$\left[p_N = \dfrac{1}{N}\left\{\left[\dfrac{N}{k}\right]\right\}\right.$ where $[z]$ is the highest integer contained in z, because $\left[\dfrac{N}{k}\right]$ is the number of digits divided by k. $\lim\limits_{N\to\infty} p_N = \dfrac{1}{k}\Bigg]$

42. *A number 'a' is chosen at random from the set $1, \cdots, N$. Find the probability p_N that when divided by the integer $r \geq 1$, 'a' will have a remainder q.*

['a' must be one of the following integers : $q, q+r, q+2r, \cdots$. Hence $p_N = \frac{1}{N}$ $\left\{\left[\frac{N-q}{r}+1\right]\right\}$, where $[z]$ is the highest integer contained in z. $\lim_{N\to\infty} p_N = \frac{1}{r}$.]

43. *A and B each chooses an integer between 1 and m (both inclusive) independently of each other. Find the probability that two numbers chosen do not differ by more than $n(< m)$.*

[Let x, y denote numbers chosen by A, B respectively $(1 \leq x, y \leq m)$. We want to find $P(|x-y| \leq n) = 1 - P(|x-y| > n)$.

Suppose $y > x$. Then $(y - x) > n$ may occur in the following ways :

If $x = 1$, y must be $n+2, n+3, \cdots, m$ [i.e., $(m-n-1)$ choices of B],

$x = 2$, y ,, ,, $n+3, n+4, \cdots, m$ [i.e., $(m-n-2)$,, ,, B],

... ...

$x = (m-n-1)$, y ,, ,, m [i.e., 1 choice of B].

Hence total number of choices of A and B such that $y - x > n$ is $\sum_{t=1}^{m-n-1} (m-n-t)$ $= \frac{(m-n)(m-n-1)}{2}$. Similar is the case for $x - y > n$. Hence $P\{|x-y| > n\}$ $= \frac{(m-n)(m-n-1)}{m^2}$]

44. *The numbers x_1 and x_2 are chosen from the set of numbers $\{1, \cdots, N\}$ without replacement. Find $P\{x_1 > x_2\}$. When selecting three numbers, find the probability that the second number lies between the first and the third.* $\left[\frac{1}{2}, \frac{1}{3}\right]$

45. *A number x is chosen at random from the numbers $\{1, \cdots, N\}$. Find $\lim_{N\to\infty} p_N$ where p_N is the probability that $x^2 - 1$ is divisible by 10.*

[We must have $x^2 = 10l + 1$ $(l = 0, 1, \cdots)$ and hence

$$x = 10m + 1 \quad \text{or} \quad 10m + 9, \qquad (m = 0, 1, \cdots), \tag{i}$$

Let $N = 10k + r$, $k = 0, 1, 2, \cdots$ and $0 \leq r \leq 9$.

(a) Suppose $N = 10k$. Number of x's satisfying (i) is $2k$. Thus $p_N = \frac{1}{5}$.

(b) Suppose $N = 10k + r,\ \ 0 < r < 9$. Number of x's satisfying (i) is $2k + 1$. Thus $p_N = \dfrac{2k+1}{10k+r}$.

(c) $N = 10k + 9$. Here $p_N = \dfrac{2k+2}{10k+9}$. In any case $\lim\limits_{N\to\infty} p_N = \dfrac{1}{5}$]

46. *An integer x is chosen at random from the set $\{0, 1, \cdots, 10^p - 1\}$. Find the probability that this number is a k-digit number, $k < p + 1$.*

[Number of k-digited numbers $= (10^k - 1) - (10^{k-1} - 1) = 9 \times 10^{k-1}$. Hence probability $= 9 \times 10^{k-p-1}$]

47. *The numbers X, Y are chosen at random from a set of natural numbers $\{1, \cdots, N\}$, $N \geq 3$ with replacement. Find which is greater : $P_2 = P\{X^2 - Y^2$ is divisible by 2$\}$, $P_3 = P\{X^2 - Y^2$ is divisible by 3$\}$.*

[$X^2 - Y^2$ is divisible by 2, if both X, Y are even or odd; is divisible by 3 if both X, Y are simultaneously divisible or indivisible by 3. $P_2 = \dfrac{1}{N^2}\left\{\left[\dfrac{N}{2}\right]^2 + \left(N - \left[\dfrac{N}{2}\right]\right)^2\right\}$,

$P_3 = \dfrac{1}{N^2}\left\{\left[\dfrac{N}{3}\right]^2 + \left(N - \left[\dfrac{N}{3}\right]\right)^2\right\}$, $P_3 - P_2 = \dfrac{2}{N}\left(\left[\dfrac{N}{2}\right] - \left[\dfrac{N}{3}\right]\right)\left(1 - \dfrac{1}{N}\left[\dfrac{N}{2}\right] + \left[\dfrac{N}{3}\right]\right) > 0$

for $N \geq 3$.]

48. *One sequence is chosen at random from the set of all sequences of length n consisting of numbers 0, 1, 2. Find the probabilities of the following events : $A = \{$the sequence begins with 0$\}$, $B = \{$ the sequence contains exactly $(m + 2)$ zeroes, two of them being at the end points of the sequence$\}$, $C = \{$the sequence contains exactly m units$\}$, $D = \{$the sequence contains exactly m_o zeroes, m_1 units and m_2 twos$\}$.*

[$P(A) = \dfrac{1}{3}$; $\quad P(B) = \dbinom{n-2}{m} 2^{n-m-2}/3^n \quad (0 \leq m \leq n-2)$

$P(C) = \dbinom{n}{m} 2^{n-m}/3^n$

$P(D) = \dbinom{n}{m_o}\dbinom{n-m_o}{m_1}\dfrac{1}{3^n} = \dfrac{n!}{m_o! m_1! m_2! 3^n}, \quad m_o + m_1 + m_2 = n$]

49. *Find the probability in 5 tossings a coin turns up heads at least 3 times in succession.*

[The event occurs if there are no tails, 1 tail or 2 tails. Favourable position of tails are : for 1 tail, — 1st, 2nd, 4th, 5th; for 2 tails, (1st, 2nd), (1st, 5th), (4th, 5th). Hence probability $= \dfrac{8}{32} = \dfrac{1}{4}$]

50. *What is the probability of obtaining a given sum s of points with n dice ?*

[The total number of possible cases with 6 dice is 6^n. The total number of favourable cases is the number of solutions of the equation $a_1+a_2+\cdots+a_n = s$ where a_i represents the number turned up by the ith draw and hence $1 \le a_i \le 6$, $i = 1,\cdots,n$. This number is given by the coefficient of x^s in the expansion of

$$T = (x + x^2 + \cdots + x^6)^n,$$

(becuase coefficient of x^s is $\dfrac{n!}{b_1!\cdots b_6!}$ where $b_1 + 2b_2 + \cdots + 6b_6 = s$, b_i = no. of throws with results i, $i = 1,\cdots,6$)

Now $T = x^n(1-x^6)^n(1-x)^{-n}$;

$$x^n(1-x^6)^n = \sum_{r=0}^{n}(-1)^r\binom{n}{r}x^{n+6r} \qquad \text{(i)}$$

$$(1-x)^{-n} = \sum_{k=0}^{\infty}\binom{n+k-1}{n-1}x^k \qquad \text{(ii)}$$

Multiplying (i) and (ii), the coefficient of x^s in T is

$$\sum_{r=0}^{[(s-n)/6]}(-1)^r\binom{n}{r}\binom{s-6r-1}{n-1} = N_n \quad \text{(say)},$$

For $n = 3$, $s = 15$, $N_n = 10$. The required probability $= \dfrac{N_n}{6^n}$]

51. *An urn contains a white and b black balls. Balls are drawn one by one until only those of the same colour are left. What is the probability that they are all white?*

[All the black balls should have been exhausted. The sequence of balls drawn must contain x white balls and b black balls, $x = 0, 1, \cdots, a-1$. The probability of drawing x white balls to precede the last black ball is $\binom{x+b-1}{x}\Big/\binom{a+b}{a}$ (becuase the total number of possible cases is the number of permutations of a white and b black balls and is $\binom{a+b}{a}$. Hence probability $= \sum_{x=0}^{a-1}\binom{x+b-1}{x}\Big/\binom{a+b}{a}$.

Now $\sum_{x=0}^{a-1}\binom{x+b-1}{x} = \binom{b-1}{0} + \binom{b}{1} + \binom{b+1}{2} + \cdots + \binom{a+b-2}{a-1}$, is the term independent of x in $\left[(1+x)^{b-1} + \frac{1}{x}(1+x)^b + \frac{1}{x^2}(1+x)^{b+1} + \cdots + \frac{1}{x^{a-1}}(1+x)^{b+a-2}\right] =$ $(1+x)^{a+b-1}x^{1-a}$, which is $\binom{a+b-1}{a-1}$. Hence the probability $= \dfrac{a}{a+b}$]

52. *An urn contains N_1 white, N_2 black and N_3 red balls. Balls are successively drawn without replacement until a red ball is drawn. Find the probability that :*

(i) n_1 white and n_2 black balls are drawn.

(ii) not a single white ball is drawn.

(iii) a total of k balls are drawn.

[(i) $\dfrac{\binom{n_1+n_2}{n_1}(N_1)_{n_1}(N_2)_{n_2}N_3}{(N)_{n_1+n_2+1}}$, $N = N_1 + N_2 + N_3$ since n_1 places for white balls can be chosen out of $(n_1 + n_2)$ positions in $\binom{n_1+n_2}{n_1}$ ways.

(ii) $\displaystyle\sum_{m=0}^{N_2} \frac{(N_2)_m N_3}{(N)_{m+1}}$, (iii) $\dfrac{(N_1+N_2)_{k-1}N_3}{(N)_k}$]

53. *n indistinguishable balls are distributed without replacement among $M(> n)$ urns numbered 1 to M. Determine the probability that each urn numbered 1 to n contains exactly one ball.*

The total number of possible arrangements is $\binom{n+M-1}{n}$ by Bose–Einstein statistic, all of which are assumed equally likely. Only one arrangement satisfies the desired event. Hence probability is $\left[\binom{n+M-1}{n}\right]^{-1}$.

54. *From n balls numbered $1, 2, \cdots, n$, m balls are drawn. Let X be the largest number drawn. Find $P(X = k)$ if the sampling is done : (i) with replacement, (ii) without replacement.*

[(i) $P(X \le k) = \left(\dfrac{k}{n}\right)^m$, $P(X = k) = P(X \le k) - P(X \le k-1) = \dfrac{k^m - (k-1)^m}{n^m}$

(ii) $P(X \le k) = \binom{k}{m}\Big/\binom{n}{m}$, $P(X = k) = \binom{k-1}{m-1}\Big/\binom{n}{m}$]

55. *A man has exactly n keys, one of which fits the lock. He tries the keys one at a time, at each trial choosing at random from the keys that were not tried earlier. Find the probability that the key tried at the rth trial is the correct key.*

[Probability $= \dfrac{(n-1)_{(r-1)}}{(n)_r} = \dfrac{1}{n}$]

56. *Suppose we have r boxes. Balls are placed at random one at a time into the boxes until, for the first time, some box has two balls. Find the probability that this occurs with the nth ball.*

[$(n-1)\{(r)_{n-1}\}/r^n$, because the nth ball can be placed in any of the $(n-1)$ already occupied boxes.]

57. *A box has r balls labelled* $1, 2, \cdots, r$. $N (\leq r)$ *balls are selected at random from the box, their number noted and are then returned to the box. If the procedure is repeated r times, what is the probability that none of the original N balls are duplicated?*

$$\left[\left\{\binom{r-N}{N}\bigg/\binom{r}{N}\right\}^{r-1}\right]$$

2.7. Some Elements of Set Theory

Before proceeding further we consider some elements of set theory. For this we consider a few definitions.

(a) *Union* : Let A, B be two sets. The union of A and B, denoted as $A \cup B$ is the set of all elements contained in either A or B or both. In fig. 5(a), $A \cup B$ is the region covered by either horizontal bars or dots or both. In the terminology of events, $A \cup B$ represents the event in which at least one of A and B occurs.

Similarly the union of m sets $A_1, \cdots, A_m$, namely, $\bigcup_1^m A_i$ is the set of all elements contained in at least one of $A_1, \cdots, A_m$. In fig. 5(c), $A \cup B \cup C$ is the region covered by either horizontal bars or dots or vertical bars or by any two of them or by all. When the sets $A_1, \cdots, A_m$ are mutually disjoint we shall sometime write $\bigcup_1^m A_i = A_1 + A_2 + \cdots + A_m$. $\bigcup_{i=1}^m A_i$ represents the event that at least one of $A_1, \cdots, A_m$ occurs. $\bigcup_{i=1}^m A_i$ is also called a total event.

(b) *Intersection* : The intersection of two sets A and B, denoted as $A \cap B$ or simply AB is the set of all elements contained in both A and B. In fig 5(a), $A \cap B$ is the region covered by both horizontal bars and dots. The set $A \cap B$ denotes the event that both A and B occur.

Similarly the intersection of m sets $A_1, \cdots, A_m$, denoted as $\bigcap_{i=1}^m A_i$ or simply $A_1 A_2 \cdots A_m$ is the set of all elements contained in all of $A_1, \cdots, A_m$. In fig 5(c), ABC is the region covered by all of the three markings — vertical and horizontal bars and dots. The set $A_1 \cdots A_m$ represents the event, all of $A_1, \cdots, A_m$ occur. $\bigcap_{i=1}^m A_i$ is called a compound event. If the sets $A_1, \cdots, A_m$ are mutually disjoint, $A_i \cap A_j = \phi$ for all $i \neq j = 1, \cdots, m$.

(c) *Complementation* : The complement of set A, denoted as A^c or A' or $\bar{A}$ is the set of all elements not contained in A. Hence $A^c = \mathcal{U} - A$. In fig. 5(b), A^c is the entire shaded region. The set A^c represents the event then A does not occur.

Clearly $\mathcal{U}^c = \phi$, $\phi^c = \mathcal{U}$, $(A^c)^c = A$; $P(A^c) = 1 - P(A)$.

(d) *Subset* : A set C is a subset of a set D if every member of C is necessarily a member of D. This is denoted as $C \subseteq D$. When D contains at least one element not contained in C, C is a proper subset of D and we write $C \subset D$.

For two sets A, B clearly, $A \cap B \subset (A \cup B)$. For three sets A, B, C, $ABC \subset (AB)$, $ABC \subset A \cup (BC)$, etc.

If $C \subset D$, event D is implied by C but not vice-versa. In this case, as noted in section 2.3, $P(C) \leq P(D)$.

(e) *Difference* : The difference $A - B$ is the set of all elements contained in A but not in B. Similarly $B - A$ is the set of all elements contained in B but not in A. In fig. 5(a), $A - B$ is the region covered by bars but not by dots; $B - A$ is the region covered by dots but not by bars. Clearly $A - B \neq B - A$. Also $(A - B) \subset A, (B - A) \subset B$. Clearly $A - B$ and $B - A$ are disjoint sets.

If follows that $A - B = A \cap B^c$, $B - A = B \cap A^c$.

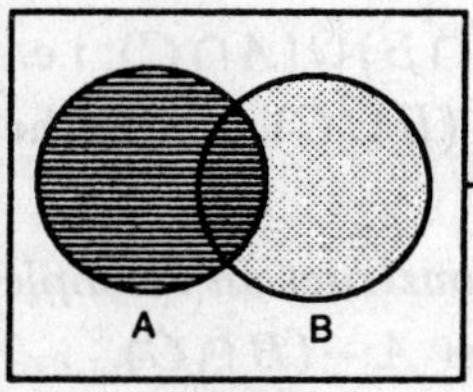

Fig. 5(a)

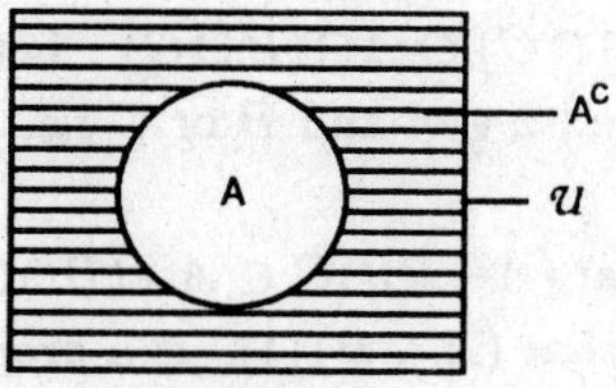

Fig. 5(b)

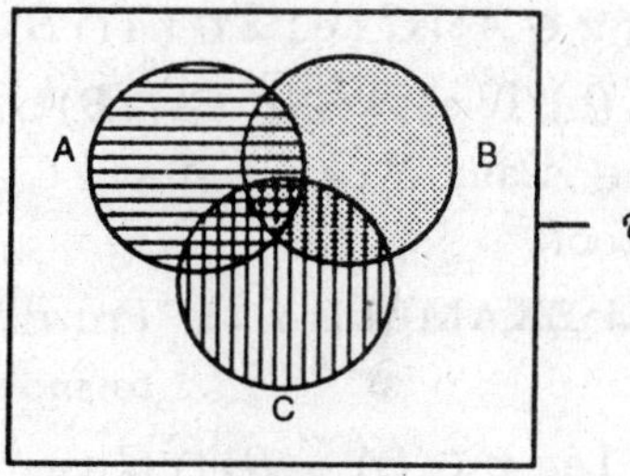

Fig. 5(c)

The operations of union and intersection satisfy the following properties :

Commutativity : $A \cup B = B \cup A; \quad A \cap B = B \cap A.$

Associativity : $A \cup (B \cup C) = (A \cup B) \cup C; \quad A \cap (B \cap C) = (A \cap B) \cap C.$

Distributivity : $A \cap (B \cup C) = (A \cap B) \cup (A \cap C);$

$$A \cup (B \cap C) = (A \cup B) \cap (A \cup C).$$

Idempotency : $A \cup A = A; \quad A \cap A = A.$

They also obey De Morgan's laws :

(a) $(A \cup B)^c = A^c \cap B^c$

For m sets, $A_i, \quad i = 1, \cdots, m$

$$\left(\bigcup_{i=1}^{m} A_i \right)^c = \bigcap_{i=1}^{m} A_i^c.$$

(b) $(A \cap B)^c = A^c \cup B^c$

For m sets, $A_i, i = 1, \cdots, m$

$$\left(\bigcap_{i=1}^{m} A_i \right)^c = \bigcup_{i=1}^{m} A_i^c.$$

❑ **EXAMPLE 2.19.** *Prove that $A \cap (B - C) \subset A - (B \cap C)$.*

Proof. Let $x \in A \cap (B - C)$; i.e., $x \in A$ and $x \in (B - C)$ i.e., $x \in A$, $x \in B$ but $x \notin B \cap C$.

Hence $x \in A - (B \cap C)$. Thus $\{x \in A \cap (B - C)\} \Rightarrow \{x \in (A - B \cap C)\}$.

Hence $A \cap (B - C) \subset A - (B \cap C)$.

To establish that $A \cap (B - C)$ is a proper subset of $A - (B \cap C)$ consider the following example. Let $A = \{1, 4, 5, 7\}$, $B = \{2, 4, 6, 7\}$, $C = \{3, 5, 6, 7\}$. Here $A \cap (B - C) = \{4\}$. $A - (B \cap C) = \{1, 4, 5\}$.

❑ **EXAMPLE 2.20.** *Prove that $A \cap (B \cup C) = (A \cap B) \cup (A \cap C)$.*

(a) To show that $A \cap (B \cup C) \subset (A \cap B) \cup (A \cap C)$: Let $x \in A \cap (B \cup C)$; i.e., $x \in A \cap B$ or $x \in A \cap C$ i.e., $x \in (A \cap B) \cup (A \cap C)$.

(b) To show that $(A \cap B) \cup (A \cap C) \subset A \cap (B \cup C)$: Let $x \in (A \cap B) \cup (A \cap C)$; i.e., $x \in A$ and B or $x \in A$ and C, i.e., $x \in A$ and B or C i.e., $x \in A \cap (B \cup C)$. Hence the proof.

❑ **EXAMPLE 2.21.** *Prove that $(A - B) \cap C \subset A - (B \cap C)$ and construct an example to show that $(A - B) \cap C$ is a proper subset $A - (B \cap C)$.*

Let $x \in (A - B) \cap C$ i.e., $x \in (A - B)$ and C $\Rightarrow$ $x \in A - B \cap C$. Hence $(A - B) \cap C \subset A - (B \cap C)$.

Let $A = \{1, 2, 3, 4\}$, $B = \{3, 4, 5, 6\}$, $C = \{1, 6, 7, 8\}$. $(A - B) \cap C = \{1\}$.$A - B \cap C = \{1, 2, 3, 4\}$.

❑ **EXAMPLE 2.22.** *Prove that $(A - B) \cap (C - D) \subset (A \cap C) - (B \cap D)$.*

Let $x \in (A - B) \cap (C - D)$ i.e., $x \in (A - B)$ and $(C - D)$ i.e., $x \in A, x \notin B$, $x \in C$, $x \notin D$ i.e., $x \in A$, $x \in C$, $x \notin B \cup D$ i.e., $x \in A \cap C$, $x \notin B \cup D$ i.e., $x \in A \cap C - B \cup D$ $\Rightarrow$ $x \in A \cap C - B \cap D$.

❑ **EXAMPLE 2.23.** *Prove that $(A - B) \cap (A - C) \subset A - (B \cap C)$.*

Let $x \in (A - B) \cap (A - C)$ i.e., $x \in A$, $x \notin B$, $x \notin C$ i.e., $x \in A$ and $x \notin B \cup C$ i.e., $x \in A - (B \cup C)$ $\Rightarrow$ $x \in A - (B \cap C)$.

Field of Events : We now construct a field of events F which has the following properties :

(i) F contains the set $\mathcal{S}$ as one of its elements.

(ii) If A, B, subsets of $\mathcal{S}$ are elements of F, so are $A \cup B, A \cap B$, $\bar{A}$ and $\bar{B}$. Thus the null set is contained in F (because $\mathcal{S} \in F$). The union, intersection and complements of a finite number of events belong to F (for further details on this concept see Chapter 3).

❑ **EXAMPLE 2.24.** *A die is thrown. The sample space $\mathcal{S} = \{E_1, \cdots, E_6\}$, E_i denoting the elementary event, roll of point i. The field F consists of $2^6 = 64$ sets, $F = \{(\phi); (E_1), \cdots, (E_6); (E_1, E_2), \cdots, (E_5, E_6); (E_1, E_2, E_3), \cdots, (E_4, E_5, E_6); (E_1, \cdots, E_4), (E_3, \cdots, E_6); (E_1, \cdots, E_5) \cdots, (E_2, \cdots, E_6); (E_1, \cdots, E_6)\}$.*

2.8. Theorems on Probability of Union of Events

We now consider some results which give probability of union of certain events $A_1, \cdots, A_n$ in terms of probabilities of events $A_1, \cdots, A_n$, probabilities of intersection of events, $A_1 A_2,\ A_1 A_3, \cdots$, etc.

• **THEOREM 2.1.** *If $A_1, \cdots, A_n$ are all mutually exclusive then probability of union of events $A_1, \cdots, A_n$ is,*

$$P\left(\bigcup_{i=1}^{n}\right) = P(A_1) + P(A_2) + \cdots + P(A_n) \tag{2.2}$$

Proof. Let M be the total number of mutually exclusive and equally likely cases in the sample space of which m_i are favourable for the event $A_i, i = 1, \cdots, n$. Since $A_1, \cdots, A_n$ are mutually exclusive events, the sets representing them are mutually disjoint. Hence the total number of cases favourbale for the event $\bigcup_{i=1}^{n} A_i \left(= \sum_{i=1}^{n} A_i\right)$ is $\sum_{i=1}^{n} m_i$. Thus

$$P\left(\bigcup_{i=1}^{n} A_i\right) = \frac{m_1 + \cdots + m_n}{M} = \frac{m_1}{M} + \cdots + \frac{m_n}{M} = P(A_1) + \cdots + P(A_n).$$

COROLLARY 2.1. *If an event A consists of n mutually exclusive forms $A_1, \cdots, A_n$ so that A happens whenever any of the these occurs and vice-versa, then $A = A_1 + \cdots + A_n$ and*

$$P(A) = P(A_1) + \cdots + P(A_n)$$

COROLLARY 2.2. *Let the event B imply A. Then the set B is a subset of A. Now A can be decomposed into two disjoint subsets B and $A - B$.*

Hence

$$P(A) = P(B) + P(A - B)$$

Thus

$$P(A - B) = P(A) - P(B),\ B \subset A \tag{2.3}$$

❑ **EXAMPLE 2.25.** *4 cards are drawn randomly from a well-shuffled pack of 52 cards. What is the probability all are black or red.*

Let A_1, A_2 be the events, all the cards are black, red respectively. The two events are mutually exclusive. Here $P(A_1) = P(A_2) = \binom{26}{13} \Big/ \binom{52}{13} = a$ (say). Hence $P(A_1 + A_2) = 2a$.

We now consider the case when $A_1, \cdots, A_n$ are not necessarily mutually exclusive events. For simplicity we introduce the following notations. Let

$$\sum_{i=1}^{n} P(A_i) = S_1, \quad \sum\sum_{i<j=1}^{n} P(A_i A_j) = S_2$$

$$\sum\sum\sum_{i<j<k=1}^{n} P(A_i A_j A_k) = S_3$$

$$\cdots\ \cdots\ \cdots\ \cdots\ \cdots\ \cdots\ \cdots$$

$$P(A_1 A_2 \cdots A_n) = S_n$$

Thus S_1 is the sum of n terms each of probability of an event A_i; S_2 is the sum of $\binom{n}{2}$ terms each of probability of joint occurrance of a pair A_i, A_j; S_3 is the sum of $\binom{n}{3}$ terms each of probability of joint occurrance of a triplet of events A_i, A_j, A_k etc. We have

• **THEOREM 2.2.** *Let $A_1, \cdots, A_n$ be events not necessarily mutually exclusive. The probability that at least one of $A_1, \cdots, A_n$ occurs is*

$$P\left(\bigcup_{i=1}^{n} A_i\right) = S_1 - S_2 + S_3 - \cdots + (-1)^{n-1} S_n \tag{2.4}$$

Proof. (By method of induction). First consider the events A_1, A_2.

Now $A_1 \cup A_2 = A_1 + (A_2 - A_1 \cap A_2)$. Hence by theorem 2.1,

$$P(A_1 \cup A_2) = P(A_1) + P(A_2 - A_1 \cap A_2)$$

Since $A_1 \cap A_2 \subset A_2$, by corollary 2.2

$$P(A_1 \cup A_2) = P(A_1) + P(A_2) - P(A_1 \cap A_2) \tag{2.5}$$

Thus the theorem is true for $n = 2$. Assume (2.4) holds for $n = t (\geq 2)$.

Thus

$$P\left(\bigcup_{i=1}^{t} A_i\right) = \sum_{i=1}^{t} P(A_i) - \sum_{i<j=1}^{t}\sum P(A_iA_j) + \sum_{i<j<k=1}^{t}\sum\sum P(A_iA_jA_k) - \cdots$$

$$+(-1)^{t-1} P(A_1 \cdots A_t) \tag{2.6}$$

Now

$$\bigcup_{i=1}^{t+1} A_i = \bigcup_{i=1}^{t} A_i + \left(A_{t+1} - \left(\bigcup_{1}^{t} A_i\right) \cap A_{t+1}\right)$$

$$= \bigcup_{i=1}^{t} A_i + \left(A_{t+1} - \bigcup_{1}^{t} (A_i \cap A_{t+1})\right)$$

Hence

$$P\left(\bigcup_{i=1}^{t+1} A_i\right) = P\left(\bigcup_{i=1}^{t} A_i\right) + P\left(A_{t+1} - \bigcup_{1}^{t} (A_i \cap A_{t+1})\right)$$

$$= P\left(\bigcup_{i=1}^{t} A_i\right) + P(A_{t+1}) - P\left(\bigcup_{1}^{t} (A_i \cap A_{t+1})\right) \tag{2.7}$$

Since, by assumption, the theorem holds for t sets,

$$P\left(\bigcup_1^t (A_i \cap A_{t+1})\right) = \sum_1^t P(A_i \cap A_{t+1}) - \sum_{i<j=1}^{t}\sum P\left\{(A_i \cap A_{t+1})(A_j \cap A_{t+1})\right\}$$

$$+\sum_{i<j<k=1}^{t}\sum\sum P\left\{(A_i \cap A_{t+1})(A_j \cap A_{t+1})(A_k \cap A_{t+1})\right\}$$

$$- \cdots + (-1)^{t-1} P\left\{(A_1 \cap A_{t+1}) \cdots (A_t \cap A_{t+1})\right\} \tag{2.8}$$

Now

$$(A_i \cap A_{t+1})(A_j \cap A_{t+1}) = A_i A_j A_{t+1}$$

$$(A_i \cap A_{t+1})(A_j \cap A_{t+1})(A_k \cap A_{t+1}) = A_i A_j A_k A_{t+1} \text{ etc.}$$

Using in (2.7) the expressions in (2.6) and (2.8), after rearrangement of terms, we get

$$P\left(\bigcup_{i=1}^{t+1} A_i\right) = \sum_{i=1}^{t+1} P(A_i) - \sum_{i<j=1}^{t+1}\sum P(A_i A_j)$$

$$+\sum_{i<j<k=1}^{t+1}\sum\sum P(A_i A_j A_k) - \cdots + (-1)^t P(A_1 \cdots A_{t+1})$$

Hence, if the theorem holds for $n = t$, it holds for $n = t + 1$. The theorem has been shown to hold for $n = 2$. Hence it holds for all integral values of n.

An alternative Proof.

To calculate $P\left(\bigcup_{i=1}^{n} A_i\right)$ one should add the probabilities of all elementary points contained in at least one $A_i (i = 1, \cdots, n)$, each point being taken only once. Consider an arbitrary point E contained in exactly m of the sets A_i, say, $A_1, A_2, \cdots, A_m$ and find its contribution in the rhs of (2.4). $P(E)$ occurs m times in S_1, since S_1 is the sum of the probabilities $P(A_1), \cdots, P(A_m)$. Similarly it occurs $\binom{m}{2}$ times out of $\binom{n}{2}$ terms in S_2, $\binom{m}{3}$ times in $S_3, \cdots$ $\binom{m}{j}$ times in $S_j, \cdots,$ $\binom{m}{m}$ times in S_m. Its contribution to $S_{m+1}, \cdots, S_n$ is zero since E occurs only in m of the sets $A_1, \cdots, A_n$. Therefore coefficient of $P(E)$ in the expression in the rhs of (2.4) is

$$\binom{m}{1} - \binom{m}{2} + \cdots \pm \binom{m}{m}$$

which equals 1. Thus each $P(E)$ occurs only once in rhs of (2.4). Hence the proof.

COROLLARY 2.3. *For n events $A_1, A_2, \cdots, A_n$*

$$P\left(\bigcap_i^n A_i\right) = \sum_i P(A_i) - \sum_{i<j}\sum P(A_i \cup A_j) + \sum_{i<j<k}\sum\sum P(A_i \cup A_j \cup A_k) - \cdots + (-1)^{n-1} P(A_1 \cup A_2 \cdots \cup A_n).$$

Proof. We have

$$P\left(\bigcap_1^n A_i\right) = 1 - P\left(\bigcup_i A_i^c\right).$$

Now

$$P\left(\bigcup_1^n A_i^c\right) = \sum P(A_i^c) - \sum_{i<j}\sum P(A_i^c A_j^c) + \sum_{i<j<k}\sum\sum P(A_i^c A_j^c A_k^c) - \cdots + (-1)^{n-1} P(A_1^c A_2^c \cdots A_n^c) \qquad \text{(i)}$$

By De Moivre's laws,

$$\cap A_i^c = (\cup A_i)^c.$$

Hence

$$P(A_i^c A_j^c) = 1 - P(A_i \cup A_j)$$
$$P(A_i^c A_j^c A_k^c) = 1 - P(A_i \cup A_j \cup A_k), \text{ etc.}$$

Hence from (i)

$$1 - P\left(\bigcap_1^n A_i\right) = \sum \{1 - P(A_i)\} - \sum_{i<j}\sum \{1 - P(A_i \cup A_j)\} + \sum_{i<j<k}\sum\sum \{1 - P(A_i \cup A_j \cup A_k)\} - \cdots + (-1)^{n-1}\left\{1 - P\left(\bigcup_1^n A_i\right)\right\}$$

$$= \binom{n}{1} - \binom{n}{2} + \binom{n}{3} + \cdots + (-1)^{n-1}\binom{n}{n} - \sum P(A_i) + \sum_{i<j}\sum P(A_i \cup A_j) - \sum_{i<j<k}\sum\sum P(A_i \cup A_j \cup A_k) + \cdots + (-1)^n P\left(\bigcup_1^n A_i\right)$$

when the result follows.

❑ **EXAMPLE 2.26.** ***Find the probability that in a bridge game at least one of the players will get a complete suit of cards.***

Let A_i be the event that player i gets a complete suit of cards.

Thus

$$P\left(\bigcup_1^4 A_i\right) = \sum_1^4 P(A_i) - \sum_{i<j=1}^4\sum P(A_i A_j) + \sum_{i<j<k=1}^4\sum\sum P(A_i A_j A_k) - P(A_1 A_2 A_3 A_4).$$

Let

$$P(A_i) = \frac{4}{\binom{52}{13}} = p_1, \qquad P(A_iA_j) = \frac{4.3}{\binom{52}{13}\binom{39}{13}} = p_2,$$

$$P(A_iA_jA_k) = \frac{4.3.2}{\binom{52}{13}\binom{39}{13}\binom{26}{13}} = p_3, \quad P(A_1A_2A_3A_4) = \frac{4!}{\binom{52}{13}\binom{39}{13}\binom{26}{13}} = p_4$$

Hence required probability$= 4p_1 - 6p_2 + 4p_3 - p_4$.

❑ **EXAMPLE 2.27.** *Matching problem : Suppose n cards marked $1, 2, \cdots, n$ are shuffled and laid out against n fixed positions, also marked $1, 2, \cdots n$. Consider the number of matches—that is, the number of times a card marked i is in position i, $i = 1, \cdots, n$. Find the probability that (a) there is at least one match (b) there are exactly r matches.*

(a) We first find the probability there will be at least one match. Let A_i be the event that a match takes place at the ith place.

Thus A_iA_j is the event that there is a match at the ith position and a match at the *jth* position, etc. Now

$$P(A_i) = \frac{(n-1)!}{n!} = \frac{1}{n}$$

because if the entry at the ith position is kept fixed at i, the remaining positions can be filled up in $(n-1)!$ ways out of the total number of $n!$ equally likely ways of distributing the cards.

Similarly,

$$P(A_iA_j) = \frac{(n-2)!}{n!} = \frac{1}{n(n-1)},$$

$$\cdots\cdots\cdots\cdots\cdots\cdots\cdots\cdots$$

$$P(A_{i_1}, \cdots A_{i_r}) = \frac{1}{n(n-1)\cdots(n-r+1)}.$$

Hence

$$P\left(\bigcup_{i=1}^{n} A_i\right) = S_1 - S_2 + S_3 - \cdots + (-1)^{n-1}S_n$$

$$= 1 - \frac{1}{2!} + \frac{1}{3!} - \cdots + (-1)^{n-1}\frac{1}{n!} = P_1(\text{say}).$$

$1 - P_1 = p_n$ (say) is the first $(n+1)$ terms in the expansion of e^{-1}. Thus, the probability of no matches in n positions, $p_n \to e^{-1} = 0.36788$ as $n \to \infty$. Hence $P_1 \to 0.63212$ as $n \to \infty$. We have $|p_n - e^{-1}| \leq \frac{1}{(n+1)!}$.

The probability p_n is almost independent of n.

(b) Let N_r, n be the number of sample points with exactly r matches out of n positions. Required probability $= \frac{N_r, n}{n!}$

To evaluate N_r, n let us assume first that the r positions are fixed, Number of sample points with no matches in the remaining $(n-r)$ positions is $N_o, (n-r)$. Again the r positions can be chosen in $\binom{n}{r}$ ways. Thus $N_r, n = \binom{n}{r} N_o, (n-r)$. Hence the probability is

$$\frac{\binom{n}{r} N_o, (n-r)}{n!} = \frac{N_o, (n-r)}{r!(n-r)!} = \frac{p_{n-r}}{r!}$$

$$= \frac{1}{r!}\left[1 - \frac{1}{11} + \frac{1}{2!} - \cdots + (-1)^{n-r}\frac{1}{(n-r)!}\right].$$

• **THEOREM 2.3.** *(Boole's inequality). Given $n(>1)$ events $A_1, \cdots, A_n$*

$$P\left(\bigcup_{i=1}^{n} A_i\right) \leq \sum_{i=1}^{n} P(A_i) \tag{2.9}$$

Proof. Consider first A_1 and A_2. Now

$$A_1 \cup A_2 = A_1 + (A_2 - A_1),$$

the events $A_1, A_2 - A_1$ being disjoint. Hence

$$P(A_1 \cup A_2) = P(A_1) + P(A_2 - A_1)$$
$$\leq P(A_1) + P(A_2)$$

since $(A_2 - A_1) \subset A_2$. Thus the inequality is proved for $n = 2$. Now

$$\bigcup_{i=1}^{n} A_i = \left(\bigcup_{i=1}^{n-1} A_i\right) \bigcup A_n$$

$$\Rightarrow \quad P\left(\bigcup_{i=1}^{n} A_i\right) \leq P\left(\bigcup_{i=1}^{n-1} A_i\right) + P(A_n) \leq P\left(\bigcup_{i=1}^{n-2} A_i\right) + P(A_{n-1}) + P(A_n)$$

$$\cdots\cdots\cdots\cdots\cdots\cdots\cdots\cdots$$

$$\leq \sum_{i=1}^{n} P(A_i).$$

COROLLARY 2.4. *Given n events $A_1, \cdots, A_n$*

$$P\left(\bigcap_{i=1}^{n} A_i\right) \geq 1 - \sum_{i=1}^{n} P(A_i^c) = \sum_{i=1}^{n} P(A_i) - (n-1)$$

Proof. Since $\bigcap_{i=1}^{n} A_i = \left(\bigcup_{i=1}^{n} A_i^c\right)^c$

$$P\left(\bigcap_{i=1}^{n} A_i\right) = 1 - P\left(\bigcup_{i=1}^{n} A_i^c\right) \geq 1 - \sum_{i=1}^{n} P(A_i^c)$$

by (2.9). The second expression follows by putting $P(A_i^c) = 1 - P(A_i)$.

• **THEOREM 2.4.** *(Bonferroni's inequality). Given $n(>1)$ events $A_1, \cdots, A_n$,*

$$\sum_{i=1}^{n} P(A_i) - \sum_{i<j=1}^{n}\sum P(A_iA_j) \leq P\left(\bigcup_{i=1}^{n} A_i\right) \leq \sum_{i=1}^{n} P(A_i) \tag{2.10}$$

We only prove the left side. The right side follows by (2.9). For $n = 2$.

$$P(A_1) + P(A_2) - P(A_1A_2) = P(A_1 \cup A_2)$$

and thus the inequality is true for $n = 2$ (by equality). For $n = 3$

$$P\left(\bigcup_{i=1}^{3} A_i\right) = \sum_{i=1}^{3} P(A_i) - \sum_{i<j=1}^{3}\sum P(A_iA_j) + P(A_1A_2A_3)$$

and the result holds. Assuming that it holds for m events $(3 < m \leq n-1)$, we show that it holds for $m+1$.

$$\begin{aligned}
P\left(\bigcup_{i=1}^{m+1} A_i\right) &= P\left(\bigcup_{i=1}^{m} A_i \bigcup A_{m+1}\right) \\
&= P\left(\bigcup_{i=1}^{m} A_i\right) + P(A_{m+1}) - P\left\{A_{m+1} \bigcap \left(\bigcup_{i=1}^{m} A_i\right)\right\} \\
&\geq \left(\sum_{i=1}^{m} P(A_i) - \sum_{i<j=1}^{m}\sum P(A_iA_j) + P(A_{m+1}) - P\left(\bigcup_{i=1}^{m}(A_i \cap A_{m+1})\right)\right) \\
&\geq \sum_{i=1}^{m+1} P(A_i) - \sum_{i<j=1}^{m}\sum P(A_iA_j) - \sum_{i=1}^{m} P(A_iA_{m+1})
\end{aligned}$$

(by Boole's inequalities).

$$= \sum_{i=1}^{m+1} P(A_i) - \sum_{i<j=1}^{m+1}\sum P(A_iA_j).$$

• **THEOREM 2.5.** *The probability that exctly r of the events $A_1, \cdots, A_n$ occur is, $(1 \leq r \leq n)$,*

$$P_{[r]} = S_r - \binom{r+1}{r} S_{r+1} + \binom{r+2}{r} S_{r+2} - \cdots \pm \binom{n}{r} S_n \tag{2.11}$$

Proof. To calculate $P_{[r]}$ we should add the probabilities of all points which occur in exactly r of the sets $A_1, \cdots, A_n$, each point taken only once. Let E be an arbitrary point contained in exactly m of the events $A_1, \cdots, A_n$. We calculate its contribution to the expression in the rhs of (2.11).

If $m < r, P(E)$ does not contribute. If $m = r$, $P(E)$ occurs only in S_m and its coefficient in the rhs of (2.11) is 1. For $m > r$, coefficient of $P(E)$ in the rsh of (2.11) is

$$\binom{m}{r} - \binom{r+1}{r}\binom{m}{r+1} + \binom{r+2}{r}\binom{m}{r+2} \cdots \pm \binom{m}{r}\binom{m}{m} \tag{2.12}$$

since $P(E)$ occurs $\binom{m}{r}$ times in S_r, $\binom{m}{r+1}$ times in S_{r+1}, etc.

Now (2.12)

$$= \binom{m}{r}\left\{1 - \binom{m-r}{1} + \binom{m-r}{2} - \cdots \pm \binom{m-r}{m-r}\right\} = 0.$$

Hence the proof.

COROLLARY 2.5. *The probability that at least r of the n events $A_1, \cdots, A_n$ occur is*

$$P_{[r]} + P_{[r+1]} + \cdots + P_{[n]}.$$

❑ **EXAMPLE 2.28.** *Considering the matching problem of example 2.27, probabiity of exactly r match is*

$$\frac{1}{r!} - \binom{r+1}{r}\frac{1}{(r+1)!} + \cdots \pm \binom{n}{r}\frac{1}{n!} = \frac{1}{r!}\left[1 - \frac{1}{1!} + \frac{1}{2!} - \cdots \pm \frac{1}{(n-r)!}\right]$$

as obtained earlier. Similarly probability of at least r matches can be obtained.

2.9. Conditional Probability : Theorem of Compound Probability

Suppose in the sample space of a random experiment there are N mutually exclusive, exhaustive and equally likely cases. Of these, N_A are favourable to event A. Out of these N_A cases N_{AB} cases are favourable to B. Thus N_{AB} cases are favourable to both A and B. (Fig. 6.). The proportion N_{AB}/N_A is the proportion of cases among the cases favourable to A which are favourable to B and is defined as the conditional probability of occurrence of B given that the event A has occurred. This is denoted as $P(B/A)$. The proportion N_{AB}/N is the probability of joint occurrence of the events A and B and is $P(AB)$.

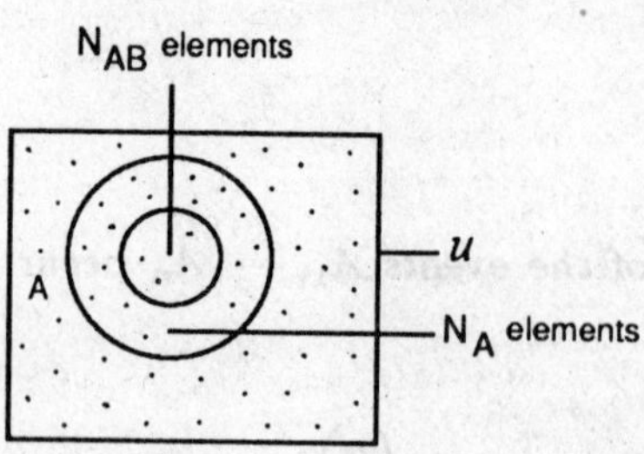

Fig. 6

Now $\quad \dfrac{N_{AB}}{N} = \left(\dfrac{N_A}{N}\right)\left(\dfrac{N_{AB}}{N_A}\right)$ i.e. $P(AB) = P(A)P(B/A)$ (2.13)

Hence we have

• **THEOREM 2.6.** *The probability of the simultaneous occurrence of two events A and B is equal to the probability of A multiplied by the conditional probability of B given that A has occurred (it is also equal to the probability of B multiplied by the conditional probability of A given that B has occurred).*

Clearly (2.13) can also be written as

$$P(AB) = P(B)P(B/A) \tag{2.14}$$

Whether one should use (2.13) or (2.14) depends on the physical conditions of the experiment,–whether it is possible for A to occur before B or otherwise. From (2.13) we have the conditional probability

$$P(B/A) = \frac{P(AB)}{P(A)} \tag{2.15}$$

provided $P(A) > 0$. The conditional probability $P(B/A)$ is not defined if $P(A) = 0$.

COROLLARY 2.6. *For three events A_1, A_2, A_3,*

$$\begin{aligned} P(A_1A_2A_3) &= P(A_1A_2)P(A_3|A_1A_2) \\ &= P(A_1)P(A_2|A_1)P(A_3|A_1A_2) \quad \text{provided } P(A_1A_2) > 0. \end{aligned}$$

Similarly for four events A_1, A_2, A_3, A_4

$$\begin{aligned} P(A_1A_2A_3A_4) &= P(A_1A_2A_3)P(A_4|A_1A_2A_3) \\ &= P(A_1)P(A_2|A_1)P(A_3|A_1A_2)P(A_4|A_1A_2A_3), \end{aligned}$$

provided $P(A_1A_2A_3) > 0$.

In general for m events $A_1, \cdots, A_m$,

$$P(A_1 \cdots A_m) = P(A_1)P(A_2|A_1) \cdots P(A_m|A_1 \cdots A_{m-1}) \tag{2.16}$$

provided $P(A_1 \cdots A_{m-1}) > 0$.

❑ **EXAMPLE 2.29.** *An urn contains 6 red and 4 black balls. Two balls are drawn without replacement. What is the probability that the second ball is red if it is known that the first is red.*

Let A and B be the events that the first and the second ball are red respectively. Hence $P(AB) = \binom{6}{2} \Big/ \binom{10}{2} = \dfrac{1}{3}$. $P(A) = \dfrac{6}{10}$. Thus $P(B|A) = \dfrac{5}{9}$.

❑ **EXAMPLE 2.30.** *Two unbiased dice are thrown. Find the conditional probability that two fives occur if it is known that the total is divisible by 5 ?*

Let A be the event the total is divisible by 5, B be the event two fives occur; then AB is the event two fives occur. $P(A) = \frac{7}{36}$, $P(AB) = \frac{1}{36}$. Hence $P(B|A) = \frac{1}{7}$.

❑ **EXAMPLE 2.31.** *A die is loaded in such a way that the probability of a given number turning up is proportional to that number (e.g. a 6 is twice as probable as 3). What is the probability of a 3, given that an odd number is rolled ?*

Here, probability of getting number i is $\frac{i}{21}$, since $\sum_{1}^{6} j = 21$. Let A be the event, an odd number turns up, B the event 3 is obtained. Thus $P(B|A) = P(AB)/P(A) = \frac{1}{3}$.

• **THEOREM 2.7.** *Let $P(AC) > 0$. Then*

$$P(AB|C) = P(A|C)P(B|AC). \tag{2.17}$$

Proof. We notice that the conditional probabilities involved in (2.17) are all defined, since $AC \subset C$, $P(AC) \le P(C)$ and thus $P(C) > 0$, because $P(AC) > 0$.

$$\text{Now} \quad P(AB|C) = \frac{P(ABC)}{P(C)} = \frac{P(B|AC)P(AC)}{P(C)} = P(B|AC)P(A|C).$$

We now consider the theorem of total probability.

• **THEOREM 2.8.** *Suppose the event A can occur only along with the event B. Suppose also B can occur only in m mutually exclusive ways $B_1, \cdots, B_m$. Then*

$$P(A) = \sum_{i=1}^{m} P(B_i)P(A|B_i) \tag{2.18}$$

provided $P(B_i) > 0$, $i = 1, \cdots, m$.

Proof. The events $AB_1, \cdots, AB_m$ are mutually exclusive. Thus $P(A) = P(AB) = \sum_{i=1}^{m} P(AB_i) = \sum_{i=1}^{m} P(B_i)P(A|B_i)$. (Actually, (2.18) remains valid even if $P(B_i) = 0$ for some i, when its contribution to the rhs of (2.18) is zero.)

COROLLARY 2.7. *For any event B which can occur along (not necessarily always) with A, $P(A) = P(AB) + P(AB^c)$.*

❑ **EXAMPLE 2.32.** *An urn contains a white balls and b black balls; another contains c white balls and d black balls. One ball is transferred from the first into the second and then one ball is drawn from the later. What is the probability that it is a white ball ?*

The ball transferred from the first urn may be a black (B) one or a white (W) one. Thus the possible events are BW, WW. Now

$$P(BW) = \frac{bc}{(a+b)(c+d+1)}, \quad P(WW) = \frac{a(c+1)}{(a+b)(c+d+1)}.$$

Hence the required probability $= P(BW) + P(WW) = \dfrac{ac+bc+a}{(a+b)(c+d+1)}$.

☐ **EXAMPLE 2.33.** *Two players play a game as follows. Taking turns, they draw the balls out of an urn containing 'a' wihite and 'b' black balls, one ball at a time. He who extracts the first white one wins the game. What is the probability that the player who starts the game will win the game.*

Let A, B be two players, A starting the game. Probability of A to win $=P(W) + P(BBW) + P(BBBBW) + \cdots$

$$= \frac{a}{a+b}\left[1+\frac{b(b-1)}{(a+b-1)(a+b-2)}+\frac{b(b-1)(b-2)(b-3)}{(a+b-1)(a+b-2)(a+b-3)(a+b-4)}+\cdots\right].$$

2.10. Independence of Events

We have seen in the previous section that $P(B|A)$ does not generally equal $P(B)$. Thus the information whether A has happened changes the probability of occurrence of B. If, however, $P(B|A) = P(B)$, probabilities of B does not depend on whether A has or has not happened. In this case events A, B are said to be stochastically independent. Hence we have the following definition.

Independence of a pair of events : Two events A and B are said to be independent if the conditional probability of occurrence of B given the occurrence of A is equal to the unconditional probability of B i.e., if $P(B|A) = P(B)$, provided $P(B|A)$ is defined.

Thus for independent events A and B,

$$P(AB) = P(A)P(B|A) = P(A)P(B).$$

It follows that

$$P(B|A) = P(B) \Leftrightarrow P(A|B) = P(A),$$

because

$$P(A)P(B|A) = P(AB) = P(B)P(A|B).$$

Thus the independence of A and B implies that the unconditional probability of either event is equal to its conditional probability given the other event.

Events A and B are stochastically independent if and only if

$$P(AB) = P(A)P(B). \tag{2.19}$$

The definition of independence of events can easily be extended to any finite number of events. Given n events $A_1, \cdots, A_n$, $P(A_i|A_1 \cdots A_{i-1}A_{i+1} \cdots A_n)$ denotes the conditional probability of A_i given that the events $A_1, \cdots, A_{i-1}, A_{i+1}, \cdots, A_n$ have occurred.

$$P(A_i|A_1 \cdots A_{i-1}A_{i+1} \cdots A_n) = \frac{P(A_1 \cdots A_n)}{P(A_1 \cdots A_{i-1}A_{i+1} \cdots A_n)}$$

if $P(A_1 \cdots A_{i-1}A_{i+1} \cdots A_n) > 0$.

Independence of a number of events : n events $A_1, \cdots, A_n$ are said to be independen[t] if the conditional probabilities of any one of these events, say A_i, given one or more o[f] the remaining events, is equal to the unconditional probability of A_i.

• **THEOREM 2.9.** *(Compound probability for independent events) If $A_1, \cdots A_n$ ar[e] independent events,*

$$P(A_1 \cdots A_n) = P(A_1)P(A_2) \cdots P(A_n).$$

Proof. We have, for any set of events $A_1, \cdots, A_n$

$$P(A_1 \cdots A_n) = P(A_1)P(A_2|A_1) \cdots P(A_n|A_1 \cdots A_{n-1}).$$

Since $A_1, \cdots, A_n$ are independent we have

$$P(A_2|A_1) = P(A_2), \cdots, P(A_n|A_1 \cdots A_{n-1}) = P(A_n).$$

The theorem follows.

Three events A, B, C are independent, by definition, if

$$P(A|BC) = P(A|B) = P(A|C) = P(A)$$
$$P(B|AC) = P(B|A) = P(B|C) = P(B)$$
$$P(C|AB) = P(C|A) = P(C|B) = P(C).$$

It is easy to verify three events, A, B, C are stochastically independent if and only i[f] they are pairwise independent i.e. if

$$\left.\begin{array}{ll} P(AB) = P(A)P(B), \ P(AC) = P(A)P(C), \ P(BC) = P(B)P(C) & \\ \text{and if also} \quad P(ABC) = P(A)\,P(B)\,P(C) & \end{array}\right\} \quad (2.20)$$

Similarly m events, $A_1, \cdots, A_m$ are stochastically independent if the following con[-]ditions hold.

$$\begin{array}{ll} P(A_iA_j) = P(A_i)\,P(A_j), & i < j = 1, \cdots, m \\ P(A_iA_jA_k) = P(A_i)\,P(A_j)\,P(A_k), & i < j < k = 1, \cdots, m \\ \cdots\cdots\cdots\cdots & \\ P(A_1 \cdots A_m) = P(A_1)\,P(A_2) \cdots P(A_m). & \end{array} \quad (2.21)$$

The conditions (2.21) are $\binom{m}{2} + \binom{m}{3} + \cdots + \binom{m}{m} = 2^m - m - 1$ in number.

❑ **EXAMPLE 2.34.** *Consider families with 3 children. Sample space, here, con[-]sists of 8 points $\{(bbb), (bbg), (bgb), (bgg), (gbb), (gbg), (ggb), (ggg)\}$, where '$b$' stands for a boy and '$g$' for a girl. Assum[e] the cases are all equally likely. Let A be the event that the fam[-]ily has children of both sexes and B the event three is at mos[t] one girl. Examine if A and B are stochastically independent.*

Here $P(A) = \frac{3}{4}$, $P(B) = \frac{1}{2}$. The event AB means the family has exactly one girl. $P(AB) = \frac{3}{8} = P(A)P(B)$. Thus A, B are independent.

❑ **EXAMPLE 2.35.** *The events A, B are independent. Determine whether the events (a) A and $\bar{B}$ (b) $\bar{A}$ and $\bar{B}$ are independent.*

(a) We have $(A\bar{B}) \cup (AB) = A$; also $A\bar{B}$ and AB are disjoint sets. Thus $P(A) = P(A\bar{B}) + P(AB)$.

$$\text{Now } P(A)\,P(\bar{B}) = P(A)\big(1 - P(B)\big) = P(A) - P(AB)$$
$$= P(A\bar{B}).$$

Thus, $A, \bar{B}$ are independent,

(b) We have $(A\bar{B}) \cup (\bar{A}\bar{B}) = \bar{B}$; Also $P(\bar{B}) = P(A\bar{B}) + P(\bar{A}\bar{B})$.

$$\text{Now } P(\bar{A})\,P(\bar{B}) = P(\bar{B})\{1 - P(A)\} = P(\bar{B}) - P(\bar{B}A) \text{ by (a)}$$
$$= P(\bar{A}\bar{B}).$$

Hence $\bar{A}, \bar{B}$ are independent.

Similarly $\bar{A}$ and B are independent.

❑ **EXAMPLE 2.36.** *Two dice are tossed. Let A denote the event of an odd total, B that of an ace on the first die, C that of having a total of seven. Are A, B, C mutually independent ?*

Here $P(A) = \frac{1}{2}$, $P(B) = \frac{1}{6}$, $P(C) = \frac{1}{6}$, $P(AB) = \frac{1}{12}$, $P(AC) = \frac{1}{6}$, $P(BC) = \frac{1}{36}$, $P(ABC) = \frac{1}{36}$. Thus A, B, C, are not mutually independent.

❑ **EXAMPLE 2.37.** *A die is thrown twice; Let x_1, x_2 be the face values turned up. Consider the following events :*

$A_1 = \{x_1 \text{ is divisible by 2, } x_2 \text{ is divisible by 3}\}$,

$A_2 = \{x_1 \text{ is divisible by 3, } x_2 \text{ is divisible by 2}\}$

$A_3 = \{x_1 \text{ is divisible by } x_2\}$

$A_4 = \{x_2 \text{ is divisible by } x_1\}$

$A_5 = \{x_1 + x_2 \text{ is divisible by 2}\}$

$A_6 = \{x_1 + x_2 \text{ is divisible by 3}\}$

Find all pairs A_i, A_j and triplets A_i, A_j, A_k of mutually independent events.

The pair A_i, A_j where $(i, j) \in \{1, 2, 5, 6\}$ are independent, The events A_1, A_5, A_6 are stochastically independent. The events A_2, A_5, A_6 are stochastically independent.

❑ **EXAMPLE 2.38.** *This is an example of pairwise independent events that are no independent. An urn contains four tickets bearing number 1234, 2341, 3412 and 4123 and one ticket is drawn. Let the events A, B, C be as follows :*

A=1st digit of the ticket drawn is 1 or 4.

B=2nd digit of the ticket drawn is 2 or 4.

C=3rd digit of the ticket drawn is 3 or 4.

Thus $P(A) = P(B) = P(C) = \frac{1}{2}, P(AB) = P(AC) = P(BC) = \frac{1}{4},\ P(ABC) = \frac{1}{4}$

Obviously A,B,C are pairwise independent but $P(ABC) \neq P(A)P(B)P(C)$. Hence A,B,C are not independent.

❑ **EXAMPLE 2.39.** *A total of n shells is fired at a target. The probability of the ith cell hitting the target is p_i, $i = 1, \cdots, n$. Find the probability that at least two cells out of n find the target.*

Required probability

$$P = 1 - \text{Prob.[at most one cell hits the target]}$$

$$= 1 - \text{Prob.[no cell hits the target]-Prob.[exactly one cell hits the target]}$$

$$= 1 - P_1 - P_2 \text{ (say).}$$

$$P_1 = q_1 q_2 \cdots q_n, \quad q_i = 1 - p_i, \quad i = 1, \cdots, n$$

$$P_2 = q_1 \cdots q_n \left[\frac{p_1}{q_1} + \frac{p_2}{q_2} + \cdots + \frac{p_n}{q_n}\right]$$

$$\text{Hence } P = 1 - \prod_{i=1}^{n} q_i \left\{1 + \sum_{k=1}^{n} \frac{p_k}{q_k}\right\}$$

❑ **EXAMPLE 2.40.** *In each of a set of games it is 2 to 1 in favour of the winner of the previous game. What is the chance that the player who wins the first game shall win at least 3 of the next 4 games.*

Required probability = prob. of winning 3 games + prob. of winning 4 games

$$= \binom{4}{3}\left(\frac{2}{3}\right)^3\left(\frac{1}{3}\right) + \left(\frac{2}{3}\right)^4 = \frac{16}{27},$$

since 3 games to be won can be chosen in $\binom{4}{3}$ ways, probability of a win is $\frac{2}{3}$, a failure is $\frac{1}{3}$ and the trials are independent.

2.11. Bayes Theorem

Suppose an event A can occur if and only if one of the hypotheses $B_1, \cdots B_k$ is true. The probability $P(B_i)$ of occurrence of B_i is known for each $i, i = 1, \cdots k$. Also known is the conditional probability $P(A|B_i)$ of occurrence of A given that B_i has already occurred, $i = 1, \cdots, k$. We want to find the conditional probability $P(B_i|A)$ of occurrence of B_i given that A has already occurred. This is given in Bayes theorem that follows.

To fix the idea suppose that a scientist observes a certain event A. He considers that A can happen only if one of the hypotheses $B_1, \cdots, B_k$ holds good. Before observing A, he assigns certain probabilities $P(B_1), \cdots, P(B_k)$ of these hypotheses to be true. He also knows the conditional probability $P(A|B_i)$ of occurrence of A when B_i is true. Now that he has observed the event A, how is he going to change his probabilities to different hypotheses $B_1, \cdots, B_k$? This is obtained by the conditional probability $P(B_i|A)$.

Generally, $P(B_i)$ and $P(B_i|A)$ will not be the same. Thus the occurrence of A generally changes one's assignment of probabilities to different hypotheses. The probabilities $P(B_i)$ which are assigned to B_i without any reference to A are called ' a priori' probabilities, $i = 1, \cdots, k$. The probabilities $P(B_i|A)$ which are calculated after A has been observed are called 'a posteriori' probabilities, $i = 1, \cdots, k$. Our main interest lies here in the hypotheses $B_1, \cdots, B_k$.

THEOREM 2.10. *Bayes theorem. Let an event A occur only if one of the hypotheses* $B_1, \cdots B_k$ *is true. Known are the probabilities* $P(B_1), \cdots, P(B_k)$ *of occurrence of* $B_1, \cdots, B_k$ *respectively. The conditional probabilities* $P(A|B_i)$, $i = 1, \cdots, k$, *are also known. The posterior probability* $P(B_i|A)$ *is given by*

$$P(B_i|A) = \frac{P(B_i)\, P(A|B_i)}{\sum_{j=1}^{k} P(B_j)\, P(A|B_j)}$$

provided at least one $P(B_t) > 0$, $t = 1, \cdots, k$.

Proof. We have

$$P(AB_i) = P(B_i)\, P(A|B_i) = P(A)\, P(B_i|A)$$

Hence

$$P(B_i|A) = \frac{P(B_i)\, P(A|B_i)}{P(A)}$$

Now

$$P(A) = P(AB_1) + P(AB_2) + \cdots + P(AB_k) = \sum_{j=1}^{k} P(B_j)P(A|B_j) \quad \text{by (2.18).}$$

Hence the theorem.

EXAMPLE 2.41. *The first of three urns contains 7 white and 10 black balls, the second contains 5 white and 12 black balls and the third contains 17 white balls (and no black ball). A person chooses an urn at random and draws a ball from it. The ball is white. Find the probabilities that the ball came from (i) the first (ii) the second (iii) the third urn.*

Let H_i be the hypothesis that the ith urn was chosen and E be the event a white ball is drawn.

$$P(H_i) = \frac{1}{3},\ i = 1, 2, 3,\ P(E|H_1) = \frac{7}{17},\ P(E|H_2) = \frac{5}{17},\ P(E|H_3) = 1$$

$$P(H_1|E) = \frac{\frac{7}{51}}{\frac{7}{51} + \frac{5}{51} + \frac{1}{3}} = \frac{7}{29},\ P(H_2|E) = \frac{5}{29},\ P(H_3|E) = \frac{17}{29}.$$

❑ **EXAMPLE 2.42.** *Two production lines manufacture the same type of items. In a given time line 1 turns out n_1 items of which n_1p_1 are defectives; in the same time, line 2 turns out n_2 items of which n_2p_2 are defectives. Suppose a unit is selected at random from the combined lot produced by the two lines. Let D be the event of a defective item, A the event the unit was produced by line 1 and B the event it was produced by line 2. Determine $P(A|D)$, $P(B|D)$.*

$$P(A) = \frac{n_1}{N},\quad P(B) = \frac{n_2}{N} \quad \text{where} \quad N = n_1 + n_2$$

$$P(D|A) = p_1, P(D|B) = p_2$$

Hence $$P(A|D) = \frac{P(A)\ P(D|A)}{P(A)\ P(D|A) + P(B)P(D|B)} = \frac{n_1p_1}{n_1p_1 + n_2p_2}$$

Similarly, $$P(B|D) = \frac{n_2p_2}{n_1p_1 + n_2p_2}$$

If $n_1p_1 > n_2p_2;\quad p(A|D) > P(B|D).$

2.12. Binomial Probability

Consider an experiment having only two outcomes, success (S) and failures (F) occurring with probabilities p and $q(= 1 - p)$ respectively. Such a trial is called a Bernoulli trial. Consider n such experiments conducted independently. We have thus n independent experiments each having two outcomes S and F with probabilities p and q respectively. An example is that of throwing a fair die n times, turning up a 6 being reckoned as S and any other point a F when $p = \frac{1}{6}$, $q = \frac{5}{6}$. We are interested in finding the probability of having r successes out of n such trials. The probability is

$$p_r = \binom{n}{r} p^r q^{n-r} \tag{2.22}$$

because the r trials to result in successes may be chosen in $\binom{n}{r}$ ways. Again any sequence of S's and F's with r S's and $(n-r)$ F's has the probability $p^r q^{n-r}$, because of independence of trials. The probability (2.22) is called the binomial probability.

❑ **EXAMPLE 2.43.** *A die is tossed 8 times. What is the probability of having 2 sixes ? at least 2 sixes ?*

Here probability of success is $\frac{1}{6}$ and that of a failure is $\frac{5}{6}$. Hence probability of having 2 sixes is $\binom{6}{2}\left(\frac{1}{6}\right)^2\left(\frac{5}{6}\right)^4 = .20094$ and at least 2 six is $\sum_{x=2}^{6}\binom{6}{x}\left(\frac{1}{6}\right)^x\left(\frac{5}{6}\right)^{6-x}$

2.13. Limitations of Classical Definition

Though the classical definition of probability is easy to understand it suffers from some serious limitations as follows.

It is limited to situations in which there is only a finite number of possible outcomes. Consider an experiment in which trials are performed until a particular event occurs, say, tossing a coin until a head occurs (exemple 2.3). Here, as has been seen, the sample space contains a countably infinite number of points rendering N infinite. Thus the probability of the event cannot be defined by classical definition. In practice such an experiment will terminate in a finite number of trials, but there is no 'a priori' assurance that this will happen.

It is desirable, both for theoretical and practical reasons, to extend the theory to situations where there is a continum of possibilities. Suppose a physical variable like the height of an individual, the value of electric current in a wire etc. are observed. Here each of the continum of possible values of these variables is to be regarded as a possible outcome. The number of such outcomes being infinite in number, probability of an event, say, the height of a man lies between 5′2″ and 5′3″, can not be defined in classical theory.

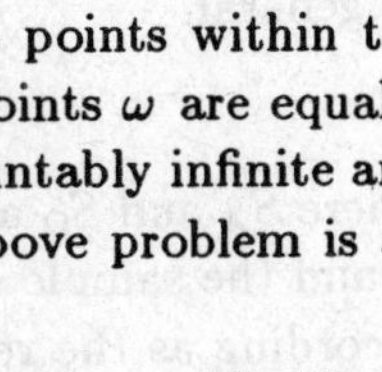

Fig. 7

Consider, again, the following problem. A point object is thrown at random on the space Ω. We want to find the probability that the object lies within the subspace A in Ω (Fig. 7). Here the sample sapce Ω is a domain on a plane and the elementary events ω are points within the domain. The conditions of the experiment are such that all the points ω are equally likely. Hence both the total number of points in Ω and A are uncountably infinite and as such probability can not be defined in clasical theory. (The above problem is an example of geometric probability dealt with in section 2.15).

The classical definition is based on the assumption that the cases are equally likely. This assumption may not be fullfilled in many cases. A simple example is the loaded die. For a die which is asymmetrical in mass or shape, it is not intuitively expected that each side has the same likelihood of turning up. The classical definition fails to answer the question like what is the probability of obtaining a six.

In the classical definition the value of probability comes necessarily as a rational number.

2.14. Statistical or Empirical Definition

We saw in section 1.2 that the experiments conducted under identical conditions a large number of times show a statistical regularity, namely, the relative frequency of an outcome in several sets of sequences of trials is more or less constant provided each set consists of a large number of trials. The rate of convergence of relative frequencies to this particular value increases rapidly as the number of trials increases. This constant value may be taken as the probability of the outcome. The basic requirement (assumption) for this definition is that the experiments must be conducted under identical conditions and the number of trials must be large.

Definition Let $f_n(A)$ be the number of times in which an event A, the outcome of an experiment, occurs in a series of n repetitions of the trial conducted under identical conditions. The relative frequency of A is $r_n(A) = fn(A)/n$. The probability of the event A is defined as

$$P(A) = \lim_{n\to\infty} \frac{f_n(A)}{n}$$

provided the limit exists and is unique.

We note that even if the conditions of an experiment is such that its elementary events are not equally likely, probability can be defined in the statistical sense, though, however, it remains undefined in the classical sense.

2.15. Geometric Probability

Geometrical approach to the calculation of probabilities is employed when the sample space Ω includes an uncountable set of elementary events ω and none of them is more likely to occur than the other. Suppose as in Fig. 7 the sample space Ω is a domain in a plane and the elementary events ω are points within Ω. If an event A is represented by the event A within Ω, so that all ω belonging to the region A are favourable to the event A, then probability of A is

$$P(A) = \frac{\text{area of subdomain } A}{\text{area of domain } \Omega}. \tag{2.23}$$

In general,

$$\text{Probability of an event } A \quad = \frac{S_A}{S_\Omega} \tag{2.24}$$

where S_A and S_Ω are measures of the specified parts of the region representing the event A and the sample space Ω respectively. The measure may be length, area, volume etc. according as the region is in one, two, three,$\cdots$ dimensions.

Formula (2.23) is a generalisation of classical formula (2.1) to an uncountable set of elementary events. The symmetry of the experiments' conditions with respect to the elementary outcomes ω is usually formulated by the assumption of randomness.

The theory of geometric probabilities is often criticised for arbitrariness in determining the probability of events (see example 2.52; exercise 67, section 2.16). Many authors point out that for an infinite number of outcomes the probability cannot be determined objectively i.e. independently of the mode of computation.

❑ **EXAMPLE 2.44.** *A straight line of unit length, say, (interval) (0,1) is divided into three intervals by choosing two points at random. What is the probability that the three line segments form a triangle ?*

Let x, y be the abscissas of any two points chosen at random on (0,1). A set of necessary and sufficient conditions for the three segments to form a triangle is that the length of any one of the segments be less than the sum of the other two. Hence we should have either

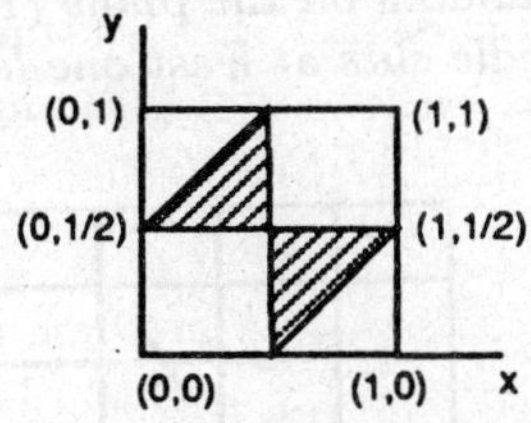

Fig. 8

$$0 < x < \frac{1}{2} < y < 1 \quad \text{and} \quad y - x < \frac{1}{2}$$

$$\text{or} \quad 0 < y < \frac{1}{2} < x < 1 \quad \text{and} \quad x - y < \frac{1}{2}$$

This is represented by the shaded area in the figure. Hence the probability is $\frac{1}{4}$.

❑ **EXAMPLE 2.45.** *Buffon's needle problem. A plane is ruled with paralled straight lines at distance L from each other. A needle of length $l < L$ is thrown at random on the plane. (Fig. 9(a)). Find the probability that it will hit one of the lines.*

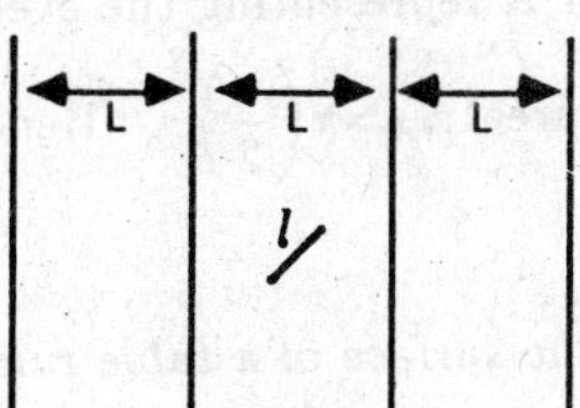

Fig. 9(a)

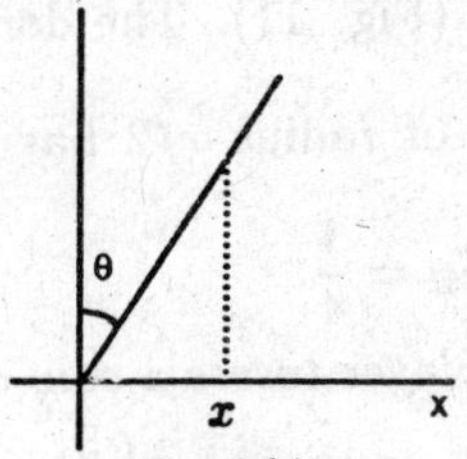

Fig. 9(b)

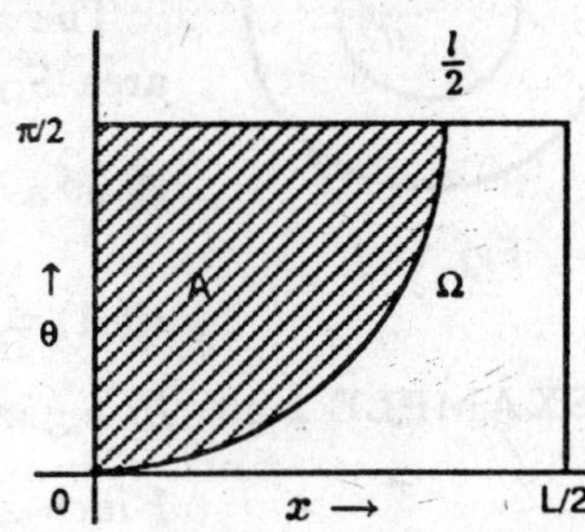

Fig. 9(c)

We characterise the outcome of the experiment by two numbers : the abscissa x of the centre of the needle with respect to the nearest line on the left and by the angle θ the needle makes with the direction of the lines (Fig. 9(b)). Since the needle is thrown at random all values of x and θ are equiprobable. Without any loss of generality we consider only the possibilities of the needle hitting the nearest line on the left when $0 \le x \le \frac{L}{2}$ and $0 \le \theta \le \frac{\pi}{2}$ (The probability is same for hitting the nearest line on the right). The sample space Ω is thus a rectangle of area $S_\Omega = \frac{L\pi}{4}$ (Fig. 9(c)). The needle will hit the line if $x < \frac{l}{2}\sin\theta$. We are thus interested in the event $A = \left\{x < \frac{l}{2}\sin\theta\right\}$. Area S_A of $A = \int_0^{\pi/2} \frac{l}{2}\sin\theta\, d\,\theta = \frac{l}{2}$. Hence $P(A) = \frac{S_A}{S_\Omega} =$

$$\frac{2l}{\pi L} = \frac{2 \text{ length of the needle}}{\text{circumference of a circle of radius } \frac{L}{2}}.$$

[This is an interesting problem, perhaps, because it suggests a relation between a pure chance experiment and a famous number π. If we take a graph paper ruled by parallel lines 1 inch apart and a needle of length 1 inch and keep track of the fraction of times the needle crosses a line, when thrown randomly on the graph paper, π may be estimated as about 2/(proportion of crosses).]

❑ **EXAMPLE 2.46.** *A plane is ruled in rectangles with sides L and M. A needle of length $l (l < L, l < M)$ is thrown at random on the plane (Fig. 10). Find the probability that the needle cuts at least one line.*

Let A, B be the events that the needle cuts a vertical line and a horizontal line respectively. The events A, B are independent. Hence

$$P(A \cup B) = P(A) + P(B) - P(A)P(B).$$

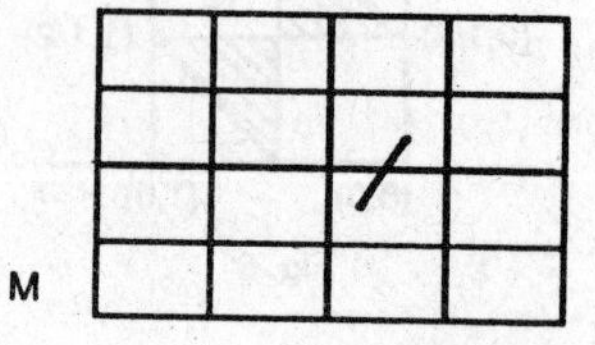

Fig. 10

On the basis of Buffon's problem, $P(A) = \dfrac{2l}{\pi L}$, $P(B) = \dfrac{2l}{\pi M}$. Hence $P(A \cup B) = \dfrac{2l}{\pi L} + \dfrac{2l}{\pi L} - \dfrac{4l^2}{\pi^2 LM}$.

❑ **EXAMPLE 2.47.** *A point image of an object M is equally likely to appear anywhere within a circular radar screen of radius r. Find the probability of the event A that the distance ρ of M from the centre of the screen is smaller than $\dfrac{r}{2}$.*

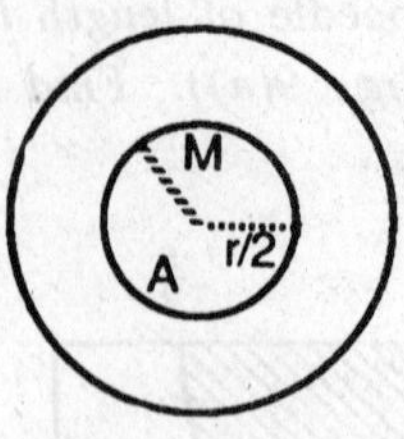

Fig. 11

The sample space Ω is the interior of the circle of radius r of area $S_\Omega = \pi r^2$ (Fig. 11). The domain A representing the event A is a circle of radius $r/2$ having area $S_A = \pi \left(\dfrac{r}{2}\right)^2$. Hence

$$P(A) = S_A / S_\Omega = \frac{1}{4}.$$

❑ **EXAMPLE 2.48.** *In a game a player tosses a coin on the surface of a table ruled 1-inch square. If the coin $\left(\dfrac{3}{4} \text{ inch in diameter}\right)$ falls entirely inside a square, the player receives a prize; otherwise he loses the coin. If the coin lands on the table, what is his chance to win ?*

We assume that the centre of the coin may fall on any region of the table with equal probability. Since the coin is $\dfrac{3''}{8}$ in radius its centre must not lie within $\dfrac{3''}{8}$ from any edge for the coin to fall entirely inside a square (Fig. 12). This restriction generates a squre A of side $\dfrac{1''}{4}$ within which the centre of the coin must lie for the required event to happen. Hence

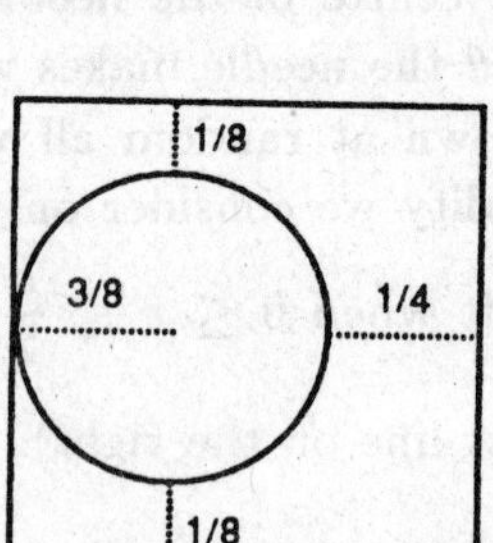

Fig. 12

$$\text{probability} = \frac{\text{area of square } A}{\text{area of the table}} = \frac{1}{16}.$$

We have ignored here the thickness of the lines marking the squre.

❑ **EXAMPLE 2.49.** *What is the prob. that the quadratic equation $x^2 + 2bx + c = 0$ has real roots ?*

Suppose b is any point chosen randomly within the interval $(-B, B)$. Similarly c is any random point in the interval $(-B, B)$. We plot b along x-axis and c along y-axis.

For the quadratic to have real roots we must have $b^2 - c \geq 0$. In the figure 13, the shaded portion represents the region with $b^2 - c > 0$.

The area of the unshaded region is $\frac{4}{3} B^{3/2}$ $(B \geq 1)$ and hence probability of getting complex roots is $\frac{4}{3} B^{3/2} \Big/ 4B^2 = \frac{1}{3\sqrt{B}}$. Now as $B \to \infty$, $\frac{1}{\sqrt{B}} \to 0$ and hence probability that the roots are real tends to 1.

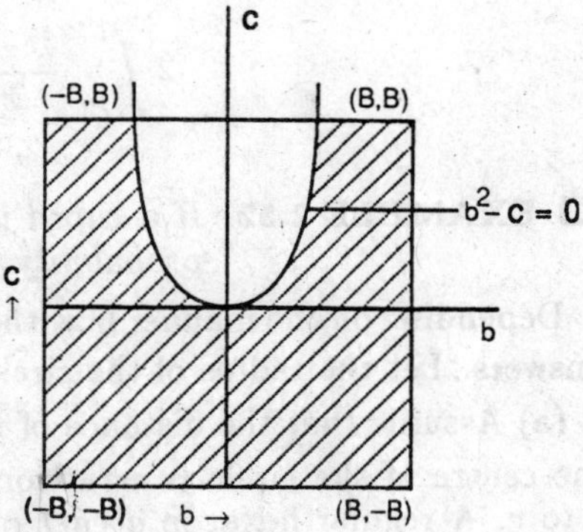

Fig. 13

❑ **EXAMPLE 2.50.** *Two persons A and B come to the club between 6 P.M. and 7 P.M. at random points of time and each stays for 10 minutes. What is the chance they will meet ?*

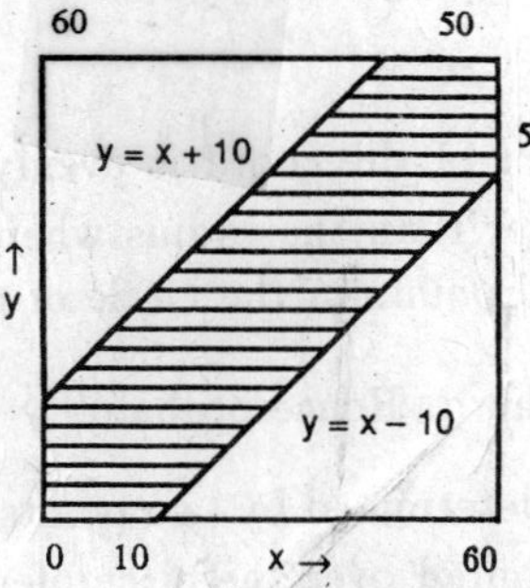

Fig. 14

Let A and B arrive respectively after x and y minutes from 6 P.M. They will meet if $(x - 10) \leq y \leq (x + 10)$ i.e. if (x, y) falls in the shaded region (Fig. 14). Area of the unshaded portion is 2500. Hence the probability they meet is $\frac{11}{36}$.

❑ **EXAMPLE 2.51.** *If a stick is broken at random in two parts, (a) what is the average length of the smaller part ? (b) what is the average ratio of the smaller length to the larger ?*

(a) The breaking point may be anywhere on the stick. If the breaking point is on the left half, the smaller half is on the left; its average size is half of that half i.e. $\frac{1}{4}$ of the total length. Similar argument applies when the breaking point is on the right half.

(b) Suppose that the point falls on the right-hand half. Then $(1 - x)/x$ is the ratio of the smaller length to the larger, assuming the stick is of unit length and x is the length of the breaking point from the left end. Since x is a random point in the interval $\left(\frac{1}{2}, 1\right)$ $\left(\text{and since a similar argument applies if } x \in \left(0, \frac{1}{2}\right)\right)$ the average value is

$$2\int_{1/2}^{1} \frac{1-x}{x} dx = 2\log_e 2 - 1 \simeq .386.$$

❑ **EXAMPLE 2.52.** *If a chord is selected at random on a fixed circle what is the probability that its length exceeds the radius of the circle ?*

Depending on the manner how the term 'random' is explained, we have three different answers. Let the radius of the circle be r.

(a) Assume that the distance of the chord from the centre of the circle is a random value within 0 to r. A regular hexagon *abcdef* of side r can be inscribed in the circle (Fig. 15). Any chord lying within this polygon will have length greater than r. Hence the required prob. $= \frac{h}{r}$, h being the distance of ab from O. Now $h = r\sqrt{\frac{3}{2}}$. Hence

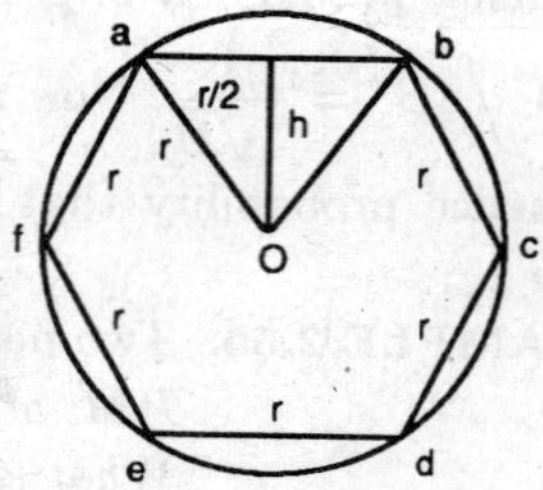

Fig. 15

probability $= \frac{\sqrt{3}}{2} \simeq 0.866.$

(b) Assume that the midpoint of the chord is evenly distributed over the interior of the circle. The chord is longer than the radius when the midpoint of the chord is within h of the centre. Thus all points in the circle of radius h with O as centre can serve as the mid-points of the chord. Hence probability $= \frac{\pi h^2}{\pi r^2} = \frac{3}{4}$.

(c) Assume that the chord is determined by two points chosen so that their positions are independently evenly distributed over the circumference of the circle. Assuming that the first point falls at a, the second point must fall on the arc *fab* for the chord to be shorter than radius. The length of *fab* $= \frac{1}{3}$ of the circumference. Hence probability $= \frac{2}{3}$.

2.16. Exercise

1. (a) *Verify the following relations for sets A, B, C.*

(i) $(A \cup B \cup C)' = A'B'C'$

(ii) $A' \cup B' \cup C' = (ABC)'$

(iii) $(A \cup B) \cap C = AC \cup BC$

(iv) $(A \cap C) \cap (B \cap C) = ABC$

(v) $(A \cup B) - (A \cap B) = (A - B) \cup (B - A)$

(vi) $A \cap (B - C) = (A \cap B) - (A \cap C)$

(vii) $(A \cup B)' \cap C = C - C \cap (A \cup B)$

(viii) $(A \cup B) - C \subset A \cup (B - C)$

(ix) $A \cap (B \cup C) \subset (A \cap B) \cup C$

(b) *Verify the correctness or otherwise of the statements :*

(a) $(A \cap B) - C = A \cap B \cap C$ [incorrect]

(b) $(A \cup B) - C = (A - C) \cup (B - C)$ [correct]

(c) $(A \cup B)'C = (A' \cap C) \cup (B' \cap C)$ [incorrect]

(d) $(A \cap B) \cap (A' \cap C) = \phi$ [correct]

(e) $(A - B) \cup B = A$ [incorrect]

2. *Let A, B, C be three arbitrary events. Find expressions for the following events :*

(i) all of A, B, C occur [ABC]

(ii) at least one occurs [$A \cup B \cup C$]

(iii) at least two occur [$(AB) \cup (AC) \cup (BC)$]

(iv) only one and no more occurs [$(AB'C') \cup (A'BC') \cup (A'B'C)$]

(v) not more than two occurs [$(ABC)'$]

(vi) both A and B but not C occurs [ABC']

(vii) none occurs [$A'B'C'$]

3. *A box contains three balls numbered 1, 2 and 3. The balls are drawn from the box one at a time without replacement and with equal probability. Find the probability that the ordinal number of at least one ball coincides with its own number.*

[Let A_i be the event that the ball marked i is drawn at the ith draw. $P(A_i) = \frac{2!}{3!} = \frac{1}{3}$, $P(A_iA_j) = \frac{1}{6}$. $P(A_1 \cup A_2 \cup A_3) = \frac{2}{3}$]

4. *Six cards are drawn from an ordinary pack with replacement. What is the probability that each of the four suits will be represented at least once ?*

[Required probability $= 1 - P(B)$, B being the event that at least one suit is not represented. $P(B) = 4\left(\frac{3}{4}\right)^6 - 6\left(\frac{1}{2}\right)^6 + 4\left(\frac{1}{4}\right)^6 = .61914$]

5. *A person writes 6 letters and 6 envelopes. If the letters are placed in the envelopes at random, what is the chance that not more than one letter is placed in the correct envelope ?*

[$P(0 \text{ match}) = \frac{1}{2!} - \frac{1}{3!} + \cdots + \frac{1}{6!} = .36805$

$P(1 \text{ match}) = \frac{1}{2!} - \frac{1}{3!} + \frac{1}{4!} - \frac{1}{5!} = .36944$. Hence probability $= .73750$]

6. *Find the probability that a hand of 13 cards from a standard pack contains the ace and the king of at least one suit.*

[Let A_i denote the event that the ace and the king of suit i is selected, $i = 1, \cdots, 4$.

$$P\left(\bigcup_1^4 A_i\right) = S_1 - S_2 + S_3 - S_4$$

$$= \frac{1}{\binom{52}{13}}\left[\binom{4}{1}\binom{50}{11} - \binom{4}{2}\binom{48}{9} + \binom{4}{3}\binom{46}{7} - \binom{44}{5}\right]$$

$$= \frac{1}{\binom{52}{13}}\sum_{k=1}^{4}(-1)^{k-1}\binom{4}{k}\binom{52-2k}{13-2k}\Bigg]$$

7. *In a survey report it is stated that 65% of male college students in Calcutta like soccer, 34% like hockey and 41% cricket, 10% both soccer and hockey, 17% both soccer and cricket and 15% both hockey and cricket, 6% all three. Show that the statement, as it stands, must be incorrect.*

[Probability that a randomly selected person likes at least one of the three $= P(S) + P(H) + P(C) - P(SC) - P(SH) - P(HC) + P(SHC) = 1.05 > 1$]

8. *A company manufacturing cornflakes put a card numbered 1, 2, $\cdots$, r at random in each packings, all numbers being equally likely to occur. If $n(> r)$ packages of cornflakes are purchased, find the probability there are at least one complete set of cards.*

[Let A_i be the event that the number i is represented, $i = 1, \cdots, r$.

$$\bigcap_{i=1}^{r} A_i = \left(\bigcup_{i=1}^{r} A_i^c\right)^c.$$

$$P\left(\bigcap_1^r A_i\right) = 1 - P\left(\bigcup_1^r A_i^c\right)$$

$$= 1 - \sum_i P(A_i^c) + \sum_{i<j}\sum P(A_i^c A_j^c) - \sum_{i<j<k}\sum\sum P(A_i^c A_j^c A_k^c) + \cdots$$

$$+ (-1)^r P(A_1^c \cdots A_r^c).$$

$$= 1 - \binom{r}{1}\left(\frac{r-1}{r}\right)^n + \binom{r}{2}\left(\frac{r-2}{r}\right)^n - \cdots + (-1)^{r-1} r\left(\frac{1}{r}\right)^n.$$

$$= \sum_{k=0}^{r}(-1)^k\binom{r}{k}\left(\frac{r-k}{r}\right)^n = \frac{\Delta^r 0^n}{r^n}$$

when $\sum_{k=0}^{r}(-1)^k\binom{r}{k}(r-k)^n = \Delta^r 0^n = \Delta^r x^n\Big|_{x=0}$, Δ being the operator defined by $\Delta f(x) = f(x+1) - f(x)$.

For $n = 10$, $r = 6$, probability $\simeq .27181$.]

9. *n balls are distributed among m cells ($n > m$). Find the probability that : (i) no cell remains empty, (ii) k cells remain empty.*

[(i) Let A_i be the event that the ith cell is empty. $P(A_i) = \left(\frac{m-1}{m}\right)^n$, $P(A_iA_j) = \left(\frac{m-2}{m}\right)^n$, etc.

The required probability $= 1 - P\left(\bigcup_1^m A_i\right)$

$$= 1 - m\left(\frac{m-1}{m}\right)^n + \binom{m}{2}\left(\frac{m-2}{m}\right)^n - \cdots + (-1)^{m-1} m\left(\frac{1}{m}\right)^n$$

$$= \sum_{k=0}^{m}(-1)^k\binom{m}{k}\left(\frac{m-k}{m}\right)^n = \frac{\Delta^m 0^n}{m^n} = p_{n,m} \quad \text{(say)}.$$

(ii) If n balls are distributed among m cells, let $N_{n,m}$ be the number of sample points leaving no cell empty. Thus $p_{n,m} = \frac{N_{n,m}}{m^n}$. Suppose first the k cells to remain empty are specified. The remaining $(m-k)$ cells must remain occupied. The number of sample points in the distribution of n balls in $(m-k)$ cells, in which all these cells remain occupied is $N_{n,m-k}$. Again the k cells may be chosen in $\binom{m}{k}$ ways. Hence probability =

$$\frac{\binom{m}{k}N_{n,m-k}}{m^n} = \binom{m}{k}\left(\frac{m-k}{m}\right)^n p_{n,m-k} = \binom{m}{k}\frac{\Delta^{m-k}0^n}{m^n}.\Bigg]$$

10. *(Coupon collector's problem) In every box of a certain breakfast cereal is a picture of a baseball player. A full set is represented by m players. Find the probability that n boxes of cereals must be purchased to obtain a full set of pictures.*

[Let Z denote the number of boxes that must be purchased to obtain a full set of m pictures. If the pictures are taken as cells, boxes as balls, the problem reduces to finding the number of balls that must be distributed so that all the m cells remain occupied. By the notation of problem 9,

$$P(Z \le n) = p_{n,m}.$$

$$\therefore \quad P(Z=n) = P(Z \le n) - P(Z \le n-1) = p_{n,m} - p_{n-1,m}$$
$$= \sum_{k=1}^{m-1} (-1)^{k-1} \binom{m-1}{k-1} \left(\frac{m-k}{m}\right)^{n-1} \Bigg]$$

11. *A die is cast repeatedly until each of the six faces appears at least once. Find the probability that it must be cast exactly 10 times.*

[Proceed as in exercise 10, Probability = .06998]

12. *A sample of r individuals are taken from a population of n people with replacement. Find the probability p_r that m specified individuals are selected. Show that if $n \to \infty$ and $r \to \infty$ so that $\frac{r}{n} \to p$, a constant, $p_r \to (1-e^{-p})^m$.*

[A_i : the event that the ith individual in the set of m specified individuals is not selected.

$$p_r = 1 - P\left(\bigcup_1^m A_i\right) = 1 - \binom{m}{1}\left(\frac{n-1}{n}\right)^r + \binom{m}{2}\left(\frac{n-2}{n}\right)^r - \cdots + (-1)^m \left(\frac{n-m}{n}\right)^r$$
$$= \sum_{k=0}^{m} (-1)^k \binom{m}{k} \left(1 - \frac{k}{n}\right)^r.$$

Now $\left(1 - \frac{kp}{r}\right)^r \to e^{-kp}$ as $r \to \infty$ $\therefore$ $p_r \to \sum_{k=0}^{m} (-1)^k \binom{m}{k} e^{-kp} = (1-e^{-p})^m$ as $n \to \infty$, $r \to \infty$.]

13. *In a lottery m tickets are drawn at a time out of the total number of n tickets and returned before the next drawing is made. What is the probability that in k drawings each of the numbers $1, 2, \cdots, n$ will appear at least once ?*

[A_i : the event, number i does not occur. The required probability $= 1 - P\left(\bigcup_1^n A_i\right)$.

$$\text{Now} \qquad P(A_i) = \left[\binom{n-1}{m} \Big/ \binom{n}{m}\right]^k = \left(1 - \frac{m}{n}\right)^k.$$
$$P(A_i A_j) = \left[\left(1 - \frac{m}{n}\right)\left(1 - \frac{m}{n-1}\right)\right]^k, \text{ etc.} \Bigg]$$

14. *A number 'a' is chosen at random from the numbers $\{1, \cdots, N\}$. Find the probability p_N that (a) the number 'a' is not divisible either by a_1 or by a_2 where a_1, a_2 are fixed natural coprime numbers, (b) the number 'a' is not divisible by either of the numbers $a_1, \cdots, a_k$ where a_i are natural pairwise coprime numbers. Find $\lim_{N \to \infty} p_N$ in case (a) and (b).*

[(a) $p_N = 1 - P\{\text{'a' is divisible at least by one of } a_1 \text{ or } a_2\} = 1 - q_N$ (say).

$q_N = P_1 + P_2 - P_3$ where P_i = probability 'a' is divisible by a_i, $(i = 1, 2)$, P_3 = probability a is divisible by both. Thus

$$p_N = 1 - \frac{1}{N}\left\{\left[\frac{N}{a_1}\right]+\left[\frac{N}{a_2}\right]-\left[\frac{N}{a_1 a_2}\right]\right\}, \quad \lim_{N\to\infty} p_N \left(1-\frac{1}{a_1}\right)\left(1-\frac{1}{a_2}\right).$$

(b) $$p_N = 1 - \frac{1}{N}\left\{\sum_1^k \left[\frac{N}{a_i}\right] - \sum\sum_{i<j=1}^{k}\left[\frac{N}{a_i a_j}\right] + \sum\sum\sum_{i<j<l}^{k}\left[\frac{N}{a_i a_j a_l}\right] + \cdots + (-1)^{k-1}\left[\frac{N}{a_i \cdots a_k}\right]\right\}$$

$$\lim_{N\to\infty} p_N = \prod_{i=1}^{k}\left(1-\frac{1}{a_i}\right)\Bigg]$$

15. *A man tosses 2 fair dice. What is the conditional probability that the sum of the two dice will be 7, given that : (i) the sum is odd, (ii) the sum is greater then 6, (iii) the outcome of the first die was odd, (iv) the outcome of the second die was even, (v) the outcome of at least one of the dice was odd, (vi) the 2 dice had the same outcomes, (vii) the 2 dice had different outcomes ?*

$\left[\text{(i) } \frac{1}{3}, \text{ (ii) } \frac{2}{7}, \text{ (iii) } \frac{1}{6}, \text{ (iv) } \frac{1}{6}, \text{ (v) } \frac{2}{9}, \text{ (vi) } 0, \text{ (vii) } \frac{1}{5}\right].$

16. *One card is drawn at random from 100 cards numbered 00, 01, ..., 99. Suppose x_1 and x_2 are the sum and the product respectively of the digits on the selected card. Find $P\{x_1 = i | x_2 = 0\}$.* $\left[\text{Probability} = \frac{1}{19} \text{ for } i = 0; = \frac{2}{19} \text{ for } i = 1, \cdots, 9\right].$

17. *A die is loaded in such a way that the probability of a given number turning up is proportional to that number (eg. a 4 is twice as probable as a 2). (i) What is the probability of a 5, given that an odd number turns up; (ii) What is the probability of rolling an even number, given that a number less then 5 turns up.* $\left[\frac{5}{9}, \frac{3}{5}\right].$

18. *In an office 30% of employees have scooters and 25% have cars. Among those who have scooters, 90% do not have cars. What is the probability that an employee has a car given that he does not have a scooter.*

[We have $P(\bar{C}|S) = .9$, $P(S) = .3$; hence $P(\bar{C}S) = .27$. Now $P(\bar{C}S) + P(\bar{C}\bar{S}) = P(\bar{C}) = .75$;

$\therefore$ $P(\bar{C}\bar{S}) = .48$. Also $P(\bar{C}\bar{S}) + P(C\bar{S}) = P(\bar{S}) = .70$.

$\therefore$ $P(C\bar{S}) = .22$. Thus $P(C|\bar{S}) = .31429$].

19. *From an urn containing M white and (N − M) black balls, all the balls are drawn one by one and without replacement. Find the probabilities : (i) A_k = {the kth* balls is white}, (ii) $A_{k,l}$ = {the kth ball and the lth ball are white}, (iii) $B_{k,l}$ = {the kth ball is black and the lth ball is white}.

[Consider any two sequences, say, $BBWWB\cdots, WBBWB\cdots$

$$P(BBWWB\cdots) = \frac{(N-M)(N-M-1)M(M-1)(N-M-2)\cdots}{N(N-1)(N-2)(N-3)(N-4)\cdots}$$

$$P(WBBWB\cdots) = \frac{M(N-M)(N-M-1)(M-1)(N-M-2)\cdots}{N(N-1)(N-2)(N-3)(N-4)\cdots}$$

Thus all sequences are equally likely. Hence $P(A_k) = P(A_1) = \frac{M}{N}$, $P(A_{kl}) = P(A_{12}) = \frac{M(M-1)}{N(N-1)}$, $P(B_{kl}) = P(B_{12}) = \frac{(N-M)M}{N(N-1)}$].

20. *Three newspapers A, B and C are published in a certain city. A survey shows that for the adult population 20% read A, 16% read B and 14% read C, 8% read both A and B, 5% read both A and C, 4% read both B and C and 2% read all three. If an adult is chosen at random, find the probability that : (i) he reads none of the papers, (ii) he reads only one of these papers and (iii) he reads at least A and B if it is known that he reads at least one paper.*

[(i) .65, (ii) .22, (iii) Let H be the event that he reads at least one paper. $P(H) = .35$. Let E be the event that he reads at least A and B. $P(E) = .08$. $P(H|E) = 1$. $P(E|H) = \frac{8}{35}$].

21. *(Polya's urn model). An urn contains initially b black and r red balls. A ball is drawn and then replaced with c additional balls of the same colour. The process is repeated. Let B_k be the event that the kth ball drawn is black and R_k the event that the kth ball drawn is red. Suppose on any draw it is equally likely that any ball in the urn will be drawn. Determine: (a) $P(R_2|R_1)$ (b) $P(B_1B_2B_3)$. Show by induction the probability of selecting a red ball at any trial is $\frac{r}{(b+r)}$.*

[(a) $\frac{r+c}{b+r+c}$ (b) $\frac{b(b+c)(b+2c)}{(b+r)(b+r+c)(b+r+2c)}$]

22. *A symmetric die is thrown as long as necessary for a six to turn up. Find the conditional probability that more than three throws will be necessary given that the first throw did not show up a six.* $\left[\left(\frac{5}{6}\right)^2\right]$

23. *Two die are tossed. Let A_e = {the face value of the first die is divisible by e}, B_e = {the face value of the second die is divisible by e}, C_e = {the sum total of face values of both dice is divisible by e}. Find whether the following pairs of events are independent : (i) A_l, B_k for any k, l, (ii) A_2, C_2, (iii) A_4, C_4.*

[(i) independent for any k, l, (ii) independent, $P(A_2) = P(C_2) = \frac{1}{2}$, $P(A_2C_2) = \frac{1}{4}$, (iii) dependent, $P(A_4) = \frac{1}{6}$; $P(C_4) = \frac{1}{4}$, $P(A_4C_4) = \frac{1}{36}$]

24. *A pair of die is thrown once. Let A be the event that the first die has a 1 on it, B be the event that the second die has a 6 on it and C be the event that the sum is 7. Are A, B and C mutually independent ?* [No]

25. *Let the events $A_1, \cdots, A_n$ be mutually independent. If $P(A_i) = p_i$, what is the probability that : (i) none of the events will occur, (ii) at least one of the events will occur, (iii) at most one of the events will occur.*

$$\left[\text{(i)}\ \prod_{i=1}^{n}(1-p_i) = Q \text{ (say)}, \quad \text{(ii)}\ 1-Q, \quad \text{(iii)}\ \prod_{k=1}^{n}(1-p_k)\left[1+\sum_{i=1}^{n}\frac{p_i}{1-p_i}\right].\right]$$

26. *The probability p of malfunction of an electric circuit on any operation is about 10^{-5}. what is the probability of at least one malfunction in 10^5 independent operations?*

[F_i = event of a malfunction at the ith operation, $P(F_i) = p$. $P\left(\bigcup_{i=1}^{n} F_i\right) = 1 - P\left(\bigcap_{i=1}^{n} F_i^c\right) = 1-(1-p)^n \simeq 1-e^{-np}$ for small $p = 1-e^{-1} \simeq .63$.]

27. *(The paradox of de Méré). Show that it is more probable to get at least one ace with four dice than at least one double ace in 24 throws of two dice.*

[Probability of getting no ace with 4 dice $= \left(\frac{5}{6}\right)^4 = P_1$ (say). Probability of getting no double ace with 24 throws of two dice $= \left(\frac{35}{36}\right)^4 = P_2$ (say). Now $1 - P_1 = .51775 > 1 - P_2 = .49140$]

28. *Let the probability that the weather on any day is of the same type (rain or no rain) as the previous day be p. Let P_1 be the probability of rain on the first day of the year. What is the probability P_n of rain on the nth day ?*

$$\begin{aligned}\Big[P_n &= P_{n-1}p + (1-P_{n-1})(1-p) = (2p-1)P_{n-1} + 1 - p \\ &= (2p-1)\{(2p-1)P_{n-2} + (1-p)\} + (1-p) = \cdots\cdots\cdots \\ &= (2p-1)^{n-1}P_1 + (1-p)\{(2p-1)^{n-2} + (2p-1)^{n-3} + \cdots + 1\} \\ &= (2p-1)^{n-1}P_1 + \frac{1}{2}\{1-(2p-1)^{n-1}\} = \frac{1}{2}\{1+(2P_1-1)(2p-1)^{n-1}\}\Big]\end{aligned}$$

29. *Three prisoners A, B and C with apparently equally good records have applied for parole. The parole board has decided to release two of the three and the prisoners know this but not which two. A warder friend of prisoner A knows who are to be released. Prisoner A realises that it would be unethical to ask the warder if he, A is to be released, but think of asking for the name of one prisoner other than himself who is to be released. He thinks that before he asks, his chance of release are $\frac{2}{3}$. He thinks that if the warder says, "B will be released" his own chances have now gone down to $\frac{1}{2}$ because either A and B or B and C are to be released. So he decides not to reduce his chance by asking. Are his calculations justified?*

[His probability of being released, before asking the warder is $P(A) = \frac{2}{3}$. On his asking the warder the following results may occur :

A and B released and the warder says B : probability is $\frac{1}{3}$ (event D_1)

B and C " " " " " B : probability is $\frac{1}{6}$ (event D_2)

B and C " " " " " C : probability is $\frac{1}{6}$ (event D_3)

A and C " " " " " C : probability is $\frac{1}{3}$ (event D_4).

Hence probability of A's release given that the warder says B is $P(D_1)/[P(D_1) + P(D_2)] = \frac{2}{3}$. The chances remain the same.]

30. *The probability that a family will have k children is αp^k when $k = 1, \cdots, n$, the probability that one will have no child being $1 - \alpha \sum_{k=1}^{\infty} p^k$. Assuming that the ratio at birth is 1 : 1, (a) show that the probability that a family will have $m(\geq 1)$ sons is $2\alpha p^m/(2-p)^{m+1}$. (b) Given that a family has at least one boy, what is the probability that there are two or more?*

[(a) Let B_n be the event that a family has n children and A_k the event the family has k boys.

$$P(A_m) = \sum_{n=m}^{\infty} P(B_n)P(A_m|B_n) = \sum_{n=m}^{\infty} \alpha p^n \binom{n}{m}\left(\frac{1}{2}\right)^n$$

$$= \alpha \sum_{n=m}^{\infty} \binom{n}{m}\left(\frac{p}{2}\right)^n = \alpha\left(\frac{p}{2}\right)^m \sum_{n=m}^{\infty} \binom{n}{m}\left(\frac{p}{2}\right)^{n-m}$$

$$= \alpha\left(\frac{p}{2}\right)^m \left(1 - \frac{p}{2}\right)^{m+1} = \frac{2\alpha p^m}{(2-p)^{m+1}}$$

$$\left[\because \sum_{n=m}^{\infty} \binom{n}{m}\left(\frac{p}{2}\right)^{n-m} = 1 + (m+1)\frac{p}{2} + \binom{m+2}{2}\left(\frac{p}{2}\right)^2 + \cdots = \left(1 - \frac{p}{2}\right)^{m+1}\right].$$

(b) $$P\left(\bigcup_{k=2}^{\infty} A_k \,\Big|\, \bigcup_{k=1}^{\infty} A_k\right) = \frac{P\left\{\left(\bigcup_{2}^{\infty} A_k\right) \cap \left(\bigcup_{k=1}^{\infty} A_k\right)\right\}}{P\left\{\bigcup_{1}^{\infty} A_k\right\}} = \frac{P\left(\bigcup_{2}^{\infty} A_k\right)}{P\left(\bigcup_{1}^{\infty} A_k\right)}$$

$$= \frac{\sum_{2}^{\infty} P(A_k)}{\sum_{1}^{\infty} P(A_k)}.$$

Now $$\sum_{2}^{\infty} P(A_k) = \frac{2\alpha}{(2-p)} \sum_{2}^{\infty} \left(\frac{p}{2-p}\right)^k = \frac{\alpha p^2}{(2-p)^2(1-p)},$$

$$\sum_{1}^{\infty} P(A_k) = \frac{\alpha p}{(2-p)(1-p)}.$$

Hence probability $= \dfrac{p}{2-p}$].

31. *Consider three loaded dice. Die A is certain to throw a 3. Die B is twice as likely to show a 2 as to show a 5 and will not show any other number. Die C is twice as likely to show a 4 as a 1 and will not show any other number. The three dice are rolled once. Find the probability that : (i) A shows a larger number than B, (ii) B shows a larger number than C, (iii) C shows a larger number than A, (iv) A shows a larger number than B and B shows a larger number than C.*

[(i) $\frac{2}{3}$, (ii) $P(B_2C_1 + B_5C_4 + B_5C_1) = \frac{5}{9}$, (iii) $P(A_3C_4) = \frac{2}{3}$,

(iv) $P(A_3B_2C_1) = \frac{2}{9}$]

32. *A and B alternatively throw a die. He who throws the first 6 wins the game. If A has the first throw what is the probability of his winning the game ?*

[Prob. $= \frac{1}{6} + \left(\frac{5}{6}\right)^2 \times \frac{1}{6} + \left(\frac{5}{6}\right)^4 \times \frac{1}{6} + \cdots = \frac{6}{11}$, since A may win at the first or third or fifth,$\cdots$ throw and trials are independent]

33. *Two players, in turn, draw balls (without replacement) from an urn containing M white and N black balls. The person who is to draw a white ball first wins the game. Find the probability that the person who draws a ball first would be the first to draw a white ball (i) when $N = 3$, $M = 1$ (ii) $N = 4$, $M = 1$ (iii) $N = 6$, $M = 2$.*

[(i) $P(W) + P(BBW) = \frac{3}{4}$, (ii) $P(W) + P(BBW) + P(BBBBW) = \frac{3}{5}$, (iii) $\frac{4}{7}$.]

34. *Three players A, B, C, in turn draw a ball from an urn with 10 white and 10 black balls, taking one ball at a time. He who extracts the first white ball wins the* *game. Supposing they start in the order A, B, C find the probability for each of them to win the game.*

$$\left[P(A) = \frac{10}{20} + \frac{10 \cdot 9 \cdot 10}{20 \cdot 19 \cdot 18} + \frac{10 \cdot 9 \cdot 8 \cdot 7 \cdot 10}{20 \cdot 19 \cdot 18 \cdot 17 \cdot 16} + \cdots = 0.56584,\right.$$

$$\left. P(B) = 0.29144, \quad P(C) = 0.14272\right]$$

35. *Two players play a game as follows. Taking turns, they draw the balls, one by one, out of an urn containing a white balls and b black balls. He who extracts the first white one wins the game. What is the probability that the player who starts will win the game.*

[Let A, B be two players, A starting the game. Probability of A to win=(W) + $P(BBW)+P(BBBBW)+\cdots = \frac{a}{a+b}\left(1 + \sum_{r=2,4,6,\cdots} \frac{(b)_r}{(a+b-1)_r}\right)$]

36. *A player tosses a coin and is to score one point for every head turned up and two for every tail. He is to play on unitl his score reaches or passes n. Find p_n, the probability of attaining exactly n.*

[The player attains n if : (i) he tosses a head after attaining $(n-1)$ points or (ii) he tosses a tail after attaining $(n-2)$ points. Hence the difference equation is $p_n = \frac{1}{2}p_{n-1} + \frac{1}{2}p_{n-2}$.

Thus

$$p_n - p_{n-1} = -\frac{1}{2}(p_{n-1} - p_{n-2}) = \left(-\frac{1}{2}\right)^2 (p_{n-2} - p_{n-3}) = \cdots = \left(-\frac{1}{2}\right)^{n-2} (p_2 - p_1).$$

$$\therefore \quad p_n = p_{n-1} + (-1)^n \frac{1}{2^n} \quad \left(\because \; p_1 = \frac{1}{2}, \; p_2 = \frac{3}{4}\right) = p_{n-1} + (-1)^n \frac{1}{2^n}\left(\frac{2}{3} + \frac{1}{3}\right)$$

i.e. $p_n - \frac{1}{3}(-1)^n \cdot \frac{1}{2^n} = p_{n-1} - \frac{1}{3}(-1)^{n-1} \cdot \frac{1}{2^{n-1}}$.

Thus $p_n - \frac{1}{3}(-1)^n \cdot \frac{1}{2^n} = p_1 - \frac{1}{3}(-1) \cdot \frac{1}{2}$ when $p_n = \frac{1}{3}\left(2 + (-1)^n \cdot \frac{1}{2^n}\right)$]

37. *To win a prize A has to win at least two tennis sets in a row in a three-set series to be played with B and C alternatively—BCB or CBC, according to his choice. C is a better player than B. What series should A choose ?*

[Let f and c be the probabilities of A for winning in a set with B and C respectively, $f > c$. Probabilities of winning the prize for different choices of series by A are :

Series 1	B	C	B	Probability	Series 2	C	B	C	Probability
results	W	W	W	fcf	results	W	W	W	cfc
	W	W	L	$fc(1-f)$		W	W	L	$cf(1-c)$
	L	W	W	$(1-f)cf$		L	W	W	$(1-c)fc$
			Total	$fc(2-f)$				Total	$fc(2-c)$.

Since $f > c$, A should choose the series CBC.]

38. *An urn contains 15 white and 5 black balls, another urn contains 7 white and 8 black balls. Two balls are transferred from the first urn and placed in the second urn and then one ball is taken from the later. What is the probability thac it is a white ball ?*

[Three possible cases are BBW, BWW, WWW where the first two letters stand for the balls transferred to the second urn and the third for the ball drawn from this urn.

$$P(BBW) = \binom{5}{2}7\Big/\binom{20}{2}17 = .02167, \quad P(BWW) = .18576, \quad P(WWW) = .29257$$

The required probability = .5]

39. *Three urns contain respectively 1 white, 2 black balls, 2 white, 1 black balls, 2 white, 2 black balls. One ball is transferred from the first into the second. Then one from the second to third; finally one ball is drawn from the third urn. What is the probability of its being white ?*

[$P(BBW) = \frac{2}{15}$, $P(BWW) = \frac{1}{5}$, $P(WBW) = \frac{1}{30}$, $P(WWW) = \frac{3}{20}$, the required probability $= \frac{31}{60}$]

40. *An urn contains n white and m black balls, a second urn contains N white and M black balls. A ball is transferred randomly from the first to the second urn and then a ball randomly from the second to the first urn. A ball is then selected randomly from the first urn. Wht is the probability it is white.*

[$P(BWW) = \frac{m}{m+n} \cdot \frac{N}{M+N+1} \cdot \frac{n+1}{m+n}$. The required probability $= P(BWW) + P(BBW) + P(WBW) + P(WWW) = \frac{n}{m+n} + \frac{mN - nM}{(m+n)^2(N+M+1)}$]

41. *Each of the 'n' urns contains 'a' white and 'b' black balls. A ball is selected randomly from the first urn and then transferred into the second, another one from the second to the third and so on. Finally, a ball is drawn at random from the nth urn. What is the probability for it to be white ? How does the probability change when the first ball transferred is known to be white ?*

[Let the urn be numbered $1, 2, \cdots, n$ and p_r the probability of drawing a white ball from the rth urn. Then

$$p_r = p_{r-1}\frac{a+1}{a+b+1} + (1-p_{r-1})\frac{a}{a+b+1}, \quad r = 2, 3, \cdots, n.$$

We have $p_1 = \dfrac{a}{a+b}$.

$$p_2 = \frac{a(a+1)}{(a+b)(a+b+1)} + \frac{ba}{(a+b)(a+b+1)} = \frac{a}{a+b}.$$

Thus $p_n = \dfrac{a}{a+b}$.

In the second case, $p_1 = 1$, $p_2 = \dfrac{a+1}{a+b+1} = \dfrac{a}{a+b} + \dfrac{b}{(a+b)(a+b+1)}$, $p_3 = \dfrac{(a+1)^2}{(a+b+1)^2} + \dfrac{ab}{(a+b+1)^2} = \dfrac{a}{a+b} + \dfrac{b}{(a+b)(a+b+1)^2}$.

Proceeding in this case, $p_n = \dfrac{a}{a+b} + \dfrac{b}{(a+b)(a+b+1)^{n-1}} \to \dfrac{a}{a+b}$ as $n \to \infty$]

42. *Two urns contain respectively a white and b black balls and b white and a black balls. A series of random drawings is made as follows : (i) each time only one ball is drawn and it is returned to the same urn from which it was obtained, (ii) if the ball drawn is white, the next drawing is made from the first urn; otherwise the next drawing is made from the second urn.*

Assuming that the first ball drawn is drawn from the first urn, find the probability p_n that the nth ball drawn is white and obtain its limiting value as $n \to \infty$.

[We have

$$p_n = p_{n-1}\frac{a}{a+b} + (1-p_{n-1})\frac{b}{a+b} = \frac{a-b}{a+b}p_{n-1} + \frac{b}{a+b}$$

$$= \alpha + \beta p_{n-1} \quad \text{(say)} \quad \text{i.e.} \quad \frac{p_n}{\beta^n} = \frac{\alpha}{\beta^n} + \frac{p_{n-1}}{\beta^{n-1}}.$$

This is a difference equation in p_n. Replacing n by $n-1, n-2, \cdots, 2$ and adding

$$\frac{p_n}{\beta^n} = \alpha\left(\frac{1}{\beta^n} + \cdots + \frac{1}{\beta^2}\right) + \frac{p_1}{\beta}.$$

$$\text{or} \quad p_n = \frac{\alpha(1-\beta^{n-1})}{1-\beta} + p_1\beta^{n-1} = \frac{1}{2} + \frac{1}{2}\left(\frac{a-b}{a+b}\right)^n \quad (\because p_1 = 1)$$

$$\to \frac{1}{2} \text{ as } n \to \infty]$$

43. *An urn contains a white and b black balls. After a ball is drawn, it is to be returned to the urn if it is white; if it is black it is replaced by a white ball (from another urn). Show that the probability of drawing a white ball after the above operation has been repeated n times is*

$$p_n = 1 - \frac{b}{a+b}\left[1 - \frac{1}{a+b}\right]^n.$$

[Let A_r denote the expected number of white balls after r operations (for definition of expectation see chapter 4). Then

$$A_{r+1} = A_r + 1\left(1 - \frac{A_r}{a+b}\right) = A_r\left(1 - \frac{1}{a+b}\right) + 1,\ A_0 = a.$$

$$\text{Thus}\quad A_n = (a+b) - b\left(1 - \frac{1}{a+b}\right)^n \quad \text{(by repeated application)}$$

$$\text{Now}\quad p_n = \frac{A_n}{a+b}.\text{]}$$

44. *Two urns contain respectively (a white(w), b black(b)), (c white, d black) balls. One ball is taken from the first urn and transferred into the second and simultaneously one ball is taken from the second urn and transferred into the first. Find the probabilities p_n of drawing a white ball from the first urn after this operation has been repeated n times.*

[Let A_r, C_r denote the expected number of white balls in the first and the second urn respectively, after the operation has been repeated r times. $A_r + C_r = a + c$, $A_0 = a$, $C_0 = c$.

$$A_{r+1} = A_r + (\text{prob. } a \text{ white ball is transferred from the second urn})$$

$$- (\text{prob. } a \text{ white ball is transferred from the first urn.})$$

$$= A_r + \frac{C_r}{c+d} - \frac{A_r}{a+b},$$

$$\text{when}\quad p_n = \frac{A_n}{a+b} = \frac{a+c}{a+b+c+d} + \frac{ad-bc}{(a+b)(a+b+c+d)}\left[1 - \frac{1}{a+b} - \frac{1}{c+d}\right]^n$$

$$\to \frac{a+c}{a+b+c+d} \quad \text{as} \quad n \to \infty\ \text{]}$$

45. *The game of craps is played as follows. The player throws two dice and the player wins at once if the total for the first throw is 7 or 11, loses at once if it is 2,3 or 12. Any other throw is called his point. If the first throw is a point, the player* *throws the dice repeatedly until he either wins by throwing his point again or loses by throwing a 7. Find the player's chance of win.*

[At the first throw he may win with probability $P(7)+P(11)=\frac{8}{36}\simeq .22222$ or lose with probability $P(2)+P(3)+P(12)=\frac{4}{36}$.

Probability of making a point at the first throw are as follows : $P(4)=P(10)=\frac{3}{36}$, $P(5)=P(9)=\frac{4}{36}$, $P(6)=P(8)=\frac{5}{36}$; also $P(7)=\frac{6}{36}$.

Probabilities of a win when initial point is $4=P_4$ (say) $=\left(\frac{3}{36}\right)^2+\left(\frac{27}{36}\right)\left(\frac{3}{36}\right)^2+\left(\frac{27}{36}\right)^2\left(\frac{3}{36}\right)^2+\cdots=\frac{1}{36}=P$ (say) since he may score a 4 in the second throw or score a point different from 4 or 7 in the second throw and score 4 in third throw, etc. Similarly $P_5=\frac{2}{45}$, $P_6=\frac{25}{396}$

Hence probability of a win by throwing a point

$$=(P_4+P_5+P_6+P_8+P_9+P_{10})=2\left(\frac{1}{36}+\frac{2}{45}+\frac{25}{396}\right)\simeq 0.27076.$$

Probability of a win $\simeq .49293$.]

46. *A three-man jury has two members each of whom independently has probability p of making the correct decision and a third member who flips a coin for each decision. The decision of the majority becomes final. A one-man jury has probability p of making the correct decision. Which jury has the better probability of making the correct decision ?*

[The serious jurors agree on correct decision in $p\cdot p=p^2$ proportion of cases. For these cases the vote of the other jurors does not count.

In the other correct decisions by the three men-jury, the probability that the serious jurors disagree is $p(1-p)+(1-p)p=2p(1-p)$. In half of these cases the other jury vote for a correct decision. Hence in such cases the probability of a correct decision is $p(1-p)$.

Thus the total probability of a correct decision by the three-men jury is $p^2+p(1-p)=p$. Hence both the jury have the same probability of a correct decision].

47. *Using a probabilistic argument, prove the identity* $(A>a)$,

$$1+\frac{A-a}{A-1}+\frac{(A-a)_{(2)}}{((A-1))_{(2)}}+\cdots+\frac{(A-a)!}{((A-1))_{(A-a-1)}\cdot a}=\frac{A}{a}$$

[Suppose an urn contains A balls of which a are white. The balls are drawn at random without replacement. The probability that a white ball will be drawn ultimately is one. Thus

$$\frac{a}{A}+\frac{(A-a)a}{A(A-1)}+\frac{(A-a)(A-a-1)a}{A(A-1)(A-1)}+\cdots+\frac{(A-a)!}{A(A-1)\cdots(a+1)}=1\Big]$$

48. *There are two boxes with parts of the same type. The first contains 'a' sound parts and 'b' defective ones and the second contains 'c' sound parts and 'd' defecive ones. A box is selected at random and one part is drawn from it and is found to be sound. What is the probability that box 2 was selected?* $\left[\left(\frac{c}{c+d}\right)\Big/\left(\frac{a}{a+b}+\frac{c}{c+d}\right)\right]$

49. *The contents of three vessels 1,2 and 3 are as follows : 2 white, 1 red and 3 black balls; 1 white, 3 red and 2 black balls; 3 white, 1 red and 2 black balls. A vessel is chosen at random and from it two balls are drawn at random. The two balls are one red and one white. What is the probability they come from the second vessel ?*

[Let B_i be the event that the ith vessel is chosen. $P(B_1|A)=\frac{1}{4}$, $P(B_2|A)=$

$P(B_3|A)=\frac{3}{8}$]

50. *In a doll factory, machines M_1, M_2 and M_3 manufacture respectively 45%, 25% and 30% of total output. Of these outputs 6%, 8% and 3% are defectives. A doll is chosen at random from the produce and found to be defective. What are the probabilities that it was manufactured by M_1, M_2, M_3 ?* [.482, .357, .161]

51. *Three out of 10 students taking a test are excellently prepared, four are well-prepared, two are adequately prepared and one is poorly prepared. There are 20 questions on the test paper. A student who was excellently prepared can answer all the 20 questions, a student who was well prepared can answer 16 questions, a student who was adequately prepared can answer 10 questions and a student who was poorly prepared can answer only 5 questions. A student was called at random and answered three arbitrarily chosen questions. Find the prob. that the student was (i) excellently prepared (ii) poorly prepared.*

[$P(E)=.3, P(W)=.4, P(A)=.2, P(P)=1$. Let H be the event that he answered 3 randomly chosen questions

$P(H|E)=1, P(H|W)=\binom{16}{3}\Big/\binom{20}{3}=.49123$

$P(H|A)=\binom{10}{3}\Big/\binom{20}{3}=.10526, P(H|P)=\binom{5}{3}\Big/\binom{20}{3}=.00877$

$$P(E|H) = \frac{P(E)P(H|E)}{P(E)P(H|E) + P(W)P(H|W) + P(A)P(H|A) + P(P)P(H|P)}$$
$$= .57868$$
$$P(P|H) \simeq .002$$
]

52. *An urn contains n balls, each of different colour, of which one is white. Two independent observers, each with probability 0.1 of telling the truth, assert that a ball drawn at random from the urn is white. Prove that the probability that the ball is, in fact, white is* $\frac{n-1}{n+80}$. *Also show that if* $n < 20$, *this probability is less than the probability that at least one of the observers is telling the truth.*

[Let A, B be two observers. Denote balls by $a_1, \cdots, a_n$; a_1 is the white ball.

$$P(A \text{ says } a_k|a_k) = P(B \text{ says } a_k|a_k) = \frac{1}{10},$$

$$P(A \text{ says } a_r|a_s) = P(B \text{ says } a_r|a_s) = \frac{9}{10} \cdot \frac{1}{n-1}$$

Hence, $P(a_1|A, B \text{ say } a_1) = \dfrac{P(a_1, (A, B \text{ say } a_1))}{P(A, B \text{ say } a_1)}$

$$= \frac{P(A, B \text{ say } a_1|a_1)\, P(a_1)}{\sum_{s=1}^{n} P(A, B \text{ say } a_1|a_s)\, P(a_s)}$$

$$= \frac{\left(\frac{1}{10}\right)^2 \cdot \frac{1}{n}}{\left(\frac{1}{10}\right)^2 \cdot \frac{1}{n} + (n-1)\left(\frac{9}{10}\right)^2 \left(\frac{1}{n-1}\right)^2 \cdot \frac{1}{n}} = \frac{n-1}{n+80} \to 1 \text{ as } n \to \infty$$

Again, prob. of at least one of A and B telling the truth$=1 - \frac{81}{100} = \frac{19}{100} > \frac{n-1}{n+80}$ for $n < 20$.]

53. *A safety device is designed to have a high conditional probability of operating when there is a failure (dangerous condition) and a high conditional probability of not operating when a failure does not occur. For a particular band of safety device both these probabilities are 0.98. Given that a dangerous condition occurs with probability .001, find the conditional probability that there was a failure when the safety device worked.*

[Let W : the event the safety device works.

F : there is a failure (dangerous condition)

Then $P(F) = .001,\ P(W|F) = 0.98\ P(W^c|F^c) = 0.98$

$$P(F|W) = \frac{P(F)\,P(W|F)}{P(F)\,P(W|F) + P(F^c)\,P(W^c|F^c)} = .04676\Big]$$

54. *A noisy communication channel is transmitting a message which is a sequence of 0's and 1's, a 1 being sent with probability p. A transmitted symbol may or may not be perturbed into the opposite symbol in the process of transmission. It is known that a 1 is converted into 0 with probability p_1 and a 0 is converted into 1 with probability p_2 during the process of transmission. Calculate*

(i) the conditional prob. that a 1 was sent given that a '0' is received at the reception.

(ii) the conditional prob. that a 0 was sent given that a '1' is received at the reception.

[Let A stand for an event 1 and B for 0. Thus $P(A) = p$, $P(B) = 1 - p$. It is given

$$P(B|A) = p_1,\quad P(A|A) = 1 - p,\quad P(A|B) = p_2,\quad P(B|B) = 1 - p_2$$

$$\text{(i)}\quad P(A|B) = \frac{pp_1}{pp_1 + (1-p)(1-p_2)},\qquad \text{(ii)}\quad P(B|A) = \frac{(1-p)p_2}{(1-p)p_2 + p(1-p_1)}.$$

55. *Let A be the event of a correct choice from among n alternatives. Let B be the event of knowing prior to the selection which of the alternatives is correct. Suppose $P(B) = p$, $0 < p < 1$. (a) Determine $P(B|A)$, (b) Show that $P(B|A) \geq P(B)$ and $P(B|A)$ increases with increasing n for fixed p.*

$$\Big[\ P(B) = p,\ P(B^c) = 1 - p, P(A|B) = 1, P(A|B^c) = \frac{1}{n}$$

$$P(B|A) = \frac{P(B)P(A|B)}{P(B)P(A|B) + P(B^c)P(A|B^c)} = \frac{p}{p + (1-p)\cdot\dfrac{1}{n}}$$

$$= \frac{p}{\dfrac{n-1}{n}p + \dfrac{1}{n}} > p,\quad \text{because}\quad \frac{n-1}{n}p + \frac{1}{n} < 1.$$

$$\text{Also}\quad P_{n+1}(B|A) - P_n(B|A) = \frac{(n+1)p}{np+1} - \frac{np}{(n-1)p+1}$$

$$= \frac{p(1-p)}{(np+1)\Big\{(n-1)p+1\Big\}} > 0.\Big]$$

56. *A multiplex transmission system has n channels. The channels are used independently and at any time there is a probability p that any given channel will be in* *use. What is the probability that exactly k of the channels will be in use? What is the probability that not more than k are in use at any given time ?*

$$\left[\text{(i)} \binom{n}{k} p^k (1-p)^{n-k} \quad \text{(ii)} \sum_{r=0}^{k} \binom{n}{r} p^r (1-p)^{n-r}\right]$$

57. *Five random digits are chosen. What is the probability that at least three of them will be the same ?* $\left[\sum_{r=3}^{5} \binom{5}{r} \left(\frac{1}{10}\right)^r \left(\frac{9}{10}\right)^{5-r}\right]$

58. *A coin is tossed until both head and tail appear at least once. Let z be the number of throws required. Find $P(z > n)$.*

$\left[z > n \text{ means the first } n \text{ throws are either all heads or all tails. } P(z > n) = \frac{2}{2^n} = \frac{1}{2^{n-1}}\right]$

59. *An unbiased coin is tossed 2n times. Find the probability $p(2n)$, that there will be an equal number of heads and tails. Prove that $p(2n)$ is a decreasing function of n.*

$$\left[p(2n) = \binom{2n}{n}\left(\frac{1}{2}\right)^n ; \quad \frac{p(2n+2)}{p(2n)} = \frac{2n+1}{2n+2} < 1\right]$$

60. *A game is played with dice as follows. A player rolls a pair of (fair) dice 5 times. He scores a 'hit' at any given throw in the sequence if he throws either a 7 or 11. He wins if he scores an even number of hits (zero counted as an even number) in the sequence of 5 throws. (a) What is the probability p of scoring a hit on any given throw ? (b) What is the probability that the dice thrower wins ?*

$$\left[\text{(a)} \quad p = \frac{2}{9}, \qquad \text{(b)} \quad (1-p)^5 + \binom{5}{2} p^2 (1-p)^3 + \binom{5}{4} p^4 (1-p)\right]$$

61. *A unbiased die is tossed $(r + s)$ times. Find the probability of getting $r(> s)$ consecutive sixes.*

[As $r > s$, only one sequence of r consecutive sixes can occur. Let p_a be the probability of getting r consecutive sixes in $(r+a)$ tosses. Hence $p_s = p_{s-1}+$ probability of getting a 6 at the $(r+s)$th throw immediately after obtaining $(r-1)$ 6's preceded by a number other than 6. $= p_{s-1} + \left(\frac{1}{6}\right)^r \cdot \frac{5}{6}$.

Similarly,

$$p_{s-1} = p_{s-2} + \left(\frac{1}{6}\right)^r \frac{5}{6}, \quad \cdots\cdots\cdots\cdots, \quad p_1 = p_0 + \left(\frac{1}{6}\right)^r \frac{5}{6}$$

Adding $\quad p_s = p_0 + \dfrac{5s}{6^{r+1}} = \dfrac{1}{6^r} + \dfrac{5s}{6^{r+1}} = \dfrac{6+5s}{6^{r+1}}$]

62. *A and B have, respectively, n + 1 and n coins. If they toss their coins simultaneously, what is the probability that (a) A will have more heads than B ? (b) A and B will have equal number of heads ? (c) B will have more heads than A ?*

[(a) Let P_n be the probability of A having more heads than B.

$$P_n = \sum_{x=1}^{n+1}\sum_{a=0}^{n}\binom{n+1}{x+a}\binom{n}{a}\frac{1}{2^{n+1}}.$$

Now considering the coefficient of t^x in

$$(1+t)^{n+1}\left(1+\frac{1}{t}\right)^n = \frac{(1+t)^{2n+1}}{t^n}, \sum_{a=0}^{n}\binom{n+1}{x+a}\binom{n}{a} = \binom{2n+1}{n+x}.$$

$$\text{Hence}\quad P_n = \frac{\sum_{x=1}^{n+1}\binom{2n+1}{n+x}}{2^{2n+1}} = \frac{\frac{1}{2}\cdot 2^{n+1}}{2^{n+1}} = \frac{1}{2}.$$

(b) The probability $= \dfrac{1}{2^{2n+1}}\displaystyle\sum_{x=1}^{n}\binom{n+1}{x}\binom{n}{x}$

$$= \frac{1}{2^{2n+1}}\sum_{x=1}^{n}\binom{n+1}{x}\binom{n}{n-x} = \frac{\binom{2n+1}{n}}{2^{2n+1}} = Q_n \quad \text{(say)}$$

(c) The probability $= 1 - P_n - Q_n = \dfrac{1}{2} - Q_n$]

63. *A and B play a series of games with Rs. a and Rs. b respectively. The stake is Rs. 1 on each game and no game can be drawn. If the probability of A winning a game is p (a constant), find the initial probability of A's ultimate win (winning B's total capital). Show that if b is infinitely large, (i) A is sure to be ruined for $p = \frac{1}{2}$ (ii) A's initial probability of ruin is $\frac{1}{2}$ if $p = 2^{\frac{1}{a}}/(1+2^{\frac{1}{a}})$.*

[Let p_n be the probability of A's ultimate win when his capital is n. Then

$$p_n = pp_{n+1} + (1-p)p_{n-1}, \tag{i}$$

because, when his capital is n, if he wins a game (prob. p) his capital will be $n+1$ (prob. of ultimate win at that time is p_{n+1}) and if he loses his capital will be $n-1$. Initial assumptions : $p_0 = 0$, $p_{a+b} = 1$. From (i)

$$p_{n+1} - p_n = \frac{1-p}{p}(p_n - p_{n-1}) = \left(\frac{1-p}{p}\right)^n p_1 \quad \text{(by repeated application)} \tag{ii}$$

$$\therefore\quad p_{n+1} = p_1\left[1 - \left(\frac{1-p}{p}\right)^{n+1}\right]\frac{p}{2p-1}$$

Using $p_{a+b} = 1$, we have

$$p_n = \frac{1-\left(\frac{1-p}{p}\right)^n}{1-\left(\frac{1-p}{p}\right)^{a+b}}.$$

Initial probability of A's win is $p_a = \dfrac{p^a-(1-p)^a}{p^{a+b}-(1-p)^{a+b}} \cdot p^b$. (iii)

Initial probability of A's ruin $=1-p_a$.

For $p=\frac{1}{2}$, $p_a = \dfrac{a}{a+b}$ (from (i), by similar analysis) $\to 0$ as $b \to \infty$.

For $p \neq \frac{1}{2}$ and $b \to \infty$, (from (iii)), $p_a = \frac{1}{2}$ when $p = \dfrac{2^{1/a}}{(1+2)^{1/a}}$.]

64. *The trains R-passanger and C-express arrive at B-junction on their way to Calcutta between 8-50 A.M. and 9-15 A.M. and each of the trains wait for the other for at most 10 minutes. Supposing both the trains arrive at junction at random moments between 8-50 A.M. and 9-15 A.M., what is the probability that one person travelling by the passanger on an urgent business will be able to get the express train at the junction to reach Calcutta in time ?* [Proceed as in example 2.50. Ans. 9/25]

65. *Two points x, y are tossed at random and independently onto a line segment of length L. What is the probability that the distance between the two points does not exceed l ?*

[Suppose x falls in the interval $(0, L)$ of the x-axis, y in the interval $(0, L)$ of the y-axis. The required probability is the probability that a point (x, y) tossed at random into the square will fall in the region bounded by the lines $y = l + x$ and $y = -l + x$.

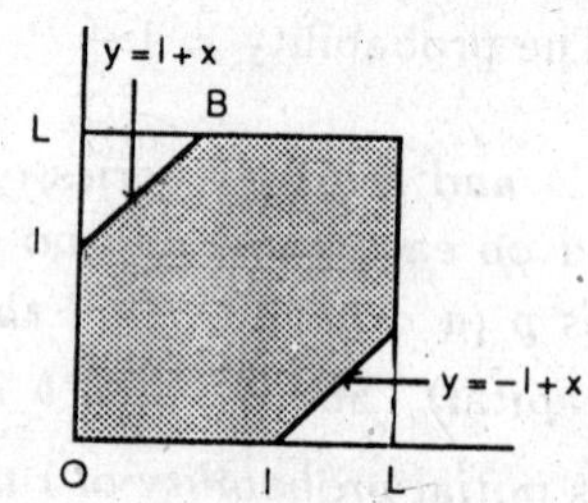

Hence probability $= \dfrac{\text{area of the shaded region}}{L^2}$

$$= \frac{2Ll - l^2}{L^2} = \frac{l(2L-l)}{L^2}\Bigg]$$

66. *A rod of length l is broken into three parts. What is the probability that a triangle can be formed from these parts ?*

[Let the length of the three parts be $a, b-a, l-b (a < b < l)$. A set of necessary and sufficient conditions for the three segments to form a triangle is that the length of any of the segments be less than the sum of the other two. Hence we must have

$$\begin{array}{lcr} a+(b-a) > l-b & \text{or} & b > (l/2) \\ (b-a)+(l-b) > a & \text{or} & a < (l/2) \\ (l-b)+a > b-a & \text{or} & b-a < (l/2). \end{array}$$

Hence $0 \leq a \leq \frac{l}{2}$ and $\frac{l}{2} < b \leq \frac{l}{2} + a$.

$$\text{The required probability} = \left(\int_0^{l/2}\int_{l/2}^{(l/2)+a} da\,db\right)\Big/\left(\int_0^{l}\int_0^{a} da\,db\right) = \frac{1}{4}$$

67. *(Bertrand's paradox). A chord is chosen at random in a circle of radius r. What is the probability its length will exceed the length of the side of the equilateral triangle inscribed in the circle ?*

Solution 1. Consider any chord. Draw a diameter perpendicular to it. It is seen that only chords that intersect the diameter between $\frac{1}{4}$ and $\frac{3}{4}$ of its length exceeds the side of the regular triangle inscribed in the circle. Hence probability is $\frac{1}{4}$ [fig. (16.a)].

Solution 2. Fix one end of the chord on the circle, say A, in advance. The tangent to the circle at A and the two sides of the regular triangle with vertex at A form three 60° angles. Clearly, chords falling in the middle angle satisfy the conditions of the event. Hence probability is $\frac{1}{3}$ [fig. (16.b)].

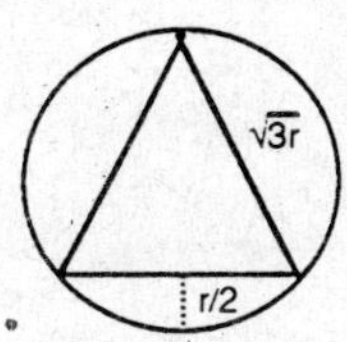

Fig. (16.0)

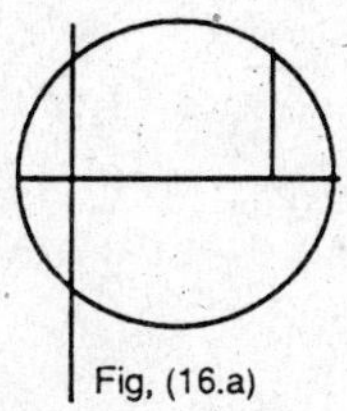
Fig, (16.a)

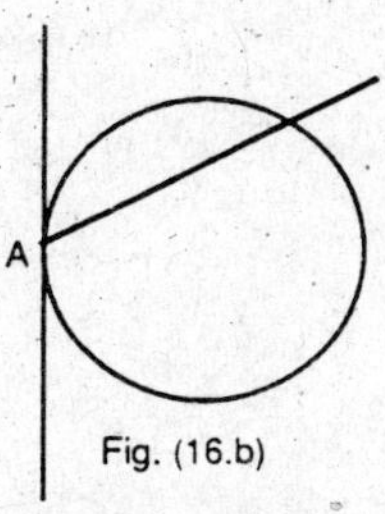

Fig. (16.b)

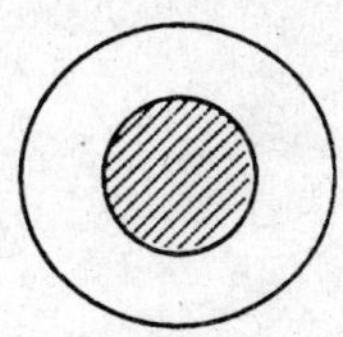
Fig. (16.c)

Solution 3. To define the position of the chord it is sufficient to specify its mid point. The chord will satisfy the conditions of the problem if its midpoint lies within distance $\frac{r}{2}$ from the centre i.e., within a circle concentric with the given one but with radius $\frac{r}{2}$ [fig (16.c)]. Hence probability $= \frac{1}{4}$.

3 AXIOMATIC APPROACH

3.1. Introduction

We have noted in the previous chapter the short-comings of the classical definition. The vagueness of the classical definition sometimes led to paradoxical results (example: Bertrand's paradox). We shall consider here axiomatic approach to the theory of probability.

The axioms are propositions that are regarded as true and are not proved within the framework of the subject. All other propositions of the theory are derived from the accepted axioms in a logical fashion. The axioms are formulated on the basis of a prolonged accumulation of facts and their logical analysis. The problem of an axiomatic construction of probability theory was first posed and solved by S. N. Bernstein in the first quarter of this century. A. N. Kolmogorov (1933) considered a different approach which closely relates the probability theory with the set theory and the metric theory of functions. We shall follow here Kolmogorov's approach.

We first consider some elements of algebra of sets, set functions and measure.

3.2. Some Elements of Theory of Measure

The union and intersection of a countable number of sets $A_1, A_2, \cdots$ are written as $\bigcup_{i=1}^{\infty} A_i$ and $\bigcap_{i=1}^{\infty} A_i$ respectively.

Given n sets $A_1, \cdots, A_n$ we can write

$$\bigcup_{i=1}^{n} A_i = A_1 + A_1'A_2 + A_1'A_2'A_3 + \cdots + A_1' \cdots A_{n-1}'A_n \tag{3.1}$$

as the sum of n disjoint sets $A_1, A_1'A_2, A_1'A_2'A_3, \cdots$ etc. For a countably infinite number of sets $A_1, A_2, \cdots$

$$\bigcup_{i=1}^{\infty} A_i = A_1 + A_1'A_2 + A_1'A_2'A_3 + \cdots$$

Definition 3.1. Sequence and limits of sets. To every integer $n = 1, 2, \cdots$ we assign a set. The ordered class of sets $A_1, A_2, \cdots$ is a sequence $\{A_n\}$ of sets.

$\{A_n\}$ is monotonically increasing if $A_n \subseteq A_{n+1}$. In this case $\bigcup_{k=1}^{n} A_k = A_n$; $\bigcup_{k=1}^{\infty} A_k = A$ is called the limit of $\{A_n\}$. Symbolically $A_n \uparrow A$.

$\{A_n\}$ is monotonically decreasing if $A_n \supseteq A_{n+1}$. Here $\bigcap_{k=1}^{n} A_k = A_n$; $\bigcap_{k=1}^{\infty} A_k = A$ is called the limit of $\{A_n\}$. symbolically $A_n \downarrow A$.

Now we define limit of arbitrary sequence of sets. For any sequence $\{A_n\}$ define

$$B_n = \inf_{k \geq n} A_k = \bigcap_{k=n}^{\infty} A_k$$

$$= \{w : w \text{ belong to all } A_k \text{ except perhaps for } A_1, \cdots, A_{n-1}\}.$$

$$C_n = \sup_{k \geq n} A_k = \bigcup_{k=n}^{\infty} A_k$$

$$= \{w : w \text{ belong to at least one of } A_n, A_{n+1}, \cdots\}.$$

Thus B_n is a monotonically increasing sequence with limit $B = \bigcup_{n=1}^{\infty} \bigcap_{k=n}^{\infty} A_k = \liminf A_k = \underline{\lim} A_n$. B is the set of all points which belong to almost all A_n (all but any finite number of sets). C_n is a monotonically decreasing sequence with limit $C = \bigcap_{n=1}^{\infty} \bigcup_{k=n}^{\infty} A_k$ $= \limsup A_k = \overline{\lim} A_n$. C is the set of all those points which belongs to infinitely many A_n.

Since every point which belongs to almost all A_n belongs to infinitely many A_n, $\underline{\lim} A_n \subseteq \overline{\lim} A_n$. If $\underline{\lim} A_n = \overline{\lim} A_n = A$ (say), the limit of $\{A_n\}$ is said to exist and A is called the limit of $\{A_n\}$.

Definition 3.2. Real valued point function on a set. A real valued point function f defined on a set S is a rule that associates with each element x of S a real number $f(x)$ called the value of f for the point x. The set S is called the domain of f and the set of values $\{f(x); x \in S\}$ is called the range of f. We denote the set containing the range of f as R.

If for each x, $f(x)$ is different, the function is one-to-one (1:1). If two or more x's have the same value $f(x)$, the function is many-to -one. An example of a 2-1 function is $f(x) = x^2$.

Thus a function f which maps S to R (we denote it as $S \xrightarrow{f} R$) assigns to each $x \in S$, a value $f(x) \in R$. $f(x)$ is called the image of x under f and x is called the pre-image of $f(x)$.

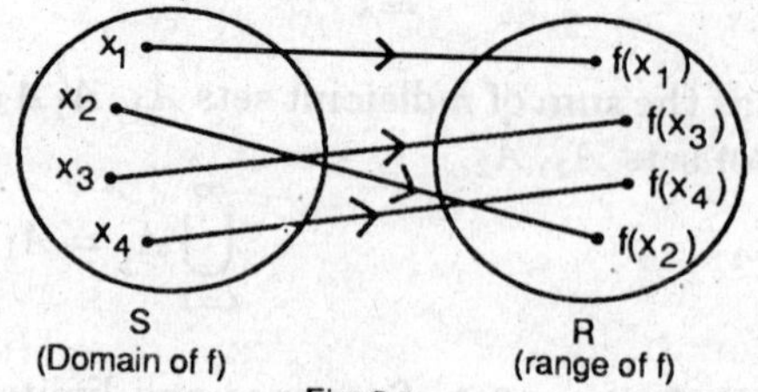

Fig. 3.1

If $R = \{f(x) : x \in S\}$, f is said to be a mapping from S onto R. If $\{f(x) : x \in S\} \in R$, f is said to be a mapping from S into R.

Real-valued point functions f_1 and f_2 are equal if $f_1(x) = f_2(x) \forall x \in S$. The function f is a numerical function if $R = R^1$, the real line, $\{-\infty \leq y \leq \infty\}$.

Definition 3.3. Inverse function of a point function on a set. If the set $B \in R$, then $\{x : f(x) \in B\}$ is called the inverse image of B under f and is denoted as $f^{-1}(B)$. Thus with each point function f (with domain S and range R) there is a set function f^{-1} whose domain is R and range is S. f^{-1} is called the inverse function of f. The following relations hold :

$$f^{-1}\left(\bigcap_k B_k\right) = \bigcap_k f^{-1}(B_k), \quad f^{-1}\left(\bigcup_k B_k\right) = \bigcup_k f^{-1}(B_k), \quad f^{-1}(B') = (f^{-1}(B))'.$$

Definition 3.4. Class of sets. A class of sets is a collection of sets. The elements of this class are sets. An example is $\mathcal{C} = \left\{\{a\}, \{b\}, \{a, b\}\right\}$.

Definition 3.5. Real valued set function. Let $\mathcal{C}$ be a class of sets. A rule f that associates with each element set C of $\mathcal{C}$ a real number $f(C)$ is called a real valued set function, defined on $\mathcal{C}$. The class $\mathcal{C}$ is called the domain of f, the set of values $\{f(C), C \in \mathcal{C}\}$ is called the range of f.

Definition 3.6. Finitely additive class of sets. A class $\mathcal{C}$ of sets is called finitely additive if it has the following properties : (i) $\phi \in \mathcal{C}$ (ii) if $C \in \mathcal{C}$ then $C' \in \mathcal{C}$ (iii) if C_i, $i = 1, \cdots, n \in \mathcal{C}$, then $\bigcup_{i=1}^{n} C_i \in \mathcal{C}$. It follows that $\bigcap_{i=1}^{n} C_i \in \mathcal{C}$. Clearly the universal set $\in \mathcal{C}$. Such a class is also called a Boolean field of sets (or a Boolean algebra).

Definition 3.7. Completely additive class of sets. A class $\mathcal{C}$ of sets is completely additive if it has the following properties : (i) $\phi \in \mathcal{C}$ (ii) if $C \in \mathcal{C}$, then $C' \in \mathcal{C}$ (iii) if $C_i, i = 1, 2, \cdots \in \mathcal{C}$ then $\bigcup_{i=1}^{\infty} C_i \in \mathcal{C}$. It follows that $\bigcap_{i=1}^{\infty} C_i \in \mathcal{C}$. Such a class is also called a Borel field or a σ-field (or a σ-algebra) of sets.

Given a class $\mathcal{C}$ of sets, the smallest (minimal) (Boolean field) σ-field containining $\mathcal{C}$ is the intersection of all (Boolean fields) σ-fields containing $\mathcal{C}$. There is one and only one minimal (Boolean field) σ-field containing $\mathcal{C}$. This is the minimal (Boolean field) borel field of subsets of $\mathcal{C}$. This field can be generated out of an initial subclass of sets in $\mathcal{C}$.

If $S = R^1$, the real line, $\mathcal{C}$ is the class of all real half-open intervals $(a, b]$, minimal σ-field containing $\mathcal{C}$ is the borel field $\mathcal{B}_1$ of subsets of the real line and its sets are borel sets of R^1. The same borel field is obtained if $\mathcal{C}$ is taken as the class of all intervals of the type (a, b) or $[a, b)$ or $[a, b]$.

If $\mathcal{S} = R^n$, n-dimensional Eucledian space (the set of all n-tuples of finite real numbers), $\mathcal{C}$ is the class of all n-dimensional half-open intervals $\{(x_1, \cdots, x_n),\ a_i < x_i \leq b_i,\ i = 1, \cdots, n\}$, minimal σ-field containing $\mathcal{C}$ is $\mathcal{B}_n$ and its sets are borel sets of R^n.

Definition 3.8. Finitely additive real-valued set function. Let $\mathcal{C}$ be a finitely additive class of sets. A real valued set function μ defined on $\mathcal{C}$ is finitely additive if (i) $\mu(\phi) = 0$ (ii) if $C_i \in \mathcal{C}$, $i = 1, \cdots, n$ are disjoint for any n, then $\mu\left(\bigcup_{i=1}^{n} C_i\right) = \sum_{i=1}^{n} \mu(C_i)$.

Definition 3.9. Completely additive real-valued set function. Let $\mathcal{C}$ be a σ-field of sets. A real-valued set function μ defined on $\mathcal{C}$ is called completely additive if (i) $\mu(\phi) = 0$ (ii) if $C_i \in \mathcal{C}$, $i = 1, 2, \cdots$ are disjoint then $\mu\left(\bigcup_{i=1}^{\infty} C_i\right) = \sum_{i=1}^{\infty} \mu(C_i)$.

Definition 3.10. Measure and measurable sets. A non-negative, additive, real-valued set function defined on an additive class of sets is called a measure. Thus if $\mathcal{C}$ is an additive class of subsets of a set $\mathcal{U}$ (universal set), a real-valued set function defined on $\mathcal{C}$ is called a measure if

(i) $\mu(C) \geq 0 \quad \forall C \in \mathcal{C}$, (ii) $\mu(\phi) = 0$

(iii) if $C_i \in \mathcal{C}$ $(i = 1, 2, \cdots)$ are disjoint such that $\bigcup_{1}^{\infty} C_i \in \mathcal{C}$ then $\mu\left(\bigcup_{1}^{\infty} C_i\right) = \sum_{1}^{\infty} \mu(C_i)$

The measure μ is finitely additive if (iii) is valid for finite number of disjoint sets; it is completely additive if (iii) also extends to a countable number of disjoint sets. The sets C are called measurable sets w.r.t. measure μ. The measure μ is called bounded if $\mu(\mathcal{U}) < \infty$.

Definition 3.11. Lebesgue measure and Lebesgue measurable sets. On the borel field $\mathcal{B}_1$ of subsets of R^1 a unique finite measure can be defined such that

$$\mu[(a, b]] = b - a$$

The measure μ is called the Lebesgue measure. Sometimes one may take the σ-field obtained by adjoining to $\mathcal{B}_1$ all subsets of measure zero and here μ is defined over this larger σ-field, $\mathcal{B}'_1$ (say). The members of $\mathcal{B}'_1$ are called Lebesgue measurable sets. μ defined over $\mathcal{B}_1$ or $\mathcal{B}'_1$ is called the Lebesgue measure.

Definition 3.12. A property which is true everywhere, except for a set of measure zero is said to hold almost everywhere (abbreviated to a.e.).

In the theory of probability, the universal set corresponds to the sample space $\mathcal{S}$ underlying a random experiment and the σ-field $\mathcal{A}$ corresponds to the class of all events (subsets of $\mathcal{S}$) under consideration.

3.3. Axiomatic Definition of Probability

Let (i) $\mathcal{S}$ be any abstract set, the sample space underlying a random experiment (ii) $\mathcal{A}$ be a σ-field (σ-algebra) of subsets of $\mathcal{S}$ (events resulting from the experiment) (iii) P be a measure defined on $\mathcal{A}$ such that $P(\mathcal{S}) = 1$. Then P is called a probability measure defined on $\mathcal{A}$. The triplet $(\mathcal{S}, \mathcal{A}, P)$ is called a probability space.

[Some authors denote $(\mathcal{S}, \mathcal{A})$ or $(\mathcal{S}, \mathcal{A}, P)$ as the sample space; we shall denote $\mathcal{S}$ as the sample space and $(\mathcal{S}, \mathcal{A}, P)$ as the probability sapce].

If the set $\mathcal{S}$ contains only a finite number of points (elementery events), we have a finite sample space. If $\mathcal{S}$ contains at most a countable number of points we have a discrete sample space. If $\mathcal{S}$ contains uncountably many points, we have a uncountable sample space. In particular if $\mathcal{S} = R^k$ or some rectangles in R^k, we have a continuous sample space.

In case the sample space $\mathcal{S}$ is finite, the σ-field $\mathcal{A}$ in the axiomatic definition coincides with all subsets of $\mathcal{S}$. In case $\mathcal{S}$ is infinite, our primary interest would usually be a well-defined family $\mathcal{C}$ of events (subsets of $\mathcal{S}$), which may not form an additive system. It is usual to take the σ-field $\mathcal{A}$ in the axiomatic definition as the (unique) minimal σ-field containing $\mathcal{C}$.

❑ **EXAMPLE 3.1.** *Suppose $\mathcal{S}$ is finite containing N elementary events $w_i, i = 1, \cdots N$. We assign to each w_i an arbitrary non-negative number p_i such that $\sum_{i=1}^{N} p_i = 1$. As the σ-field we take the class of all elements in $\mathcal{S}$, the total number of such events being 2^N. If A is an event $P(A) = \sum_{w_i \in A} P(w_i)$.*

❑ **EXAMPLE 3.2.** *Suppose $\mathcal{S}$ consists of a countable number of elementary events $w_i, i = 1, 2, \cdots$. To each w_i attach a non-negative quantity p_i, $\sum_{i=1}^{\infty} p_i = 1$. As the σ-field we take the class of all events of $\mathcal{S}$. For an event A, $P(A) = \sum_{w_i \in A} P(w_i)$.*

❑ **EXAMPLE 3.3.** *Let $\mathcal{S}$ be infinite and uncountable, say, R^1. Such cases occur when the experiment consists in taking measurement of characters such as the length, breadth, height, weight of an article. Here A is the class of all borel sets of R^1. As a probability function one may take arbitrary non-negative, countably additive set function on the borel sets such that $P(\mathcal{S}) = 1$. Such a function is uniquely defined by its values for the specific interval $(-\infty, x]$ where x is any finite real number.*

The similar method applies when $\mathcal{S}$ is a real plane R^2 or the real hyperplane R^n.

It follows from the definition that we have postulated the following axioms :

(a) With each event $A \in \mathcal{A}$ there is associated a non-negative real quantity $P(A)$, called its probability.

(b) $P(\mathcal{S}) = 1$

(c) If events $A_1, A_2, \cdots$ are pairwise mutually exclusive, $P\left(\bigcup_{i=1}^{\infty} A_i\right) = \sum_{i=1}^{\infty} P(A_i)$.

It is obvious that if axiom (c) is accepted, the result in axiom (c) also holds for a finite number of disjoint events, $A_1, \cdots, A_n$ i.e.

(c)′ $P\left(\bigcup_{i=1}^{n} A_i\right) = \sum_{i=1}^{n} P(A_i)$ if $A_i A_j = \phi$, $i \neq j = 1, \cdots, n$ (a proof has been given below).

In the classical definition there was no need to postulate the properties in axioms (b) and (c)′, since these properties were proved by us. The assertion in axiom (a) was contained in the calssical definition.

From axioms (a), (b), (c)′ we get the following important properties :

(i) Probability of an impossible event (ϕ) is zero $\left[\text{because } \phi = \bigcup_{i=1}^{n} \phi, \text{ union of disjoint sets } \phi;\ P(\phi) = P(\phi) + \cdots + P(\phi), \text{ which implies } P(\phi) = 0\right]$.

(ii) For any random event A, $P(A') = 1 - P(A)$ [because, $\mathcal{S} = A + A'$; $P(\mathcal{S}) = P(A) + P(A')$.]

(iii) For any random event $A, 0 \leq P(A) \leq 1$.

(iv) If event A implies B ($A \subset B$), then
(a) $P(A) \leq P(B)$ (b) $P(B - A) = P(B) - P(A)$
[because, $B = A + (B - A)$ and $P(B) = P(A) + P(B - A)$ by (c)′]

(v) Let A and B be two arbitrary events. Now
$A \cup B = A + (B - AB)$ and $B = AB + (B - AB)$. By axiom (c)′
$P(A \cup B) = P(A) + P(B - AB)$.
$P(B) = P(AB) + P(B - AB)$.

Hence

$$P(A \cup B) = P(A) + P(B) - P(A\,B)$$
$$\leq P(A) + P(B) \qquad [\because\ P(AB) \geq 0].$$

By induction,

$$P\left(\bigcup_{i=1}^{n} A_i\right) \leq \sum_{i=1}^{n} P(A_i).$$

All the properties (i)-(v) were proved in the classical approach. We note that (c)′ follows from (c) by taking $A_i = \phi$ for $i = n+1, n+2 \cdots$ and by applying the result (i).

All the axioms (a)-(c) satisfy the properties of the relative frequencies which provide the basis for the statistical definition of probability. If $f_n(A)$ is the number of times an event A occurs in n repetitions of an experiment then $\lim_{n\to\infty} r_n(A)$ has been defined as the probability of occurrence of A where $r_n(A) = f_n(A)/n$, the relative frequency of A. Obviously $(a)' r_n(A) \geq 0$, $(b)'$ if A is a sure event $r_n(A) = 1$. Axioms $(a), (b)$ are taken simply with the idea that the probability which is the long-term relative frequency should have these properties. Also if the events $A_i (i = 1, \cdots, k)$ are mutually exclusive and occur with relative frequency $r_n(A_i)$ then the relative frequency with which any one of them occurs is

$$r_n(A_1) + r_n(A_2) + \cdots + r_n(A_n).$$

Axiom (c) represents a generalisation of this property to the probability of any one of a countable number of events.

The following theorem is of basic importance.

• **THEOREM 3.1.** *(Continuity theorem). If $\{A_n\}$ is a monotone sequence of events belonging to $\mathcal{A}$, then $P\{\lim_{n\to\infty} A_n\} = \lim\{P(A_n)\}$.*

Proof. Suppose that $\{A_n\}$ is monotonically non-decreasing. Here $\lim A_n = \bigcup_{n=1}^{\infty} A_n$.

Now

$$A_n = A_1 + (A_2 - A_1) + \cdots + (A_n - A_{n-1}) = \sum_{k=1}^{n} (A_k - A_{k-1}),$$

sum of n disjoint subsets. Hence

$$\lim A_n = \sum_{k=1}^{\infty} (A_k - A_{k-1}),$$

$$\therefore \quad P(\lim A_n) = P\left(\sum_{k=1}^{\infty} (A_k - A_{k-1})\right)$$

$$= \sum_{k=1}^{\infty} P(A_k - A_{k-1}) \quad \text{(by axiom (c))}$$

$$= \lim_{n\to\infty} \sum_{k=1}^{n} P(A_k - A_{k-1}) = \lim P(A_n)$$

Suppose now $\{A_n\}$ is monotonically non-increasing. The sequence $\{A_n^c\}$ is then monotonically non-decreasing and each A_n^c belongs to $\mathcal{A}$.

Hence

$$P(\lim A_n^c) = \lim P(A_n^c)$$

$$\text{i.e.} \quad P\left[(\lim A_n)^c\right] = \lim[1 - P(A_n)]$$

$$\left[\text{Since} \quad \lim A_n^c = \bigcup_{n=1}^{\infty} A_n^c = \left(\bigcap_{n=1}^{\infty} A_n\right)^c = (\lim A_n)^c\right]$$

$$\text{i.e.} \quad 1 - P(\lim A_n) = 1 - P \lim P(A_n)$$

$$\text{and hence} \quad P(\lim A_n) = \lim P(A_n).$$

3.4. Conditional Probability

Consider the probability space $(\mathcal{S}, \mathcal{A}, \mathcal{P})$. Let A_1, A_2 be two events in $\mathcal{A}$ such that $P(A_i) > 0$, $i = 1, 2$. The ratio

$$P(A_2|A_1) = \frac{P(A_1 \cap A_2)}{P(A_1)} \tag{3.2}$$

is called the conditional probability of the event A_2 given that A_1 occurs. When $P(A_1) = 0$, $P(A_1 \cap A_2) = 0$ and $P(A_2|A_1)$ remains undefined.

For fixed A_1 in $\mathcal{A}$ such that $P(A_1) > 0$, $(\mathcal{S}, \mathcal{A}, P(\cdot|A_1))$ is a probability space where $P(\cdot|A_1)$ is a probability measure which takes the value $P(A_2|A_1)$ on the event $A_2 \in \mathcal{A}$. To prove this it is enough to prove

• **THEOREM 3.2.** *$P(A_2|A_1)$ given in (3.2) satisfies axioms (a), (b), (c) provided* $P(A_1) > 0$.

Proof. (a) $P(A_2|A_1) \geq 0$ since $P(A_1 \cap A_2) \geq 0$

(b) $P(\mathcal{S}|A_1) = \dfrac{P(A_1)}{P(A_1)} = 1$

(c) Let $\{B_i\}$ be a sequence of disjoint events in $\mathcal{A}$. We have

$$\left\{\bigcup_{i=1}^{\infty} B_i\right\} \cap A_1 = \bigcup_{i=1}^{\infty}(B_i \cap A_1).$$

Since $\{B_i\}$ is a sequence of disjoint events in $\mathcal{A}$, $\{B_i \cap A_1\}$ is a sequence of disjoint events in $\mathcal{A}$. Hence

$$P\left[\left\{\bigcup_{i=1}^{\infty} B_i\right\} \cap A_1\right] = \sum_{i=1}^{\infty} P(B_i \cap A_1).$$

Dividing both sides by $P(A_1)$

$$P\left(\left(\bigcup_{i=1}^{\infty} B_i\right)\middle| A_1\right) = \sum_{i=1}^{\infty} P(B_i|A_1)$$

Hence the proof.

(3.2) can also be written as

$$P(A_1 \cap A_2) = P(A_1)P(A_2|A_1). \tag{3.3}$$

If $P(A_2|A_1) = P(A_2)$, then (3.3) reduces to the case of independent events A_1, A_2 and we have

$$P(A_1 \cap A_2) = P(A_1)P(A_2).$$

More generally, it can be shown that :

If $A_1, A_2, \cdots$ is a finite or countably infinite sequence of events in $\mathcal{A}$ having probability measure P such that $P(A_1) > 0$, $P(A_1 \cap A_2) > 0$, $P(A_1 \cap A_2 \cap A_3) > 0, \cdots$, then

$$P(A_1 \cap A_2 \cap \cdots) = P(A_1)P(A_2|A_1)P(A_3|A_1 \cap A_2)\cdots P(A_n|A_1 \cap A_2 \cap \cdots \cap A_{n-1})\cdots \quad (3.4)$$

In the case of mutual independence of $A_1, A_2, \cdots$, (3.4) reduces to

$$P(A_1 \cap A_2 \cap \cdots) = P(A_1)P(A_2)P(A_3)\cdots$$

• **THEOREM 3.3.** *If $A_1, A_2, \cdots$ is a finite or countably infinite sequence of disjoint events in $\mathcal{A}$ having nonzero probabilities, where $\bigcup_{i=1}^{\infty} A_i = \mathcal{S}$ (sample sapce), and if A is any event in $\mathcal{A}$ then*

$$P(A) = P(A_1)P(A|A_1) + P(A_2)P(A|A_2) + \cdots$$

Proof. Since $A_1, A_2, \cdots$ is a sequence of disjoint events in $\mathcal{A}$ and $\cup A_i = \mathcal{S}$, $A \cap A_1$, $A \cap A_2, \cdots$, is a sequence of disjoint events in $\mathcal{A}$ and $\bigcup_{i=1}^{\infty}(A \cap A_i) = A$. Hence

$$P(A) = \sum_{i=1}^{\infty} P(A \cap A_i) = \sum_{i=1}^{\infty} P(A_i)P(A|A_i).$$

Remarks 1. Theorems of the previous chapter (theorem 2.2 onwards) were proved for a finite number of events in $\mathcal{A}$. The axiomatic approach entends the validity of these theorems to countably infinite sequence of events in $\mathcal{A}$. Thus theorem 2.2 can now be modified as

• **THEOREM 3.4.** *If $\{A_i\}$ is a sequence of disjoint events in $\mathcal{A}$,*

$$P\left(\bigcup_{i=1}^{\infty} A_i\right) = \sum_{i=1}^{\infty} P(A_i) - \sum\sum_{i<j=1}^{\infty} P(A_i \cap A_j) + \sum\sum\sum_{i<j<k=1}^{\infty} P(A_i \cap A_j \cap A_k) - \cdots\cdots$$

Theorem 2.8 (theorem of total probability) is a particular case of theorem 3.3 for finite number of events. Independence of events have been discussed at length at the previous chapter.

3.5. Independent Trials

Consider two random experiments e_1, e_2 which are independent—results of e_1 do not depend on those of e_2 and vice-versa—with sample sapces $\mathcal{S}_1, \mathcal{S}_2$ respectively. The Cartesian product $\mathcal{S}$ of the sample sapces $\mathcal{S}_1$ and $\mathcal{S}_2$ is the set of all ordered pairs of elementary events $(w^{(1)}, w^{(2)})$ where $w^{(1)} \in \mathcal{S}_1$, $w^{(2)} \in \mathcal{S}_2$. We write

$$\mathcal{S} = \mathcal{S}_1 \times \mathcal{S}_2.$$

Similarly if $A^{(1)}, A^{(2)}$ are events in (subsets of) $\mathcal{S}_1, \mathcal{S}_2$ respectively, the cartesian product of $A^{(1)}$ and $A^{(2)}$ is the event $A = \left\{(w^{(1)}, w^{(2)}); (w^{(1)} \in A^{(1)}, w^{(2)} \in A^{(2)})\right\}$ and we write

$$A = A^{(1)} \times A^{(2)}.$$

The event A is called the joint occurance of $A^{(1)}$ and $A^{(2)}$. A is empty if and only if $A^{(1)}$ or $A^{(2)}$ is an empty set.

$\mathcal{S}_1, \mathcal{S}_2$ are called the components or marginal sample spaces of $\mathcal{S}$. If we take all points $(w^{(1)}, w^{(2)})$ in $\mathcal{S}$ such that $w^{(1)} \in A^{(1)}$ we obtain a cylindrical set in $\mathcal{S}$ which is the cartesian product $A^{(1)} \times \mathcal{S}_2$. Thus

$$A^{(1)} \times \mathcal{S}_2 = \left\{(w^{(1)}, w^{(2)}) : w^{(1)} \in A^{(1)}\right\}.$$

Similarly $\mathcal{S}_1 \times A^{(2)}$ is the cylindrical set $\{(w^{(1)}, w^{(2)}) : w^{(2)} \in A^{(2)}\}$. Thus $A^{(1)} \times A^{(2)}$, $A^{(1)} \times \mathcal{S}_2$, $\mathcal{S}_1 \times A^{(2)}$ are events in the sample space $\mathcal{S} = \mathcal{S}_1 \times \mathcal{S}_2$ such that the cartesian product $A^{(1)} \times A^{(2)}$ is the intersection of the cylindrical sets $A^{(1)} \times \mathcal{S}^{(2)}$ and $\mathcal{S}_1 \times A^{(2)}$.

$$A^{(1)} \times A^{(2)} = (A^{(1)} \times \mathcal{S}_2) \cap (\mathcal{S}_1 \times A^{(2)}).$$

The notion extends readily to any finite or countably infinite number of events $A^{(1)}, A^{(2)}, \cdots$ in sample spaces $\mathcal{S}_1, \mathcal{S}_2, \cdots$ respectively.

If $\mathcal{A}^{(1)}, \mathcal{A}^{(2)}$ are σ-fields of events of $\mathcal{S}_1, \mathcal{S}_2$ respectively, then the minimal borel field of subsets of $\mathcal{S}, \mathcal{A}$ can be generated out of the sets of the type $A^{(1)} \times A^{(2)}$, $A^{(1)} \in F^{(1)}$, $A^{(2)} \in F^{(2)}$ where $F^{(1)}, F^{(2)}$ are the Boolean fields generating $\mathcal{A}^{(1)}, \mathcal{A}^{(2)}$ respectively.

Let $P^{(1)}, P^{(2)}$ be probability measures on $\mathcal{S}_1, \mathcal{S}_2$ respectively. For any set $A = A^{(1)} \times A^{(2)}$ where $A^{(1)} \in \mathcal{A}^{(1)}$, $A^{(2)} \in \mathcal{A}^{(2)}$ let us assign a probability measure by the formula

$$P(A) = P^{(1)}(A^{(1)}) \times P^{(2)}(A^{(2)}). \tag{3.5}$$

In this case we say the component probability spaces $(\mathcal{S}_1, \mathcal{A}^{(1)}, P^{(1)})$ and $(\mathcal{S}_2, \mathcal{A}^{(2)}, P^{(2)})$ are statistically independent. The probability space $(\mathcal{S}, \mathcal{A}, P)$ where P is defined by (3.5) is the cartesian product of the two component probability spaces, is

$$(\mathcal{S}_1, \mathcal{A}^{(1)}, P^{(1)}) \times (\mathcal{S}_2, \mathcal{A}^{(2)}, P^{(2)}).$$

The notion extends readily to more than two probability spaces.

❑ **EXAMPLE 3.4.** *Consider n Bernoullian trials, $e_1, e_2, \cdots, e_n$. The probability space for e_i is $(S_i, \mathcal{A}^{(i)}, P^{(i)})$ where $S_i = \{S', F\}$ [S' = success, F = failure], $\mathcal{A}^{(i)} = \{\phi, \{S'\}, \{F\}, \mathcal{S}_i\}$, $P^{(i)}(S') = p$, $P^{(i)}(F) = 1-p$. The product sample sapce $\mathcal{S} = \mathcal{S}_1 \times \cdots \times S_n = \left\{(w^{(1)}, \cdots, w^{(n)}),\right.$ $\left. w^{(i)} = S' \text{ or } F, i = 1, \cdots n\right\}$ consists of 2^n points, $\mathcal{A}$ is the minimal borel field over $\mathcal{S}$, the probability measure P over $\mathcal{A}$ is such that*

$$P\Big\{(w^{(1)}, \cdots w^{(n)})\Big\} = p^r(1-p)^{n-r}$$

where $w^{(i_1)} = \cdots = w^{(i_r)} = S'$ and the remaining w's are each F.

❑ **EXAMPLE 3.5.** *A and B each draws two cards from a full pack without replacement. The probability space for A's experiment is $(\mathcal{S}_1, \mathcal{A}^{(1)}, P^{(1)})$ where $\mathcal{S}_1$ consists of $\binom{52}{2}$ elementary events $w, \mathcal{A}^{(1)}$ is the minimal σ-field of subsets of $\mathcal{S}_1$ and $P^{(1)}$ is such that $P^{(1)}(w_i) = 1\Big/\binom{52}{2}$ $\forall i = 1, \cdots, 52$. The sample space for B is identical. The joint occurence of the event $A^{(1)}$ – A has two aces, and $A^{(2)}$ – B has two queens is the event $A = A^{(1)} \times A^{(2)}$. The probability measure P on the product probability space $(\mathcal{S}, \mathcal{A}, P) = (\mathcal{S}_1, \mathcal{A}^{(1)}, P^{(1)}) \times (\mathcal{S}_2, \mathcal{A}^{(2)}, P^{(2)})$ is such that $P(E) = P^{(1)}(E_1) \times P^{(2)}(E_2)$ for each $E_1 \in \mathcal{A}^{(1)}$, $E_2 \in \mathcal{A}^{(2)}$ and $E \in \mathcal{A}$ where $E = E_1 \times E_2$. Now $P^{(1)}(A^{(1)}) = \binom{4}{2}\Big/\binom{52}{2}\Big(= \sum_{w_i \in A^{(1)}} P^{(1)}(w_i)\Big) = P^{(2)}(A^{(2)})$ and hence $P(A) = \left[\binom{4}{2}\Big/\binom{52}{2}\right]^2$.*

4 EXPECTATION OF A DISCRETE RANDOM VARIABLE

4.1. Introduction

In the previous chapters we have seen how from a given random experiment we can assign a probability function to all its possible outcomes. Calculation of probability of an event starting from the probabilities of each elementary event would often be impracticable. We introduce here the concept of a random variable.

4.2. Random Variable

Definition 4.1. A random variable (r.v.) is a real valued function defined on the sample space $\mathcal{S}$ of a random experiment (a more elaborate definition is given in the next chapter).

To fix the idea suppose $\mathcal{S}$ consists of a finite number of points $w_1, w_2, \cdots, w_m$. Let X be a point function defined on $\mathcal{S}$ such that its value on w_i is $X(w_i) = x_i, i = 1, \cdots, m$. Here X is a random variable with domain $\mathcal{S}$ and range $\{x_i, i = 1, \cdots, m\}$.

❑ **EXAMPLE 4.1.** *A coin is tossed three times. Here* $\mathcal{S} = \{w_1 = (HHH), w_2 = (HHT), w_3 = (HTH), w_4 = (HTT), w_5 = (THH), w_6 = (THT), w_7 = (TTH), w_8 = (TTT)\}$. *Let* X *denote the number of heads. Then* $X(\omega_1) = 3, X(w_2) = X(w_3) = X(w_5) = 2, X(w_4) = X(w_6) = X(w_7) = 1, X(w_8) = 0$. X *is a r.v. with domain* $\mathcal{S}$ *and range* $\{0, 1, 2, 3\}$.

❑ **EXAMPLE 4.2.** *A coin is tossed until a head appears. Let X be the number of tosses (including the toss in which H occurs) required. Here $S = \{w_1 = (H), w_2 = (TH), w_3 = (TTH), \cdots\}$ and X assumes the countably infinite number of values $1, 2, 3 \cdots$ with $X(w_1) = 1, X(w_2) = 2$, etc.*

In both the above examples X takes only the discrete values and are thus discrete random variables. Sometimes, as in the cases of observing the length of a leaf, height of a person, diameter of a product, weight of an individual, X can take any value within a certain range. X in such cases are continuous random variables (elaborate definitions are given in the next chapter).

We shall in this chapter consider in some details the properties of a discrete random variable. We shall assume throughout X takes only a finite number of values $x_1, x_2, \cdots, x_m$. All the results presented here will hold if X takes countably infinite number of values, $x_1, x_2, \cdots$.

Definition 4.2. Let X be a discrete r.v. defined on S, taking values $x_1, x_2, \cdots, x_m$ with probabilities $p_1, p_2, \cdots, p_m$. The set of values of X, together with their corresponding probabilities $\{x_i, p_i, i = 1, \cdots, m\}$ is called the probability distribution of X. We must have $p_i \geq 0, \sum_{i=1}^{m} p_i = 1$.

For example 4.1, writing $P(X = i) = p_i, p_o = \frac{1}{8}, p_1 = \frac{3}{8}, p_2 = \frac{3}{8}, p_3 = \frac{1}{8}, \{i, p_i\}$ is the probability distribution of X. For example 4.2, $p_i = \frac{1}{2^i}; \{i, p_i; i = 1, 2, \cdots\}$ is the probability distribution of X.

4.3. Expectation

Definition 4.3. Let X be a discrete r.v. with probability distribution $\{x_i, p_i, i = 1, \cdots, m\}$. The expected value of X, also called mean value of X, is defined as

$$E(X) = \sum_{i=1}^{m} x_i p_i.$$

As noted before, p_i is defined as the long-run relative frequency with which X takes the value x_i; hence $E(X)$ may be defined as the long run average value of X.

❑ **EXAMPLE 4.3.** *An unbiased die is tossed. Calculate its expected face value.*

The elementary events are $\{1\}, \cdots \{6\}$. The r.v. X, denoting the face value, takes value i with probability $\frac{1}{6}$, $i = 1, \cdots, 6$. Thus $E(X) = \frac{1}{6}\sum_{i}^{6} i = \frac{7}{2}$.

❑ **EXAMPLE 4.4.** *For example 4.2, calculate the value of the number of trials required (including the last toss in which a head has to appear).*

Here, $P(X = i) = \frac{1}{2^i}$ $(i = 1, 2, \cdots)$. Thus $E(X) = \sum_{i=1}^{\infty} \frac{i}{2^i} = \frac{1}{2} + 2 \cdot \frac{1}{2^2} + 3 \cdot \frac{1}{2^3} + \cdots = 2.$

Definition 4.4. Indicator random variable. For an event A associated with a random experiment we define a random variable $I_A(w)$ on each point w in the sample space $\mathcal{S}$ such that

$$I_A(w) = 1(0) \quad \text{if} \quad w \epsilon A \quad (w \notin A).$$

$I_A(w)$ is called the indicator r.v. (function) of the set A.

$$E\{I_A(w)\} = 1\sum_{w\epsilon A} p(w) + 0\sum_{w \notin A} p(w) = \sum_{w \epsilon A} p(w) = P(A).$$

Thus probability of an event A is the expectation of its indicator r.v. $I_A(w)$.

Indicator r.v. will often be useful in solving many problems in expectation.

Definition 4.5. Function of a r.v. and its expectation. Let X be a discrete r.v. with domain $\mathcal{S}$ and range $R_x = \{x_1, x_2, \cdots, x_m\}$. Let $H(X)$ be a function of X defined on R_x with range of values $\{h_1, h_2 \cdots, h_t\}$. Clearly $H(X)$ is also a random variable.

$$P[H(X) = h_i] = \phi_i \quad \text{(say)} = \sum_{\{x_j : H(x_j) = h_i\} = B \text{ (say)}} P(X = x_j)$$

$$= \sum_{\substack{w_k : \\ X(w_k) = x_j \epsilon B}} P\{w_k\} \geq 0.$$

It follows that $\sum_{i=1}^{t} \phi_i = 1$. Thus $\{i, \phi_i, i = 1, \cdots, t\}$ is probability distribution of $H(X)$. It expectation is

$$E[H(X)] = \sum_{i=1}^{t} h_i P[H(X) = h_i] = \sum_{i=1}^{t} h_i \phi_i$$

❑ **EXAMPLE 4.5.** *Let X be a r.v. with* $P(X_k = \pm m) = \frac{1}{2k}, m = 1, \cdots, k$. *Find* $E(X_k^2)$.

Here X_k^2 is a r.v. with $P(X_k^2 = m^2) = P(X_k = m) + P(X_k = -m) = \frac{1}{k}, m = 1, \cdots, k$.

$$E(X_k^2) = \frac{1}{k} \sum_{1}^{k} m^2 = \frac{(k+1)(2k+1)}{6}.$$

We now consider some theorems on expectation of a discrete r.v. The properties of expectation of a r.v. in the general case have been considered in details in section 6.1.

• **THEOREM 4.1.** *Let $X = c$, a constant, whatever be the outcome of the experiment. Then $E(X) = c$.*

Proof. Follows obviously since $x_i = c, i = 1, \cdots, m$.

• **THEOREM 4.2.** *Let $Y = bX$, b an arbitrary constant, be a function of X. Then $E(Y) = bE(X)$.*

Proof. $E(Y) = \sum_{i=1}^{m} b x_i p_i = b \sum_{i=1}^{m} x_i p_i = bE(X)$.

• **THEOREM 4.3.** *If X, Y are two r.v.'s defined on $\mathcal{S}$ and $Z = X + Y$ then $E(Z) = E(X) + E(Y)$.*

Proof. Let X take the values $x_1, \cdots, x_m$ and Y the values $y_1, \cdots, y_m$ on elementary events $w_1, \cdots, w_m$ with probabilities $p_1, \cdots, p_m$ respectively. Clearly Z is a r.v. taking value $x_i + y_i$ on w_i with probability p_i. Thus

$$E(Z) = \sum_{i=1}^{m}(x_i + y_i)p_i = \sum_{i=1}^{m} p_i x_i + \sum_{i=1}^{m} p_i y_i = E(X) + E(Y).$$

COROLLARY 4.1. *If $Y = a + bX$, a, b being arbitrary constants, $E(Y) = a + bE(X)$.*

Proof. $E(Y) = E(a) + E(bX)$ (Th. 4.3) $= a + bE(X)$ (Th. 4.1 and 4.2).

THEOREM 4.4. *If $X_1, \cdots, X_n$ are n r.v.'s defined on $\mathcal{S}$ and if $Z = X_1 + \cdots + X_n$ then $E(Z) = E(X_1) + \cdots + E(X_n)$.*

Proof. Let X_j take the value x_{ji} on w_i with probability p_i, $i = 1, \cdots, m; j = 1, \cdots, n$.

Z is a r.v. taking value $\sum_{j=1}^{n} x_{ji}$ with probability p_i on w_i. Hence

$$E(Z) = \sum_{i=1}^{m}\left(\sum_{j=1}^{n} x_{ji}\right)p_i = \sum_{j=1}^{n}\left(\sum_{i=1}^{m} x_{ji}p_i\right) = \sum_{j=1}^{n} E(X_j).$$

COROLLARY 4.2. *If $X_1, \cdots, X_n$ are n r.v.'s defined on $\mathcal{S}$ and if $Z = c + a_1X_1 + \cdots + a_nX_n$, $c, a_1 \cdots, a_n$ being arbitrary constants then $E(Z) = c + \sum_{i=1}^{n} a_i E(X_i)$.*

Proof. $E(Z) = E(c) + \sum_{i=1}^{n} E(a_i X_i) = c + \sum_{i=1}^{n} a_i E(X_i)$.

❑ **EXAMPLE 4.6.** *The face cards are discarded from two decks of cards. Find the expectation of the sum of the number drawn.*

Let X_i be the r.v. denoting the number drawn from the ith deck $(1 = 1, 2)$. X_i takes the values $1, 2, \cdots, 10$ each with probability $\frac{1}{10}$. Hence $E(X_i) = \frac{1}{10}\sum_{i=1}^{10} i = \frac{11}{2}$. $E(X_1 + X_2) = 11$.

❑ **EXAMPLE 4.7.** *Ten numbers are selected from 30 members $1, 2, \cdots, 30$ each with equal probability and without replacement. Find the expectation of the sum of the selected numbers.*

Let X_i be the number drawn at the ith draw $(i = 1, \cdots, 10)$. Since the numbers are drawn without replacement each of $X_1, \cdots, X_{10}$ will be different.

$$P(X_1 = i) = \frac{1}{30}, \quad i = 1, \cdots, 30.$$

$$P(X_2 = i) = \sum_{j(\neq i)=1}^{30} P(X_1 = j)P(X_2 = i|X_1 = j) = \sum_{j(\neq i)=1}^{30} \frac{1}{30} \cdot \frac{1}{29} = \frac{1}{30}, \quad i = 1, \cdots, 30.$$

It can be similarly shown $P(X_k = i) = \dfrac{1}{30}, i = 1, \cdots, 30$ and $k = 3, \cdots, 10$. Thus each of $X_1, \cdots, X_{10}$ has the same probability distribution and hence the same expectation. Now $E(X_1) = \dfrac{31}{2}$; hence $E(X_1 + \cdots + X_{10}) = 155$.

❑ **EXAMPLE 4.8.** *A box contains a white and b black balls; c balls are drawn one by one with replacement. Find the expectation of the number of white balls drawn.*

Let X_i be a random variable which takes the value 1 if the ith ball drawn is a white one and 0 if it is black. Then the number Z of white balls drawn is

$$Z = X_1 + X_2 + \cdots + X_c$$

Now

$$P(X_i = 1) = \frac{a}{a+b}, \quad P(X_i = 0) = \frac{b}{a+b}.$$

Hence $E(X_i) = \dfrac{a}{a+b}$. Thus $E(Z) = \dfrac{ac}{a+b}$.

❑ **EXAMPLE 4.9.** *n balls are distributed at random among m cells. Find the expected number of cells that remain empty.*

For the cell i define a random variabe X_i such that $X_i = 1(0)$ if the ith cell is empty (occupied). Then $X = \sum_{i=1}^{m} X_i$ is the total number of empty cells. Now

$$E(X_i) = P(X_i = 1) = \frac{(m-1)^n}{m^n}.$$

Hence $E(X) = \dfrac{m(m-1)^n}{m^n}$.

Lemme 4.1. If possible values of X are $0, 1, 2, \cdots$,

$$E(X) = P(X > 0) + P(X > 1) + P(X > 2) + \cdots \tag{4.1}$$

Proof. The rhs of (4.1)

$$= \{P(X=1) + P(X=2) + \cdots\} + \{P(X=2) + P(X=3) + \cdots\}$$
$$+ \{P(X=3) + P(X=4) + \cdots\} + \cdots$$
$$= \sum_{i=1}^{\infty} iP(X=i) = E(X).$$

❑ **EXAMPLE 4.10.** *Balls are successively distributed among m cells. Find the expected number of balls that must be distributed before all the cells are occupied.*

Let Z be the r.v. denoting the number of balls required to achieve a full occupancy. We have seen in exercise 9, section 2.16, the probability that all the m cells are occupied when n balls are distributed at random is

$$p_{n,m} = \sum_{k=0}^{m} (-1)^k \binom{m}{k} \left(\frac{m-k}{m}\right)^n$$

Hence $P(Z \leq n) = P_{n,m}$.

$$E(Z) = \sum_{n=0}^{\infty} P(Z > n) = \sum_{n=0}^{\infty}(1 - p_{n,m})$$

$$= \sum_{n=0}^{\infty}\left[\binom{m}{1}\left(\frac{m-1}{m}\right)^n - \binom{m}{2}\left(\frac{m-2}{m}\right)^n + \cdots + (-1)^m\binom{m}{m-1}\left(\frac{1}{m}\right)^n\right]$$

$$= m\left[\binom{m}{1} - \frac{1}{2}\binom{m}{2} + \frac{1}{3}\binom{m}{3} - \cdots + (-1)^m\frac{1}{m-1}\binom{m}{m-1}\right] \quad (4.2)$$

$$= m\left[1 + \frac{1}{2} + \cdots + \frac{1}{m}\right] \quad (4.3)$$

The equality of (4.2) and (4.3) follows by integrating the identity $\sum_{i=0}^{m-1}(1-t)^i = \frac{1-(1-t)^m}{t}$ over t in the range (0,1).

4.4. Variance, Covariance, Correlation Coefficient

Definition 4.6. Variance of X, denoted as $V(X)$, is a measure of dispersion or spread of X around its mean value $E(X)$.

$$V(X) = E(X - E(X))^2 = E(X^2) - (E(X))^2.$$

This is also often denoted as σ_X^2. Its positive square root σ_x is called standard deviation (s.d.) of X.

• **THEOREM 4.5.** *If $X = c$, a constant whatever be the outcome of the experiment, $V(X) = 0$.*

Proof. $X - E(X) = c - c = 0$. Thus $V(X) = 0$.

• **THEOREM 4.6.** *If $Y = cX$, c a constant, $V(Y) = c^2V(X)$.*

Proof. $V(Y) = E(cX - cE(X))^2 = c^2V(X)$.

• **THEOREM 4.7.** *If $Y = a + bX$, a, b arbitrary constant, $V(Y) = b^2V(X)$.*

Proof. $Y - E(Y) = b(X - E(X))$.

❑ **EXAMPLE 4.11.** *Find the variance of the number of points X obtained in tossing a unbiased die.*

Here $E(X^2) = \frac{1}{6}\sum_{1}^{6} i^2 = \frac{91}{6}$, $E(X) = \frac{7}{2}$. Hence $V(X) = \frac{35}{12}$.

❑ **EXAMPLE 4.12.** *A coin falls head with probability p. Find the variance of the number of heads.*

Let $X = 1(0)$ if the coin turns up a head (tail). $E(X) = 1 \cdot p + 0 \cdot (1 - p) = p$. $E(X^2) = 1^2 \cdot p = p$. $V(X) = p(1 - p)$.

Definition 4.7. Joint distribution of two r.v.s. Let X,Y be two r.v.'s, X taking the values $x_1,\cdots,x_m$ with probabilities $p_1,\cdots,p_m$ respectively and Y taking values $y_1,\cdots,y_n$ with probabilities $p'_1,\cdots,p'_n$ respectively, $\sum_{i=1}^{m} p_i = \sum_{j=1}^{n} p'_j = 1$. Clearly XY is a r.v. with possible values x_iy_j, $i = 1,\cdots,m$, $j = 1,\cdots,n$. Let p_{ij} be the probability that XY takes the value x_iy_j, $p_{ij} \geq 0$, $\sum_{i=1}^{m}\sum_{j=1}^{n} p_{ij} = 1$. The set of values $\{x_iy_j, p_{ij}\}$ is the joint probability distribution of X and Y. In case, X,Y are stochastically independent, $p_{ij} = p_ip'_j$. We have

$$p_i = \sum_{j=1}^{n} p_{ij}, \quad p'_j = \sum_{i=1}^{m} p_{ij}.$$

The set of values $\{x_i, p_i\}$, which is the probability distribution of X, is called here the marginal distribution of X. Similarly the set of values $\{y_j, p'_j\}$ is called the marginal distribution of Y.

The conditional probability that X takes the value x_i given that Y takes the value y_j is $\frac{p_{ij}}{p'_j}$. The set of values $\left\{x_i, \frac{p_{ij}}{p'_j}\right\}$ is called the conditional probability distribution of X given $Y = y_j$, $\sum_{i=1}^{m} \frac{p_{ij}}{p'_j} = 1$.

Similarly the set of values $\left\{y_j, \frac{p_{ij}}{p_i}\right\}$ gives the conditional probability dustrubution of Y given $X = x_i$. $\sum_{j=1}^{n} \frac{p_{ij}}{p_i} = 1$.

• **THEOREM 4.8.** *Let X,Y be stochastically independent r.v.'s. Then $E(XY) = E(X)\,E(Y)$.*

Proof. Let X assume value x_i with probability p_i and Y assume value y_j with probability p'_j, $i = 1\cdots,m; j = 1,\cdots,n$, $\sum_{i=1}^{m} p_i = \sum_{j=1}^{n} p'_j = 1$. Since X,Y are independent the probability that X assume value x_i and Y assume value y_j at the same time is $p_ip'_j$. Hence

$$E(XY) = \sum_{i=1}^{m}\sum_{j=1}^{n} x_iy_jp_ip'_j = \left(\sum_{i=1}^{m} x_ip_i\right)\left(\sum_{j=1}^{n} y_jp'_j\right) = E(X)E(Y).$$

COROLLARY 4.3. *If $X_1,\cdots,X_k$ are stochastically independent r.v.'s, $E(X_1\cdots X_k) = E(X_1)\cdots E(X_k)$.*

Definition 4.8. Covariance, Correlation coefficient. Covariance between two r.v.s X,Y are defined as

$$\text{Cov}(X,Y) = E(X - E(X))(Y - E(Y)) = E(XY) - E(X)E(Y) \tag{4.4}$$

The ratio

$$\rho(X,Y) = \frac{\text{Cov}(X,Y)}{\sqrt{V(X)}\sqrt{V(Y)}} \tag{4.5}$$

is called the correlation coefficient between X and Y. Both covariance and correlation coefficients are measures of association between X and Y.

Following results are easy to verify :

(i) Cov $(aX, bY) = ab$ Cov (X,Y), (ii) Cov $(X+a, Y+b) =$ Cov (X,Y), (iii) $\rho(aX+b, cY+d) = \rho(aX, cY) = (\text{sign of } ac)\ \rho(X,Y)$.

If X, Y are stochastically independent $E(XY) = E(X)E(Y)$ and hence Cov $(X,Y) = 0$ and $\rho(X,Y) = 0$. However, the converse is not necessarily true i.e. X, Y may be dependent r.v.s and yet correlation coefficient between X and Y may be zero. The following example illustrates such a situation.

❑ **EXAMPLE 4.13.** *Let U and V be two r.v.s with the same mean and the same variance. Let $X = U+V$, $Y = U-V$. Then Cov $(X,Y) = E(U^2 - V^2) - E(U-V)E(U+V) = 0$. Thus X and Y are uncorrelated though they are necessarily dependent. e.g. if U, V denote the number of points on two dice, then their sum and difference is either both even or both odd and thus are dependent.*

It follows from Cauchy-Schwartz inequality (vide section (6.3)) that

$$\text{Cov}[(X,Y)]^2 \le V(X)V(Y)$$

so that $\rho^2 \le 1$ or $|\rho| \le 1$. (Further details about ρ have been discussed in section (10.5).)

❑ **EXAMPLE 4.14.** *X, Y have joint probability distribution as follows :*

Y \ X	1	2	3	p_j
		p_{ij}		
1	1/12	2/12	1/12	4/12
2	1/12	0	1/12	2/12
3	2/12	0	4/12	6/12
p_i	4/12	2/12	6/12	1

Marginal distribution of X is $P(X=1) = \frac{4}{12}$, $P(X=2) = \frac{2}{12}$, $P(X=3) = \frac{6}{12}$. Similarly we get marginal distribution of Y.

$$\text{Here} \quad E(X) = 2\frac{1}{6} = E(Y), \quad E(X^2) = 5\frac{1}{2} = E(Y^2), \quad E(XY) = \frac{29}{6}$$

$$V(X) = V(Y) = \frac{29}{36}, \quad \text{Cov}(X,Y) = \frac{5}{36}, \quad \text{Corr. Coeff. } \rho(X,Y) = \frac{5}{29}.$$

• **THEOREM 4.9.** *If $X_1, \cdots X_n$ be n random variables with finite variances $\sigma_1^2, \cdots, \sigma_n^2$ respectively, then*

$$V(l_1X_1 + \cdots + l_nX_n) = \sum_{i=1}^{n} l_i^2\sigma_i^2 + \sum_{i\neq j=1}^{n}\sum l_il_j \text{ Cov } (X_i, X_j)$$

where $l_1, \cdots, l_n$ are arbitrary constants.

Proof.

$$V(l_1X_1 + \cdots + l_nX_n) = E(l_1X_1 + \cdots + l_nX_n - l_1E(X_1) - \cdots - l_nE(X_n))^2$$

$$= E\left(\sum_{i=1}^{n} l_i(X_i - E(X_i))\right)^2 = \sum_{i=1}^{n} l_i^2\sigma_i^2 + \sum_{i\neq j}^{n}\sum l_il_j \text{ Cov } (X_i, X_j)$$

$$= \sum_{i=1}^{n} l_i^2\sigma_i^2 + \sum_{i\neq j=1}^{n}\sum \rho_{ij}\sigma_i\sigma_j$$

where ρ_{ij} = correlation coefficient (X_i, X_j).

COROLLARY 4.4. *If* $l_1 = l_2 = \cdots = l_n = \frac{1}{n}$, *so that* $\sum_{i=1}^{n} l_iX_i = \frac{1}{n}\sum_{i=1}^{n} X_i = \bar{X}$, *and if Cov.* $(X_i, X_j) = 0$, $i \neq j = 1, \cdots, n$, $V(\bar{X}) = \sum_{i=1}^{n} \frac{\sigma_i^2}{n^2}$. *If further* $\sigma_i^2 = \sigma^2$, $i = 1, \cdots, n$, $V(\bar{X}) = \frac{\sigma^2}{n}$.

❑ **EXAMPLE 4.15.** *n trials in each of which an event A may or may not occur are performed. If X denotes the number of occurences of A, find $E(X)$, $V(X)$.*

Let $X_i = 1(0)$ if A occurs (does not occur) at the ith trial. Let $P(X_i = 1) = p_i, i = 1, \cdots, n$. Then $E(X_i) = p_i, E(X_i^2) = p_i, V(X_i) = p_i(1 - p_i)$. Cov $(X_i, X_j) = E(X_iX_j) - p_ip_j = P(X_i = 1, X_j = 1) - p_ip_j = p_{ij} - p_ip_j$ (say). Hence $V(X) = \sum_{i=1}^{n} p_i(1 - p_i) + \sum_{i\neq j=1}^{n}\sum(p_{ij} - p_ip_j)$. If $p_i = p$ for all i and $p_{ij} = P$ for $i \neq j$, $V(X) = np(1 - p) + n(n - 1)(P - p^2)$.

❑ **EXAMPLE 4.16.** *From an urn containing a white and b black balls n balls are drawn of which X balls come white. Find $E(X)$, $V(X)$.*

Probability distribution of X is

$$P(X = x) = \frac{\binom{a}{x}\binom{b}{n-x}}{\binom{a+b}{n}}$$

This is called a hypergeometric distribution with parameters a, b, n.

Let $X_i = 1(0)$ if the ith selected ball is white (black), $i = 1, \cdots, n$. The variables X_i are mutually dependent but the theorem 4.4 still applies.

$$E(X_i) = P(X_i = 1) = \frac{a}{a+b}, \quad i = 1, \cdots, n, \quad E(X) = \frac{an}{a+b}$$

$$V(X_i) = \frac{ab}{(a+b)^2}, \qquad E(X_iX_j) = \frac{(a-1)}{(a+b)(a+b-1)}$$

$$\text{Cov}(X_i, X_j) = \frac{a}{a+b}\left(\frac{a-1}{a+b-1} - \frac{a}{a+b}\right)$$

$$V(X) = \frac{nab}{(a+b)^2} + n(n-1)\frac{a}{a+b}\left(\frac{a-1}{a+b-1} - \frac{a}{a+b}\right)$$

on simplification.

❑ **EXAMPLE 4.17.** *A number of independent trials are made in each of which an event A may occur with probability p. The trials continue until the event occurs for the first time. If X is the number of trials (including the trial in which A occurs) that must be made find E(X), V(X).*

Do the same problem if the trials are continued until A occurs for the kth time and X denotes the required number of trials.

First Part. $P(X = 1) = p$, $P(X = 2) = qp, \cdots$, $P(X = k) = qp^{k-1}$, $\cdots$ where $q = 1-p$. Here X is said to have a geometric distribution with parameter p. Thus

$$E(X) = p + 2qp + 3q^2p + \cdots + kq^{k-1} + \cdots = p/(1-q)^2 = \frac{1}{p}$$

$$E(X^2) = p + 2^2qp + 3^2q^2p + \cdots + k^2q^{k-1}p + \cdots = p(1+q)/(1-q)^3$$

$\left[\text{because, } \sum_{m=1}^{\infty} mq^m = \frac{q}{(1-q)^2}; \text{ differentiating both sides wrt } q, \sum_{m=1}^{\infty} m^2q^{m-1} = (1+q)/(1-q)^3\right]$

Second Part. Let X_i be the number of trials from the $(i-1)$th to ith occurence of A (starting immediately after the $(i-1)$th occurence and ending with the ith occurence of A). Let $X = X_1 + \cdots + X_k$. Clearly each X_i has geometric distribution with parameter p. Also X_i and X_j are independent, $i \neq j = 1, \cdots, k$. Thus $E(X) = \frac{k}{p}$, $V(X) = V(X_1) + \cdots + V(X_k) = \frac{kq}{p^2}$.

❑ **EXAMPLE 4.18.** *In the familiar matching problem, where N letters are distributed at random to N addressed envelopes and a matching is said to have occured if a letter goes to the correct envelope, find the expectation and variance of number of matches.*

Let $X_i = 1(0)$ if a match occurs (does not occur) at the ith place; $X = X_1 + \cdots + X_n$ is the total number of matches. $P(X_i = 1) = \dfrac{1}{N}$, $P(X_i = 1, X_j = 1) = \dfrac{1}{N(N-1)}$, $i \neq j = 1, \cdots, N$. Thus $E(X_i) = \dfrac{1}{N}$, $V(X_i) = \dfrac{N-1}{N^2}$, Cov $(X_i, X_j) = \dfrac{1}{N^2(N-1)}$.

Hence $E(X) = 1 = V(X)$.

❑ **EXAMPLE 4.19.** *Consider a random variable X having*

$$P(X = r) = c \cdot \frac{\binom{n}{r}}{r+1}, \quad r = 0, 1, \cdots, n,$$

c being a constant. Evaluate c and find $E(X)$, $V(X)$.

We have $P(X = r) = c\dfrac{\binom{n}{r}}{r+1} = c\dfrac{\binom{n+1}{r+1}}{n+1}$. Since $\sum_{r=o}^{n} P(X=r) = 1$, $\dfrac{c}{n+1}\sum_{r=o}^{n}\binom{n+1}{r+1}$ $= 1$, when $c = \dfrac{n+1}{2^{n+1}-1}$.

$$E(X) = \frac{1}{2^{n+1}-1}\sum_{r=0}^{n} r\binom{n+1}{r+1} = \frac{(n+1)}{2^{n+1}-1}\sum_{r=0}^{n}\binom{n}{r} - 1 = \frac{(n-2)2^n + 1}{2^{n+1}-1}.$$

$$E(X^2) = \frac{1}{2^{n+1}-1}\sum_{r=0}^{n} r^2\binom{n+1}{r+1}$$

$$= \frac{1}{2^{n+1}-1}\left[n(n+1)\sum_{r=1}^{n}\binom{n-1}{r-1} - (n+1)\sum_{r=0}^{n}\binom{n}{r} + \sum_{r=o}^{n}\binom{n+1}{r+1}\right]$$

$$= \frac{2^{n-1}(n^2 - n + 2) - 1}{2^{n+1}-1}$$

$$V(X) = \frac{(n+1)2^{n-1}}{(2^{n+1}-1)^2}\left[2^{n+1} - (n+2)\right].$$

❑ **EXAMPLE 4.20.** *A population of N distinct elements is sampled with replacement. Find the expectation and variance of the number of draws required to have r distinct units in the sample.*

The first draw contains the first distinct unit. Let X_k be the number of draws after the draw in which the kth distinct unit occurs up to the draw in which the $(k+1)$th distinct unit occurs. Thus X_1 is the number of draws starting from the second draw up to the draw in which a second distinct unit occurs, etc.

The required number of draws is $1 + X_1 + X_2 + \cdots + X_{r-1} = Z$ (say), $P(X_1 = 1) = \dfrac{N-1}{N}$, $P(X_1 = 2) = \dfrac{1}{N} \cdot \dfrac{N-1}{N}$, $P(X_1 = 3) = \dfrac{1}{N^2} \cdot \dfrac{N-1}{N}$, etc. Hence

$$E(X_1) = 1.\frac{N-1}{N} + 2 \cdot \frac{1}{N} \cdot \frac{N-1}{N} + 3 \cdot \frac{1}{N^2} \cdot \frac{N-1}{N} + \cdots = \frac{N}{N-1}$$

In general $P(X_k = 1) = \frac{N-k}{N}$, $P(X_k = 2) = \frac{k}{n} \cdot \frac{N-k}{N}$, etc. $E(X_k) = \frac{N}{N-k}$, $k = 1, 2, \cdots, r-1$.

Thus,

$$E(Z) = N\left\{\frac{1}{N} + \frac{1}{N-1} + \frac{1}{N-2} + \cdots + \frac{1}{N-r+1}\right\} \simeq N \int_{N-r+1}^{N} \frac{dx}{x} \quad \text{for large } N.$$

$$= N \log \frac{N}{N-r+1}.$$

$$E(X_k^2) = \frac{N-k}{N}\left[1 + 2^2\left(\frac{k}{N}\right) + 3^2\left(\frac{k}{N}\right)^2 + 4^2\left(\frac{k}{N}\right)^3 + \cdots\right]$$

$$= \frac{N-k}{N}\left[\frac{1}{\left(1-\frac{k}{N}\right)^2} + \frac{\frac{2k}{N}}{\left(1-\frac{k}{N}\right)^3}\right] = \frac{N(N+k)}{(N-k)^2}.$$

Hence $V(X_k) = \frac{Nk}{(N-k)^2}$, $k = 1, \cdots, r-1$.

Thus

$$V(Z) = N\sum_{k=1}^{r-1} \frac{k}{(N-k)^2} = N\sum_{k=1}^{r-1}\left\{\frac{N}{(N-k)^2} - \frac{1}{N-k}\right\}$$

$$\simeq N^2 \int_{N-r+1}^{N} \frac{dt}{t^2} - N\int_{N-r+1}^{N} \frac{dt}{t} \quad \text{for large } N = \frac{N(r-1)}{N-r+1} - N\log\left(\frac{N}{N-r+1}\right).$$

4.5. Exercises

1. *In an urn there are n tickets numbered* $1, \cdots, n$. *m tickets are drawn at a time. find expectation and variance of X, where X is the sum of the numbers of m tickets drawn.*

[Let X_i denote the number on the ith selected ticket. $X = \sum_{i=1}^{m} X_i$.

$$E(X_i) = \frac{1}{n}\sum_{k=1}^{n} k = \frac{n+1}{2}, \quad E(X_i^2) = \frac{1}{n}\sum_{k=1}^{n} k^2 = \frac{(n+1)(2n+1)}{6},$$

$$i = 1, \cdots, m$$

$$E(X_iX_j) = \sum_{k \neq s = 1}^{n}\sum \frac{ks}{n(n-1)} = \frac{1}{n(n-1)}\sum_{k=1}^{n} k\left[\sum_{s(\neq k)=1}^{n} s\right]$$

$$= \frac{1}{n(n-1)}\sum_{k=1}^{n} k\left[\frac{n(n+1)}{2} - k\right] = \frac{(n+1)(3n^2-n-2)}{12(n-1)},$$

$$i \neq j = 1, \cdots, m.$$

Hence $V(X_i) = \dfrac{n^2-1}{12}$, $\quad$ Cov $(X_i, X_j) = -\dfrac{n+1}{12}$.

Thus $E(X) = \dfrac{m(n+1)}{2}$, $\quad V(X) = \dfrac{m(n+1)(n-m)}{12}$.]

2. *Coupons on cereal boxes are numbered 1,2,⋯, n and a set of one of each is required for a prize. With one coupon per box, how many boxes on an average are required to make a complete suit ?*

[The first number is a new one, chance of getting a second new number is $\dfrac{n-1}{n}$. The number of boxes required to get a second new number follows a geometric distribution having expectation $\dfrac{n}{n-1}$. The third new number requires additional $\dfrac{n}{n-2}$ boxes on an average, etc. Hence the average number of boxes required is

$$n\left[\frac{1}{n}+\frac{1}{n-1}+\frac{1}{n-2}+\cdots+1\right] \simeq n\left[\log_e n+\frac{1}{2n}+l\right]$$

where $l=$ Euler's constant, .57721⋯]

3. *A number of independent trials are carried out in each of which an event A may occur with probability p. The trials are terminated as soon as A occurs. Find $E(X)$ where X is the number of trials that must be made, if one restricts to $X \le N$.*

[Let us first consider a r.v. Y which is the number of trials that must be made if the restriction N concerning the number of trials is removed. Y has a geometric distribution, $P(Y=k) = pq^{k-1}$ (example 4.17).

Let $X = \min\{Y, N\}$, Hence

$$E(X) = \sum_{k=1}^{N-1} kP(Y=k) + N\sum_{k=N}^{\infty} P(Y=k) = p\left\{\sum_{k=1}^{N-1} kq^{k-1} + N\sum_{k=N}^{\infty} q^{k-1}\right\}$$

Now $$\sum_{k=1}^{N-1} kq^{k-1} = \sum_{k=1}^{N-1}\frac{d}{dq}(q^k) = \frac{d}{dq}\left(\frac{q-q^N}{1-q}\right) = \frac{1-Nq^{N-1}+(N-1)q^N}{(1-q)^2}$$

Again $$N\sum_{k=N}^{\infty} q^{k-1} = Nq^{N-1}\Big/(1-q).$$

Thus $$E(X) = \frac{1-q^N}{p}\ \Big]$$

4. *If $X_1, \cdots, X_n$ are independently and identically distributed positive random variables, prove that $E(S_k/S_n) = \dfrac{k}{n}$ where $S_r = \sum_{i=1}^{r} X_i$.*

[Since X_i $(i = 1, \cdots, n)$ are positive, the denominator cannot be zero. Now $E\left(S_k/S_n\right) = \sum_{i=1}^{k} E\left(\frac{X_i}{S_n}\right)$. Again, since $X_1, \cdots X_n$ are identically distributed $E\left(\frac{X_i}{S_n}\right)$ $= E\left(\frac{X_j}{S_n}\right)$, $i \neq j = 1, \cdots, n$. Now $E\left(\frac{X_1 + \cdots + X_n}{S_n}\right) = 1$; hence $E\left(\frac{X_i}{S_n}\right) = \frac{1}{n}$. Thus $E\left(\frac{S_k}{S_n}\right) = \frac{k}{n}$.]

5. *A group of four radar units scan a region in which there are five targets $T_1, \cdots, T_5$. The region is scanned for a time t. During this time interval each unit independently of the others, may detect each target with probability p_i. Find the expected number of targets that will be detected.*

[Let $X_i = 1(0)$ if the ith target is (not) detected. $P(X_i = 1) = 1 - (1 - p_i)^4$. Hence

$$E\left(\sum_{i=1}^{5} X_i\right) = \sum_{i=1}^{5}\left[1 - (1 - p_i)^4\right].$$

6. *A and B roll a pair of unbiased dice alternatively. A wins if he throws a 6 before B throws a 7, B wins if he throws a 7 before A has a 6. If A begins the game show that his probability of winning is $\frac{30}{61}$ and the expected number of trials needed for his win is approximetely 6.*

[Probability of A's win$=\sum_{r=0}^{\infty}\left(\frac{31}{36}\right)^r\left(\frac{30}{36}\right)^r\frac{5}{36} = \frac{30}{61} = p$ (say). If X denotes the required number of trials, $P(X = 2r + 1) = \frac{1}{p}\left[\left(\frac{31}{36}\right)^r\left(\frac{30}{36}\right)^r\frac{5}{36}\right]$, because we must have $\sum_{k=1}^{\infty} P(X = k) = 1$. Hence $E(X) = \frac{61}{30}\cdot\frac{5}{36}\sum_{r=0}^{\infty}(2r + 1)\left(\frac{31}{36}\right)^r\left(\frac{30}{36}\right)^r$ $= 6.08197 \simeq 6$]

7. *A straight line with $6n (n \geq 1)$ sections is divided at random into two parts A and B, A having r sections and B, $(n - r)$ sections, $r \geq 1$. If A is known to be longer than B, find expectation and variance of length of A.*

[Here $3n + 1 \leq r \leq 6n - 1$, each of these values has the probability $\frac{1}{3n - 1}$.

$$E(r) = \frac{1}{3n - 1}\sum_{3n+1}^{6n-1} r = \frac{1}{3n - 1} \times \frac{9n(3n - 1)}{2} = \frac{9n}{2}$$

$$V(r) = (9n^3 - 45n^2 - 22n - 4)/4(3n - 1).]$$

8. *A Box contains 2^n tickets among which $\binom{n}{r}$ tickets bear the number $r(= 0, 1, 2, \cdots, n)$. A group of m tickets is selected at random. Find $E(X)$ and $V(X)$ where X is the sum of the numbers on the selected tickets.*

[Let X_i be the number on the ith selected ticket. $X = \sum_{i=1}^{m} X_i$. $E(X_i) = \sum_{k=0}^{n} k\binom{n}{k} \Big/ 2^n = \frac{n}{2^n}\sum_{0}^{n-1}\binom{n-1}{j} = \frac{n}{2}$.

$$E(X_i^2) = \sum_{k=0}^{n} k^2\binom{n}{k} \Big/ 2^n = \frac{n}{2^n}\left[(1+1)^{n-1} + (n-1)(1+1)^{n-2}\right] = \frac{n(n+1)}{4}$$

$$E(X_j X_j) = \sum_{k=1}^{n} \frac{k^2\binom{n}{k}\left\{\binom{n}{k} - 1\right\}}{2^n(2^n-1)} + \sum\sum_{k \neq s=1}^{n} \frac{ks\binom{n}{k}\binom{n}{s}}{2^n(2^n-1)} \quad \text{(i)}$$

The first term

$$= \sum_{k=1}^{n} \frac{k^2\binom{n}{k}^2}{2^n(2^n-1)} - \sum_{k=1}^{n} k^2 \frac{\binom{n}{k}}{2^n(2^n-1)}$$

The second term

$$= \sum_k \frac{k\binom{n}{k}}{2^n-1}\left\{\sum_s s\frac{\binom{n}{s}}{2^n} - k\frac{\binom{n}{k}}{2^n}\right\} = \sum_k \frac{k\binom{n}{k}}{2^n-1}\left\{\frac{n}{2} - \frac{k\binom{n}{k}}{2^n}\right\}$$

$$\text{Hence} \quad (i) = \frac{n}{2}\sum_k \frac{k\binom{n}{k}}{2^n-1} - \sum_k \frac{k^2\binom{n}{k}}{2^n(2^n-1)} = \frac{1}{2^n-1}\left[n^2 \cdot 2^{n-2} - \frac{n(n+1)}{4}\right]$$

$$\text{Thus} \quad E(X^2) = \frac{mn(n+1)}{4} + \frac{m(m-1)n}{2^n-1}\left[n2^{n-2} - \frac{n+1}{4}\right]$$

$$\text{Hence Var}(X) = \frac{mn}{4} - \frac{mn(m-1)}{4(2^n-1)}.\Big]$$

9. ***In Buffon's needle problem (Example 2.45), find the expected number of hits.***

[**When $l < L$, the number of hits is either 1 or 0 with probability of a hit $p = 2l/L\pi$ so that the the expected number of hits is p.**

When the restriction $l < L$ is removed, we divide the length l of the needle into n elementary sections each of length $\frac{l}{n} < L$. Here the number of times the needle

hits the lines, $X=\sum_{i=1}^{n}X_i$ where $X_i=1(0)$ if the ith section hits (otherwise). Hence

$$E(X)=\sum_{i=1}^{n}E(X_i)=\sum_{i=1}^{n}\frac{2l}{nL\pi}=\frac{2l}{L\pi}\Bigg]$$

10. *In a vehicle insurance policy the premium is β rupees in the first year and if no claim is made in any of the first r years, the premium is $\beta\delta^r$ Rs. in the $(r+1)$th year $(r=1,2,\cdots)$.*

If in any year a claim is made the premium in that year remains unaffected; but the next year is treated as first year requiring premium β Rs and the subsequent years are treated as second year, third year, etc. for calculation of premium. Assuming that the probability that no claim is made in a year is q, calculate the expected amount of premium payable in the nth year.

[Probability that in the nth year premium paid is $\beta\delta^{n-1}$ is q^{n-1}; or $\beta\delta^{n-j-1}$ is pq^{n-j-1}. Hence expected premium in the nth year $=\beta\left[(q\delta)^{n-1}+p\sum_{r=0}^{n-2}(q\delta)^r\right]$,

$$(p=1-q)=\frac{\beta}{1-q\delta}\left[(1-q)+q^n\delta^{n-1}(1-\delta)\right]\rightarrow\frac{\beta(1-q)}{1-q\delta}\text{ as }n\rightarrow\infty.$$

11. *A box contains N objects numbered $1,2,\cdots,N$ from which n objects are drawn at random with replacement. Find $P(X=k)$, $E(X)$ and $V(X)$ when X denotes the largest number drawn.*

We have shown in exercise 54 in section 2.6

$$P(X=k)=\left(\frac{k}{N}\right)^n-\left(\frac{k-1}{N}\right)^n$$

$$E(X)=\sum_{k=1}^{N}kP(X=k)=\frac{1}{N^n}\left\{\sum_{k=1}^{N}k^{n+1}-\sum_{k=1}^{N}k(k-1)^n\right\}$$

$$=\frac{1}{N^n}\sum_{k=1}^{N}\left\{k^{n+1}-(k-1)^{n+1}-(k-1)^n\right\}$$

$$=\frac{1}{N^n}\left\{N^{n+1}-\sum_{k=1}^{N}(k-1)^n\right\}\simeq\frac{1}{N^n}\left\{N^{n+1}-\int_0^N t^n\,dt\right\}=\frac{nN}{n+1}\quad\text{for large }N$$

$$E(X^2)=\frac{1}{N^n}\sum_{k=1}^{N}k^2P(X=k)=\frac{1}{N^n}\sum_{k=1}^{N}k^{n+2}-\sum_{k=1}^{N}k^2(k-1)^n$$

$$=\frac{1}{N^n}\sum_{k=1}^{N}\left\{k^{n+2}-(k-1)^{n+2}-2(k-1)^{n+1}-(k-1)^n\right\}$$

$$\simeq\frac{1}{N^n}\left\{N^{n+2}-2\int_0^N t^{n+1}dt-\int_0^N t^n\,dt\right\}\simeq nN^2/(n+2)\quad\text{for large}N.$$

Thus

$$V(X) \simeq \frac{nN^2}{(n+1)^2(n+2)} \quad \text{for large } N.$$

12. *In a trial there are two outcomes S, F happening with probabilities p, q respectively; n such trials are performed. A man who has no real ability to guess decides to allot r outcomes as S's and $(n-r)$ outcomes as F's at random. Given that he scores point 1(0) for every correct (incorrect) guess, find the expectation and variance of his total score.*

[Let $X_i = 1(0)$ if the guess for the ith experiment is correct (incorrect). Total socre, $X = X_1 + \cdots + X_n$. $P(X_i = 1) = p \cdot \frac{r}{n} + q\left(\frac{1-r}{n}\right)$, $i = 1, \cdots, n$. $E(X_i = 1, X_j = 1) = P(X_i = 1, X_j = 1) = p^2 \frac{r(r-1)}{n(n-1)} + 2pq\frac{r}{n} \cdot \frac{(n-r)}{(n-1)} + q^2 \frac{(n-r)(n-r-1)}{n(n-1)}$. Thus $E(S_n) = np - (n-r)(p-q)$. $V(S_n) = npq$].

13. *Consider the problem 12 with $p > \frac{1}{2}$. Suppose the man allots S to a trial with probability λ and F with probability $(1-\lambda)$. Prove that the probability of a correct guess for any trial is*

$$\left[1 - p + (2p-1)\lambda\right]$$

Derive the expectation and variance of his total score.

[$P(X_i = 1) = p\lambda + (1-p)(1-\lambda) = 1 - p + (2p-1)\lambda = P$ (say); Let $X = X_1 + \cdots + X_n$. $E(X) = nP$. $V(X) = nPQ$, $Q = 1 - P$. For fixed n, $V(X)$ is maximum for $\lambda = \frac{1}{2}$].

14. *Banach's match box problem : A person carries always two match boxes, which initially contain N match-sticks each. Whenever he wants a light he selects a box at random. Find p_r, the probability that when he finds a box empty for the first time the other box contains exactly r match-sticks and verify $\sum p_r = 1$.*

Find the mean and variance of distribution of r.

[Probability that he selects N match sticks from one box and $N - r$ sticks from the other is

$$p_r = \binom{2N-r}{N}\left(\frac{1}{2}\right)^N \left(\frac{1}{2}\right)^{N-r}, \quad r = 0, 1, \cdots, N$$

$$\sum_{r=0}^{N} p_r = 2^{-N} \sum_{r=0}^{N} \text{coefficient of } x^{N-r} \text{in} \left(1 + \frac{x}{2}\right)^{2N-r}$$

$$= 2^{-N} \text{ coefficient of } x^N \text{in} \sum_{r=0}^{\infty} x^r \left(1 + \frac{x}{2}\right)^{2N-r} \quad \text{(i)}$$

Now $$\sum_{r=0}^{\infty} x^r\left(1+\frac{x}{2}\right)^{2N-r} = \frac{\left(1+\frac{x}{2}\right)^{2N+1}}{\left(1-\frac{x}{2}\right)}\left[1-\left(\frac{2x}{2+x}\right)\right]^{N+1} \quad \text{(ii)}$$

Hence (i) $= 2^{-N}$ coefficient of x^N in $\left(1+\frac{x}{2}\right)^{2N+1}\left(1-\frac{x}{2}\right)^{-1}$ (since the last term in the squared bracket of (ii) does not contribute to the coefficient of x^N).

$$= 2^{-N}\sum_{r=0}^{N}\binom{2N+1}{r}\left(\frac{1}{2}\right)^r\left(\frac{1}{2}\right)^{N-r} = 2^{-2N}\sum_{r=0}^{N}\binom{2N+1}{r}$$

Now $\sum_{r=0}^{2N+1}\binom{2N+1}{r} = 2^{2N+1}$. Hence $\sum_{r=0}^{N}\binom{2N+1}{r} = 2^{2N}$.

$\therefore \sum_{r=0}^{N} p_r = 1.$

To find μ we have

$$N-\mu = \sum_{r=0}^{N-1}(N-r)p_r = \sum_{r=0}^{N-1}(N-r)\binom{2N-r}{N-r}\frac{1}{2^{2N-r}}$$

$$= \sum_{r=0}^{N-1}(2N-r)\binom{2N-r-1}{N-r-1}\frac{1}{2^{2N-r}} = \sum_{r=0}^{N-1}\left(\frac{2N-r}{2}\right)p_{r+1}$$

$$= \frac{2N+1}{2}\sum_{r=0}^{N-1}p_{r+1} - \frac{1}{2}\sum_{r=0}^{N-1}(r+1)p_{r+1} = \frac{2N+1}{2}(1-p_o) - \frac{\mu}{2}.$$

Hence $\mu = (2N+1)p_o - 1 = (2N+1)\binom{2N}{N}\frac{1}{2^{2N}} - 1.$

To calculate $V(r)$ we have

$$E(N-r)(N-r-1) = \sum_{r=0}^{N-2}(N-r)(N-r-1)p_r$$

$$= \frac{1}{2^2}\sum_{r=0}^{N-2}(2N-r)(2N-r-1)p_{r+2}$$

$$= \frac{1}{2^2}\left[2N(2N+1)(1-2p_o) + (4N+2)(1-p_o) + E(r^2) - (4N+3)\mu\right]$$

(since $p_o = p_1$)

Thus $E(r^2) = (2N+3) - 3(2N+1)p_o$. $V(r) = (2N+2) - (1+\mu)(2+\mu).$]

15. *Consider Bose-Einstein statistic in which only distinguishable arrangements of n indistinguishable balls in m cells are considered and each is given equal probability. Find the probability distribution of X, the number of emply cells. Find $E(X)$.*

[The total number of possible arrangements is $\binom{n+m-1}{n}$. Probability that k specified cells remain empty $= \binom{n+m-k-1}{n} \Big/ \binom{n+m-1}{n}$. Let $X_i = 1(0)$ if the ith cell is empty (occupied); $X = X_1 + \cdots + X_m$ is the total number of empty cells.

Probability that at least one cell remains empty

$$= P\left[\bigcup_{i=1}^{m}(X_i = 1)\right] = S_1 - S_2 + S_3 \cdots + (-1)^{m-1} S_m = P \quad \text{(say)}.$$

With usual notation,

$$S_k = \binom{m}{k}\binom{n+m-k-1}{n} \Big/ \binom{n+m-1}{n}$$

Thus

$$P(X = 0) = 1 - P = \sum_{k=0}^{m} (-1)^k \binom{m}{k}\binom{n+m-k-1}{n} \Big/ \binom{n+m-1}{n} = p_m \quad \text{(say)}$$

Hence

$$P(X = k) = \binom{m}{k}\binom{n+m-k-1}{n} \Big/ \binom{n+m-1}{n} p_{m-k}, \quad k = 0, 1, \cdots, m-1$$

We have

$$E(X_i = 1) = \binom{n+m-2}{n} \Big/ \binom{n+m-1}{n}$$

Thus

$$E(X) = m\binom{n+m-2}{n} \Big/ \binom{n+m-1}{n} = \frac{m(m-1)}{n+m-1}\Bigg]$$

16. *For n sets $A_1, \cdots, A_n$, show that*

$$I_{\bigcup_{i=1}^{n} A_i} = \sum I_{A_i} - \sum_{i<j}\sum I_{A_iA_j} + \sum_{i<j<k}\sum\sum I_{A_iA_jA_k} + \cdots - (-1)^{n-1} I_{\bigcap_{i=1}^{n} A_i}$$

where I_B is the indicator function of the set B. Hence prove the theorem 2.2 by taking expectation of both sides.

5 PROBABILITY DISTRIBUTIONS ON R^1

5.1. Introduction

In the previous chapter we introduced the concept of a discrete random variable and its expectation, variance etc. In this chapter we examine the concept of a random variable from a measure-theoretic point of view. In section 5.3 we study the notion of probability distribution of a random variable; section 5.4 deals with some special types of random variables. Section 5.5 considers independent random variables. In section 5.6 and 5.7 we consider Lebesgue-Stieltjes integration.

5.2. Random Variables

In chapter 3 we considered the notion of a point function defined on an arbitrary set, inverse function of a point function, set function, measure and measurable sets wrt an arbitrary σ-field of sets. We consider now the concept of a measurable function.

Definition 5.1. Consider R^1, the real line and $\mathcal{B}_1$, the borel field of subsets of R^1. If a function X which maps $S \to R^1$ is such that $X^{-1}(B) \in \mathcal{C} \forall B \subset \mathcal{B}_1$, where $\mathcal{C}$ is the σ-field over S, then X is called a $\mathcal{C}$-measurable function (measurable wrt $\mathcal{C}$).

If $S = R^1$ (or a subset of R^1) when $\mathcal{C} = \mathcal{B}_1$, and if X is measurable wrt $\mathcal{B}_1$ then X is a borel measurable function. Here $X^{-1}(B) \in \mathcal{B}_1 \forall B \in \mathcal{B}_1, X^{-1}$ is defined on $\mathcal{B}_1$, having its range in $\mathcal{B}_1$.

Definition 5.2. *Random Variable* : Consider $(\mathcal{S},\mathcal{A})$, $\mathcal{S}$ being the sample space of a random experiment. A finite single-valued function X which maps $\mathcal{S}$ into R^1 is called a random variable if the inverse image under X of all borel sets in $\mathcal{B}_1$ are events i.e. if

$$X^{-1}(B) = \{w : X(w) \in B\} \in \mathcal{A}, \quad \text{for all } B \in \mathcal{B}_1,$$

i.e. if X is a $\mathcal{A}$-measurable function. Any $\mathcal{A}$-measurable function is a random variable on $(\mathcal{S},\mathcal{A})$.

The following theorem gives a necessary and sufficient condition for a function X to be a random variable.

• **THEOREM 5.1.** *X is a random variable if and only if for each $x \in R^1$, $\{w : X(w) \leq x\} \in \mathcal{A}$.*

❑ **EXAMPLE 5.1.** *Consider the examples 4.1 and 4.2.*

(a) example 4.1. here

$$X^{-1}(-\infty,x]=\begin{cases}\phi & x<0\\ \{(TTT)\}, & 0\leq x<1\\ \{(HTT),(THT),(TTH)\}, & 1\leq x<2\\ \{(HHT),(HTH),(THH)\}, & 2\leq x<3\\ \{(HHH)\}, & 3\leq x<4\\ \mathcal{S} & 4\leq x\end{cases}$$

Thus X is a random variable.

(b) example 4.2. Here

$$X^{-1}(-\infty,x]=\begin{cases}\phi & x<1\\ \{(H)\} & 1\leq x<2\\ \{(TH)\}, & 2\leq x<3\\ \{(TTH)\}, & 3\leq x<4\\ \dots\dots & \dots\dots\dots\end{cases}$$

Thus X is a random variable.

• **THEOREM 5.2.** *A constant function $X(w)$ defined on $(\mathcal{S},\mathcal{A})$ is a random variable.*

Proof. Let $X(w) = c$ (a real quantity), $w \in \mathcal{S}$. Let $B \in \mathcal{B}_1$ be any borel set. Either $c \in B$ or $c \notin B$. If $c \in B$, then $X^{-1}(B) = \mathcal{S}$. If $c \notin B, X^{-1}(B) = \phi$. Since $\phi, \mathcal{S} \in \mathcal{A}, X$ is a $\mathcal{A}$-measurable function and hence is a random variable.

• **THEOREM 5.3.** *If $X(w)$ is a random variable on $(\mathcal{S},\mathcal{A})$, $cX(w)$ is also a random variable on $(\mathcal{S},\mathcal{A})$.*

Proof. Let X be any arbitrary but fixed real number. Then $(-\infty, x] \in \mathcal{B}_1$. For $c = 0$, the theorem is trivial (by theorem 5.2). Assume $c > 0$.

$$(cX)^{-1}(-\infty,x] = \{w : cX(w) \leq x\} = \left\{w : X(w) \leq \frac{x}{c}\right\} = X^{-1}\left(-\infty, \frac{x}{c}\right] \subset \mathcal{A}$$

$$(\because \ X \text{ is a r.v.})$$

Hence $cX(w)$ is $\mathcal{A}$-measurable and hence is a random variable on $(\mathcal{S},\mathcal{A})$. The case $c < 0$ can similarly be proved.

The following theorems hold :

• **THEOREM 5.4.** *If $\{c_i\}$, $i = 1, \cdots, n$ are real numbers and $X_1(w) \cdots, X_n(w)$ are r.v.s on $(\mathcal{S}, \mathcal{A})$ then $\sum_1^n c_i X_i$ is a r.v. on $(\mathcal{S}, \mathcal{A})$.*

• **THEOREM 5.5.** *Any polynomial of a r.v. is also a r.v.*

• **THEOREM 5.6.** *If $X(w)$ and $Y(w)$ are r.v.s on $(\mathcal{S}, \mathcal{A})$ and $\{w : Y(w) = 0\} = \phi$, then $\dfrac{X(w)}{Y(w)}$ is a r.v. on $(\mathcal{S}, \mathcal{A})$.*

• **THEOREM 5.7.** *If $X(w), Y(w)$ are r.v.s on $(\mathcal{S}, \mathcal{A})$ then max $(X(w), Y(w))$ and min $(X(w), Y(w))$ are r.v.s on $(\mathcal{S}, \mathcal{A})$.*

• **THEOREM 5.8.** *The indicator function $I_A(w)$ [vide definition 4.2] is a r.v. on $(\mathcal{S}, \mathcal{A})$ if and only if $A \in \mathcal{A}$ (if A is not measurable, I_A need not be a r.v).*

Proof. Let B be any borel set in $\mathcal{B}_1$. We shall have one of the four cases : (i) $1 \in B$, $0 \in B$, (ii) $1 \in B, 0 \notin B$, (iii) $1 \notin B, 0 \in B$, (iv) $1 \notin B, 0 \notin B$. If A is a measurable set, $I_A^{-1}(B)$ is $\mathcal{S}, A, A', \phi$ respectively in cases (i), (ii), (iii) and (iv) and each of these is a $\mathcal{A}$-measurable set.

• **THEOREM 5.9.** *Let $X(w)$ be a r.v. on $(\mathcal{S}, \mathcal{A})$ and $g(X(w))$ be a borel measurable function, $g : R^1 \to R^1$. Then $g(X(w))$ is a r.v.*

Proof. Let B be any borel set.

$$g(X)^{-1}(B) = \{w : g(X(w)) \in B\} = \{w : X(w) \in g^{-1}(B)\} = X^{-1}[g^{-1}(B)].$$

Since g is borel-measurable, $g^{-1}(B)$ is a borel set. Thus $X^{-1}[g^{-1}(B)] \subset \mathcal{A}$.

COROLLARY 5.1. *If $X(w)$ is a r.v. then $g(X(w))$ is also a r.v. when g is a continuous function from R^1 to R^1.*

Proof. Continuity of g inplies g borel-measurable.

COROLLARY 5.2. *If $X(w)$ is a r.v. then $g(X(w))$ is also a r.v. when g is a monotonic function from R^1 to R^1.*

Proof. Monotonicity of g implies g is borel- measurable.

We note that the concept of probability is not required in the definition of a r.v. We now consider the probability distribution of a r.v. X.

• **THEOREM 5.10.** *The r.v. X defined on $(\mathcal{S}, \mathcal{A}, P)$ induces a probability space $(R^1, \mathcal{B}_1, P_X)$ for the r.v. X. The probability measure P_X is given by*

$$P_X(B) = P\{X^{-1}(B)\} = P\{w : X(w) \in B\}$$

for all $B \in \mathcal{B}_1$. P_X is called the probability distribution of X.

Proof. Clearly $P_X(B) \geq 0$ for all $B \in \mathcal{B}_1$. $P_X(R^1) = P\{X \in R^1\} = P(\mathcal{S}) = 1$.

Let $B_i \in \mathcal{B}_1$, $i = 1, 2, \cdots$ with $B_i \cap B_j = \phi\ (i \neq j)$. Now $X^{-1}\left(\bigcup_{i=1}^{\infty} B_i\right) = \bigcup_{i=1}^{\infty} X^{-1}(B_i)$ and $X^{-1}(B_i)$ and $X^{-1}(B_j)$ are disjoint sets in $\mathcal{A}$.

$$P_X\left(\bigcup_1^{\infty} B_i\right) = P\left\{X^{-1}\left(\bigcup_1^{\infty} B_i\right)\right\} = P\left\{\left(\bigcup_{i=1}^{\infty} X^{-1}(B_i)\right)\right\}$$

$$= \sum_{i=1}^{\infty} P\{X^{-1}(B_i)\} = \sum_{i=1}^{\infty} P_X(B_i)$$

Thus P_X satisfies all the three axioms and hence is a probability measure. $(R^1, \mathcal{B}_1, P_X)$ is , therefore, a probability space.

❑ **EXAMPLE 5.2.** *A die is tossed. Let X be a r.v. which takes the value 1 if the result is an odd number, 2 if the result is an even number.*

Here $\mathcal{S} = \{1, 2, \cdots, 6\}$, i denoting the event, i comes up, $\mathcal{A}$ the σ-field over $\mathcal{S}$. Range space of X is $\{1, 2\}$. Let $B_1 = \{1\}$, $B_2 = \{2\}$, $\mathcal{B} = \{\{1\}, \{2\}, \{1, 2\}\}$. $X^{-1}(B_1) = \{1, 3, 5\} \in \mathcal{A}$, $X^{-1}(B_2) = \{2, 4, 6\} \in \mathcal{A}$. Thus X is a r.v.

$$P_X(B_1) = P(X^{-1}(1)) = \frac{1}{2}, \quad P_X(B_2) = P(X^{-1}(2)) = \frac{1}{2}$$

The sets B_1, B_2 are disjoint and form the range space of X.

$$P_X(B_1 \cup B_2) = P_X(B_1) + P_X(B_2) = 1$$

❑ **EXAMPLE 5.3.** *Let $A_1, \cdots A_n$ be disjoint subsets with $\bigcup_1^n A_i = \mathcal{S}$; X is a function defined on $(\mathcal{S}, \mathcal{A})$ such that $X(w) = a_i$ for each $w \in A_i$ $(i = 1, \cdots, n)$. Thus one can write*

$$X(w) = \sum_{i=1}^{n} a_i I_{A_i}(w) \tag{5.1}$$

where I_{A_i} is the indicator function of set A_i. A function of the type (5.1) is called a simple function.

Consider any borel set $B \in \mathcal{B}_1$. B includes either some of a_i's, say, $a_{i_1}, \cdots, a_{i_r}$ $(r \leq n)$ or none at all. In the former case $X^{-1}(B)$ is $\bigcup_{j=1}^{r} A_{i_j}$ and in the later ϕ. If each of $A_i (i = 1, \cdots, n) \in \mathcal{A}$, X is a r.v.

❑ **EXAMPLE 5.4.** *Consider* $\{A_n, n = 1, 2, \cdots\}$, *a sequence of disjoint subsets of* $\mathcal{S}$, $\bigcup_{1}^{\infty} A_n = \mathcal{S}$, $A_n \in \mathcal{A}$ $(n = 1, 2, \cdots)$. X *is a function on* $(\mathcal{S}, \mathcal{A})$ *such that* $X(w) = a_n$ *whenever* $w \in A_n (n = 1, 2 \cdots)$. *Such a function can be written as*

$$X(w) = \sum_{1}^{\infty} a_n I_{A_n}(w) \tag{5.2}$$

where I_{A_n} is the indicator function of A_n and is called an elementary function. As in example 5.3, it follows that $X(w)$ is $\mathcal{A}$-measurable and hence a r.v.

5.3. Probability Distribution of a r.v.

We have seen the r.v. X induces a set function $P_X(x)$ on $(R^1, \mathcal{B}_1)$. We now consider a point function

$$P(X \le x) = P_X((-\infty, x]) = F_X(x) \quad \text{(say)}$$

on $(R^1, \mathcal{B}_1)$. $F_X(x)$ is called the distribution function (d.f.) or cumulative distribution function (c.d.f.) of the r.v. X.

• **THEOREM 5.11.**

(i) F is monotonically non-decreasing

(ii) $F(-\infty) = 0$

(iii) $F(\infty) = 1$

(iv) F is continuous to the right i.e. $F(x + 0) = F(x)$.

Proof. (i) Let $x < y$ be two real numbers. Denote the events $[X \le x]$, $[X \le y]$ by B_x and B_y respectively. Now $B_x \subseteq B_y$. Hence $P(B_x) \le P(B_y)$ i.e. $F_X(x) \le F_X(y)$.

(ii) Let for integral $n(> 0)$, B_n be the event $[X \le -n]$. We have $B_1 \supset B_2 \supset B_3 \cdots$ and $\lim_{n\to\infty} B_n = \phi$. Also each of $B_1, B_2 \cdots$ is a measurable event under X. Hence by the continuity theorem (Theorem 3.1) $\lim_{n\to\infty} P(B_n) = P(\phi) = 0$ i.e. $F(-\infty) = 0$.

(iii) Let for integral $n(> 0)$, B_n be the event $[X \le n]$. Then $B_1 \subset B_2 \subset B_3 \subset \cdots$ and $\lim_{n\to\infty} B_n = \mathcal{S}$. Also each of $B_1, B_2 \cdots$ is a measurable event under X. By the continuity theorem, therefore, $\lim_{n\to\infty} P(B_n) = P(\mathcal{S}) = 1$ i.e. $F(\infty) = 1$.

(iv) Let for integral $n(> 0)$, $C_n = \left[X \le x + \dfrac{1}{n}\right]$. Then $C_1 \supset C_2 \supset \cdots$ and $\lim_{n\to\infty} C_n = [X \le x]$. Hence by the continuity theorem,

$$\lim_{n\to\infty} P[C_n] = P(X \leq x) \quad \text{i.e.} \quad \lim_{n\to\infty} P\left[X \leq x + \frac{1}{n}\right] = F_X(x)$$

$$\text{i.e.} \quad \lim_{\epsilon \downarrow 0} F(x+\epsilon) = F_X(x) \quad \text{i.e.} \quad F_X(x+0) = F_X(x).$$

We now prove that $F(x-0)$ is not necessarily equal to $F_X(x)$. Consider $D_n = \left[X \leq x - \frac{1}{n}\right]$. Then $D_1 \subset D_2 \subset \cdots$ and $\lim_{n\to\infty} D_n = [X < x]$. Thus,

$$\lim_{n\to\infty} P\left[X \leq x - \frac{1}{n}\right] = P(X < x)$$

$$\text{i.e.} \quad \lim_{\epsilon\downarrow 0} F_X(x-\epsilon) = P(X < x)$$

$$\text{i.e.} \quad F_X(x-0) = P(X < x)$$

Thus $F_X(x-0)$ is not necessarily equal to $F_X(x)$ and hence $F_X(x)$ is not continuous to the left.

The quantity $F_X(x) - F_X(x-0) = P[X \leq x] - P[X < x] = P[X = x]$ is called the saltus (or jump) of the distribution function $F_X(x)$. If $P[X = x] > 0$, $F_X(x)$ has a discontinuity at x with the saltus $P(X = x)$. If $P(X = x) = 0$, $F_X(x)$ is continuous at x (see fig. 5.1).

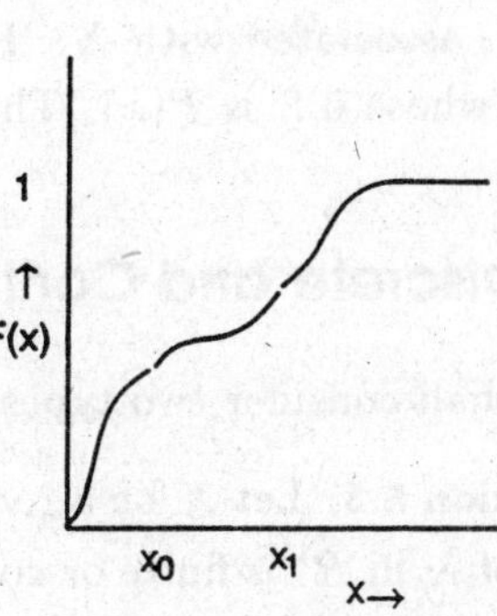

Fig. 5.1 F(x) has saltus at $x = x_0, x_1$ but is continuous at other points of x

Note : In fact, any real valued function F defined on R^1 that satisfies the conditions (i)–(iv) of theorem 5.1 is called a distribution function.

• **THEOREM 5.12.** *The points of discontinuity of F form an at most a countable set.*

Proof. Let $(a, b]$ be a finite interval with at least n discontinuity points $a < x_1 < x_2 < \cdots < x_n \leq b$. Then $F(a) < F(x_1-) < F(x_1) \leq \cdots \leq F(x_n-) < F(x_n) \leq F(b)$.

Let $p_k = F(x_k) - F(x_k-)$ (= Saltus of $F_X(x)$ at x_k), $k = 1, \cdots, n$. Then

$$\sum_{k=1}^{n} p_k \leq F(b) - F(a) \tag{5.3}$$

Let ν_ϵ be the number of discontinuity points with jump $> \epsilon$. Then

$$\sum_{k=1}^{n} p_k \geq \epsilon\nu_\epsilon + \sum{}' p_k \tag{5.4}$$

$\sum$ being summation over other points. Hence from (5.3), (5.4),

$$\epsilon \nu_\epsilon \le F(b) - F(a) \quad \text{i.e.} \quad \nu_\epsilon \le \frac{F(b) - F(a)}{\epsilon} \quad \text{(a finite quantity)}.$$

Thus for every integer n, number of discontinuity points with jump> $\frac{1}{n}$ is finite. Therefore, there are no more than a countable number of discontinuity points in any finite interval $(a, b]$. Again R^1 is union of countable number of such intervals. Hence the number of discontinuity points is at most countable.

• **THEOREM 5.13.** *(Converse of theorem 5.11) If F is a function on R^1 such that it has all the properties (i)–(iv) mentioned in theorem 5.11, then there exists a r.v. X such that $F(x)$ is the d.f. of X.*

• **THEOREM 5.14.** *Two r.v.s have the same distribution if and only if they have the same distribution functions.*

Remark 5.1. If X is a r.v. on $(\mathcal{S}, \mathcal{A}, P)$, theorem 5.11 states that $F(x) = P\{X \le x\}$ is a d.f. associated with X. Theorem 5.13 states that given a d.f. $F(x)$ there exists a r.v. X whose d.f. is $F(x)$. Thus given an r.v. there exists a d.f. and conversely.

5.4. Discrete and Continuous Random Variables

We shall consider two types of r.v.s : (i) discrete r.v. (ii) continuous r.v.

Definition 5.3. Let X be a r.v. on $(\mathcal{S}, \mathcal{A}) \to (R^1, \mathcal{B}_1)$. If the number of elements in the range of X in R^1 is finite or countably infinite then $X(w)$ is called a (one dimensional) discrete r.v.

Let $x_1, x_2, \cdots$ be the elements in the range sapce of X.

Let $p(x_i) = P[X(w) = x_i]$, $x_i \in R^1$. $p(x_i)$ is called the probability mass function (pmf) of X at the mass point x_i if

(i) $p(x_i) \ge 0$ for all $x_i \in R^1$ (ii) $\sum_{x_i \in R^1} p(x_i) = 1$.

The distribution function is given by

$$F_X(x) = P[X \le x] = \sum_{x_i \le x} p(x_i).$$

Thus F is a step function which remains constant over every interval not containing any mass point, but has at each mass point x_i a step of height p_i (figure 5.3) Fig. 5.2 shows the column diagram of pmf of a discrete r.v. X.

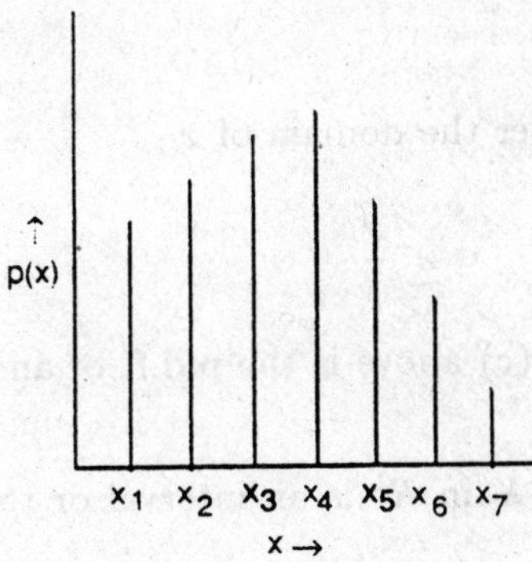

Fig. 5.2

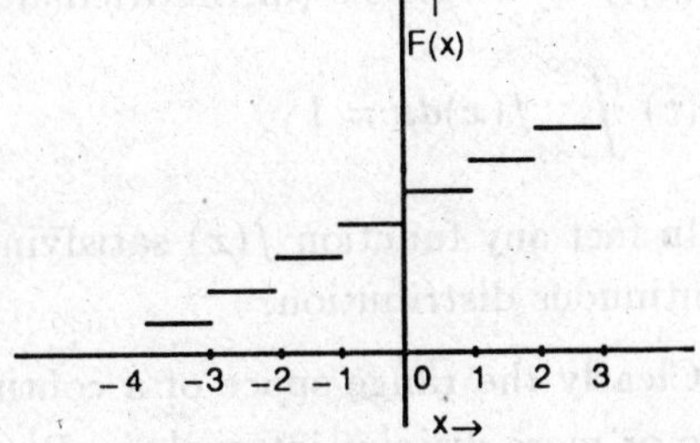

Fig. 5.3

Definition 5.4. *Continuous r.v.* Let X be a r.v. on $(\mathcal{S}, \mathcal{A}) \rightarrow (R^1, \mathcal{B}_1)$. The r.v. X is said to be of absolutely continuous (or simply of continuous) type if the distribution function F is absolutely continuous i.e. if there exists a function $F_X(x)$ (≥ 0) such that for every real number x we have

$$F_X(x) = \int_{-\infty}^{x} F_X(y)dy.$$

In this case, $\frac{dF}{dx}$ exists and $\frac{dF}{dx} = F_X(x)$ except for a set of values of X of probability 0.

The function $F_X(x)$ is called the probability density function (*p.d.f.*) of r.v. X. Since $F_X(x)$ is continuous everywhere, it has no point of jump. Thus the probability X takes the value x, $P[X = x] = 0$.

For any real number α, β

$$P[\alpha < X < \beta] = \int_{\alpha}^{\beta} f(x)dx = F(\beta) - F(\alpha).$$

$P[\alpha < X < \beta]$ represents the area under the curve of $f(x)$ between the lines $x = \alpha$ and $x = \beta$ (Fig. 5.5).

It follows (from the mean value theorem of calculus) that

$$P(x \leq X \leq x + dx) = \int_{x}^{x+dx} f(t)dt = dx f(\xi), x \leq \xi \leq x + dx \tag{5.5}$$

If dx is small, (5.5) is approximately equal to

$$f(x)dx \tag{5.6}$$

The properties of $f(x)$ are

(a) $f(x) \geq 0 \quad \forall x$

(b) f is integrable (in the Riemann sense) over the domain of x

(c) $\int_{-\infty}^{\infty} f(x)dx = 1$

In fact any function $f(x)$ satisfying (a), (b), (c) above is the p.d.f. of an absolutely continuous distribution.

Clearly the range space of a continuous r.v. X in R^1 is an interval or the union of two or more disjoint intervals in R^1.

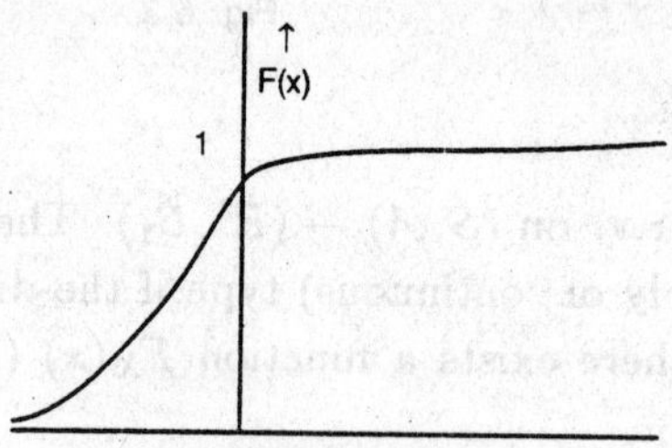

Fig. 5.4. df of a continuous x

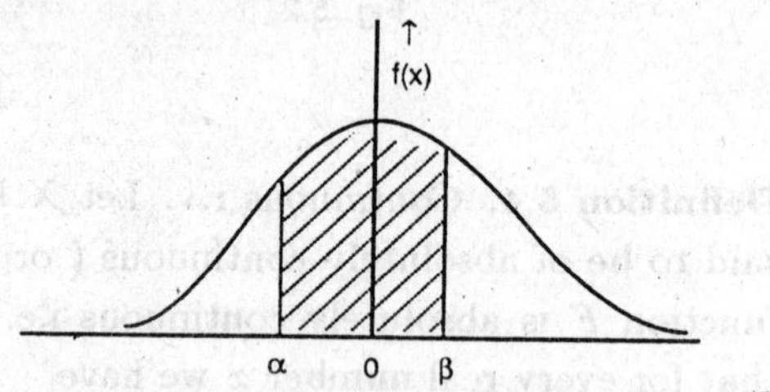

Fig. 5.5. Curve of pdf of a continuous x

Mixed distributions. Suppose the r.v. X assumes certain discrete values $x_1, \cdots, x_n$ with positive probabilities and assumes all values in an interval say, $a \leq X \leq b$ Thus X is discrete for certain range of values of X and continuous over an interval or some intervals not containing $x_1, \cdots, x_n$. The distribution of X here is of mixed type.

To each value x_i assign a non-negative number $p(x_i)$ (probability, $X = x_i$) such that $p(x_i) \geq 0$, $i = 1, \cdots, n$ and $\sum_{i=1}^{n} p(x_i) = p(< 1)$. Let f be a function defined over (a, b) such that $f(x) \geq 0$ for all $x \in (a, b)$, f is integrable and $\int_a^b f(x)dx = 1 - p$. The probability (mass or density) function defined in this way satisfies $P(-\infty \leq X \leq \infty) = 1$.

Definition 5.5. Let $0 < p < 1$. The function F defined by

$$F(t) = pG(t) + (1 - p)H(t) \tag{5.7}$$

where

$$G(t) = \sum_{x_i \leq t} p(x_i), \quad H(t) = \int_{-\infty}^{t} h(x)dx, \quad -\infty < t < \infty$$

is also a distribution function of a r.v. X and is a mixture of d.f. G of a discrete r.v. and d.f. H of a continuous r.v. The distribution of X here is of mixed type.

Such a distribution may arise if one, after making a random experiment with probability of success p and probability of failure $(1-p)$, chooses d.f. $G(t)$ if there is a success and $H(t)$ if there is a failure.

Fig. 5.1 shows the d.f. of a mixed distribution.

• **THEOREM 5.15.** *Every d.f. F can be decomposed into two parts according to*

$$F_X(x) = aF^c(x) + (1-a)F^d(x), \quad 0 \le a \le 1$$

where F^c is the d.f. of a continuous r.v. and F^d is the d.f. of a discrete r.v.

❑ **EXAMPLE 5.5.** *Let X be a continuous r.v. with p.d.f. given by*

$$f(x) = \begin{cases} ax, & 0 \le x \le 1 \\ a, & 1 \le x \le 2 \\ -ax + 3a, & 2 \le x \le 3 \\ 0 & 3 < x \end{cases}$$

Determine a and $F(x)$.

We have $\int_{-\infty}^{\infty} f(x)dx = a\int_0^1 xdx + a\int_1^2 dx + \int_2^3 (3a - ax)dx = 1$ when $a = 2$.

$$F_X(x) = \begin{cases} 0, & x < 0 \\ 2\int_0^x tdt = x^2, & 0 \le x \le 1 \\ 2\int_0^1 xdx + 2\int_1^x dt = 2x - 1, & 1 \le x \le 2 \\ 2\int_0^1 xdx + 2\int_1^2 dx + \int_2^x (6-2t)dt = 6x - x^2 - 5, & 2 \le x \le 3 \\ 1, & x < 3 \end{cases}$$

❑ **EXAMPLE 5.6.** *A number is selected at random from the interval [0, 1]. Suppose X is the square of the number selected. The c.d.f. of X is*

$$F_X(x) = \begin{cases} 0, & x < 0 \\ \sqrt{x}, & 0 \le x \le 1 \\ 1, & 1 < x \end{cases}$$

F is continuous and is differentiable everywhere except at $t = 0$ and $t = 1$.

$$f(x) = \frac{1}{2\sqrt{x}}, \quad 0 \le x \le 1$$

$$= 0 \quad \text{otherwise}$$

❑ **EXAMPLE 5.7.** *Let X have the triangular p.d.f. (Fig. 5.6(a))*

$$f(x) = \begin{cases} x, & 0 < x \le 1 \\ 2-x, & 1 \le x \le 2 \\ 0, & \text{elsewhere} \end{cases}$$

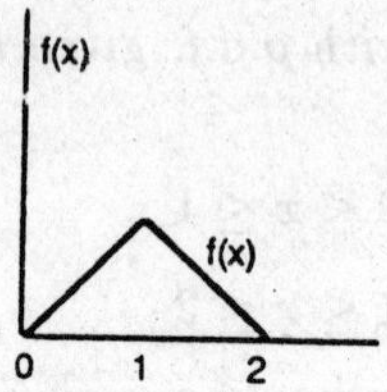

Fig. 5.6(a) Graph of f

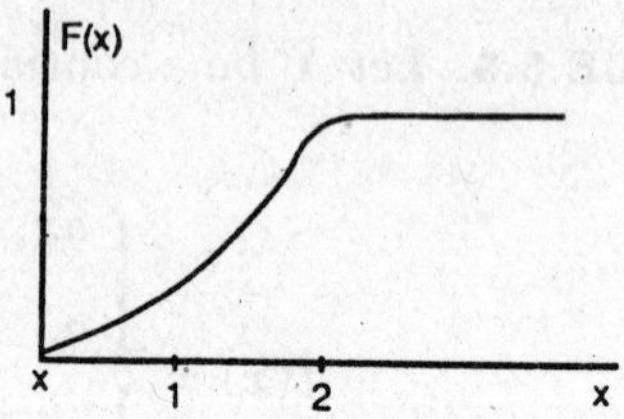

Fig. 5.6(b) Graph of F

It is seen f is a p.d.f. The d.f. is (fig. 5.6(b)) is

$$F_X(x) = \begin{cases} 0. & x \le 0 \\ \int_0^x t\,dt = \dfrac{x^2}{2}, & 0 < x \le 1 \\ \int_0^1 t\,dt + \int_1^x (2-t)\,dt = 2x - \dfrac{x^2}{2} - 1, & 1 < x \le 2 \\ 1 & x \ge 2 \end{cases}$$

❑ **EXAMPLE 5.8.** *Let for a r.v. X,*

$$F_X(x) = \begin{cases} 0, & x < 0 \\ \dfrac{x}{2}, & 0 \le x < 1 \\ \dfrac{3}{4}, & 1 \le x < 2 \\ 1, & 2 \le x \end{cases}$$

Here $F_X(x) = \dfrac{1}{2}\Big(G(x) + H(x)\Big)$ where

$$G(x) = \begin{cases} 0, & x < 1 \\ \dfrac{1}{2}, & 1 \le x < 2 \\ 1, & 2 \le x \end{cases}, \qquad H_X(x) = \begin{cases} 0, & x < 0 \\ x, & 0 \le x \le 1 \\ 1, & 1 < x \end{cases}$$

Thus X has a mixture distribution.

5.5. Independent Random Variables

Just as we defined the concept of independence between two events A and B, we may define the concept of independence between random variables. Intuitively, random variables X, Y are independent if the outcome of X in no way influences the outcome of Y and vice-versa.

❑ **EXAMPLE 5.9.** *Two persons A, B toss a die four times each. Let X_i, Y_i denote the face value of ith throw by A and B respectively. Clearly, here, (X_i, Y_j) are pairs of independent random variables, because value of X_i does not depend on the value of Y_j and vice-versa, $i, j = 1, \cdots, 4$.*

Definition 5.6. (a) Let X, Y be each one-dimensional discrete r.v. with $P(X = x_i) = p_i$ $(i = 1, 2 \cdots)$, $P(Y_j = y_j) = q_j$ $(j = 1, 2, \cdots)$. We say X, Y are independent r.v.s if and only if the joint probability $P(X = x_i, Y = y_j) = p_i q_j$ for all i, j (This case has been discussed in definition 4.7).

(b) Let X, Y be each one-dimensional continuous r.v. with p.d.f.'s $g(x)$, $h(y)$ respectively. We say, X, Y are independent r.v.s if and only if

$$P\left\{x - \frac{1}{2}dx \le X \le x + \frac{1}{2}dx, y - \frac{1}{2}dy \le Y \le y + \frac{1}{2}dy\right\}$$
$$= P\left\{x - \frac{1}{2}dx \le X \le x + \frac{1}{2}dx\right\} P\left\{y - \frac{1}{2}dy \le Y \le y + \frac{1}{2}dy\right\}$$
$$= g(x)h(y)dxdy$$

for all x, y in the domain of X, Y respectively.

The definition can similarly be extended to n varibales $X_1, \cdots, X_n$.

$X_1, \cdots, X_n$, when they are discrete, are said to be independent if and only if

$$P(X_1 = x_1, \cdots, X_n = x_n) = P(X_1 = x_1) \cdots P(X_n = x_n)$$

for all $x_1, \cdots, x_n$ in the domain of $X_1, \cdots, X_n$ respectively.

In case $X_1, \cdots, X_n$ are continuous, they are said to be independent if and only if

$$P\left\{x_1 - \frac{1}{2}dx_1 \le X_1 \le x_1 + \frac{1}{2}dx_1, \cdots, x_n - \frac{1}{2}dx_n \le X_n \le x_n + \frac{1}{2}dx_n\right\}$$
$$= P\left\{x_1 - \frac{1}{2}dx_1 \le X_1 \le x_1 + \frac{1}{2}dx_1\right\} \cdots P\left\{x_n - \frac{1}{2}dx_n \le X_n \le x_n + \frac{1}{2}dx_n\right\}$$
$$= g_1(x_1) \cdots g_n(x_n) dx_1 \cdots dx_n$$

for all $x_1, \cdots, x_n$ in the domain of $X_1, \cdots, X_n$ respectively, where $g_i(x_i)$ is the p.d.f. of X_i, $i = 1, \cdots, n$.

The concept will be discussed further in chapter 10.

5.6. Lebesgue-Stieltjes Integral

In the case of a continuous variable the d.f. is given by an integral and in the case of a discrete variable by a sum. In stating results equally true for a continuous and a discrete variable it is advantageous to have a single representation valid for both the cases. For this, we first examine a few types of integrations.

(a) *Riemann Stieltjes Integral* : Let $F_X(x)$ be a never decreasing function over $x \in [a,b]$. For any particular $x_o \in (a,b)$ both the limits $\lim_{\epsilon \downarrow 0} F(x_0+\epsilon)=F(x_0+0)$ and $\lim_{\epsilon \downarrow 0} F(x_0-\epsilon)=F(x_0-0)$ exist, since $F(x_0-\epsilon) \leq F(x_o) \leq F(x_o+\epsilon)$. If x_o is a point of continuity $F(x_o-0)=F(x_o+0)$. If $F(x_o-0) < F(x_o+0)$, x_o is a point of discontinuity (increase) of $F_X(x)$ and $F(x_o+0)-F(x_o-0)=h_o$ (say) is a measure of discontinuity at x_o [$F(x_o+0)-F(x_o)=$ right hand discontinuity; $F(x_o)-F(x_o-0)=$left hand discontinuity]. For any number of points of discontinuity, $y_o, y_1, \cdots, y_n$ the sum of discontinuities $h_o+\cdots+h_n \leq F(b)-F(a)$ and the points of discontinuities form a measurable set.

Let $g(x)$ be a continuous function in $[a,b]$. Divide the range $[a,b]$ into n subintervals by inserting points $x_1, x_2, \cdots, x_{n-1}$. In order of these arbitrarily choose points $\xi_1, \xi_2, \cdots, \xi_n$ in these intervals and form the sum

$$S=\sum_1^n g(\xi_i)[F(x_i)-F(x_{i-1})], \quad x_o=a_1, x_n=b$$

If M_i, m_i respectively are the upper and lower bound of $g(x)$ in (x_{i-1}, x_i), define upper and lower sum, respectively, as

$$S_U=\sum_1^n M_i\Big[F(x_i)-F(x_{i-1})\Big], S_L=\sum_1^n m_i\Big[F(x_i)-F(x_{i-1})\Big].$$

In the same way as for the ordinary integral (Riemann (R) integral), it is proved, when max $(x_i-x_{i-1}) \to 0$, S tends to a definite limit and is called the Riemann-Stieltjes (RS) integral of $g(x)$ wrt $F(x)$ over $[a,b]$ and is given by

$$\lim_{n\to\infty} S=\lim_{n\to\infty} S_U=\lim_{n\to\infty} S_L=\int_a^b g(x)dF(x) \tag{5.8}$$

whatever ξ_i in (x_{i-1}, x_i), provided max $(x_i-x_{i-1}) \to 0$.

The RS integral (5.8) exists in the more general case when $g(x)$ is bounded in $[a,b]$ and has at most a finite number of discontinuities r_i, provided $F(x)$ is continuous at every r_i. When $F(x)=x$, the RS integral (5.8) becomes the ordinary R-integral.

Note 1 : In case $F(x)$ is continuous everywhere in $[a, b]$ and has continuous derivative $F'(x)$ except at most in a finite number of points, it can be shown that (5.8) becomes the ordinary R-integral

$$\int_a^b g(x)F'(x)dx \tag{5.9}$$

Note 2 : If $f(x)$ is a step function with discontinuity $p_1, p_2, \cdots$ at points $x_1, x_2 \cdots$, FRS integral becomes the sum $\sum p_i g(x_i)$, which is a finite sum or absolutely convergent infinite series according as the set of points $\{x_i\}$ is finite or infinite.

Note 3 : The RS-integral extends to the case of a complex-valued function $g(x)$ and also to infinite interval (a, b) provided $g(x)$ is integrable over (a, b) wrt $F(x)$. Here

$$\lim_{n\to\infty} \sum_1^{\infty} g(\xi_i)\Big[F(x_i) - F(x_{i-1})\Big] = \int_{-\infty}^{\infty} g(x)dF(x). \tag{5.10}$$

when max $(x_i - x_{i-1}) \to 0$ as $n \to \infty$ and $a \to \infty$, $b \to \infty$ as $n \to \infty$.

(b) Lebesgue-Stieltjes Integral : Lebesgue-Stieltjes (LS) integral of $g(x)$ over a set S wrt a given (non-negative and additive set function) measure μ is similarly defined as

$$\lim_{n\to\infty} S_L = \lim_{n\to\infty} S_U = \int_S g(x)d\mu(S) \tag{5.11}$$

provided $g(x)$ is bounded for all $x \in S$ and $\mu(S) < \infty$, where S is divided into n disjoint subsets S_i, M_i, m_i being respectively the upper and lower limits of $g(x_i)$ in S_i $\left[S_U = \sum_{i=1}^{n} M_i\mu(Si),\ S_L = \sum_{i=1}^{n} m_i\mu(S_i)\right]$. Since to any measure $\mu(B)$ defined for all borel sets $B \in \mathcal{B}_1$ there exists a non-decreasing point function $f(x)$ such that $F(b) - F(a) = \mu\{(a, b]\}$, (5.11) can also be written as

$$\int_S g(x)dF(x). \tag{5.11$'$}$$

In particular when μ is a Lebesgue measure the integral (5.11) or (5.11)$'$ becomes Lebesgue-integral.

5.7. Integration of a Random Variable

Consider the probability space $(S, \mathcal{A}, P)$ and $X(w)$, a r.v. relative to $\mathcal{A}$. When $X(w)$ is a simple function (r.v.) [for an explanation, vide example 5.3], the LS-integral of $X(w)$ wrt $(S, \mathcal{A}, P)$ over S is defined by

$$\int_S X(w)dP(w) = \sum_{i=1}^{n} \ :P(A_i).$$

In general for any subset $E \in \mathcal{A}$,

$$\int_E X(w)dP(w) = \sum_{i=1}^{n} x_i P(E \cap A_i) \tag{5.12}$$

When $X(w)$ is bounded and can take infinitely many values, there exists a sequence of simple r.v.s $\{X_n(w)\}$ such that

$$\lim_{n\to\infty} X_n(w) = X(w) \tag{5.13}$$

[i.e. $|X_n(w) - X(w)| < \epsilon$ for every $\epsilon > 0$ for sufficiently large n] uniformly for all $w \in \mathcal{S}$. In this case LS-integral of $X(w)$ wrt $(\mathcal{S}, \mathcal{A}, P)$ over any $E \subset \mathcal{A}$ is defined by

$$\int_{\mathcal{S}} X(w)dP(w) = \lim_{n\to\infty} \int_{\mathcal{S}} X_n(w)dP(w) \tag{5.14}$$

and this limit is independent of the choice of the sequence $\{X_n(w)\}$ satisfying (5.13).

When $X(w)$ is not necessarily bounded, but non-negative, it is integrable if there exists a non-decreasing sequence of bounded r.v.s $\{X_n(w)\}$ such that (5.13) holds for all $w \in \mathcal{S}$ and if

$$\lim_{n\to\infty} \int_E X_n(w)dP(w) < \infty. \tag{5.15}$$

This limit is the LS-integral of $X(w)$ wrt $(S, \mathcal{A}, P)$ over $E \in \mathcal{A}$. This limit when it exists does not depend on the choice of $\{X_n(w)\}$ satisfying (5.13) and (5.15).

If $X(w)$ is arbitrary, it is integrable if it can be written as $X(w) = X_1(w) - X_2(w)$, where $X_1(w), X_2(w)$ are both non-negative integrable r.v.s and in this case

$$\int_E X(w)dP(w) = \int_E X_1(w)dP(w) - \int_E X_2(w)dP(w) \tag{5.16}$$

and its value is independent of choice of $X_1(w)$ and $X_2(w)$.

The LS- integral of a r.v. $X(w)$ wrt $(\mathcal{S}, \mathcal{A}, P)$ over $E \subset \mathcal{A}$ has the following important properties.

(i) If $X(w) = k$, a constant for all $w \in \mathcal{S}$ except possibly for a set of probablity measure zero (i.e. *a.e*)

$$\int_E X(w)dP(w) = kP(E)$$

(ii) If $\alpha \le X(w) \le \beta$ for all $w \in \mathcal{S}$ (a.e.) where α, β are constants, then

$$\alpha P(E) \le \int_E X(w)dP(w) \le \beta P(E)$$

(iii) If $X_1(w), X_2(w)$ are integrable and if α, β are constants, $\alpha X_1(w) + \beta X_2(w)$ is integrable and

$$\int_E (\alpha X_1(w) + \beta X_2(w))dP(w) = \alpha \int_E X_1(w)dP(w) + \beta \int_E X_2(w)dP(w)$$

(iv) If $X_1(w) \leq X(w) \leq X_2(w)$ for all $w \in \mathcal{S}$ (a.e.) where $X_1(w), X(w), X_2(w)$ are ingetrable, then

$$\int_E X_1(w)dP(w) \leq \int_E X(w)dP(E) \leq \int_E X_2(w)dP(w)$$

(v) $X(w)$ is integrable if and only if $|X(w)|$ is integrable and further,

$$\left|\int_E X(w)dP(w)\right| \leq \int |X(w)|dP(w)$$

(vi) If $\left\{X_n(w)\right\}$ is a sequence of r.v.s such that

$$\lim_{n\to\infty} X_n(w) = X(w) \quad \text{and} \quad \left|X_n(w)\right| \leq Y(w)$$

where $Y(w)$ is integrable, then $X(w)$ is integrable and

$$\lim_{n\to\infty} \int_E |X_n(w) - X(w)|dP(w) = 0.$$

$$\text{Also} \qquad \lim_{n\to\infty} \int_E X_n(w)dP(w) = \int_E X(w)dP(w)$$

uniformly for all $E \in \mathcal{A}$.

The readers interested in the theory of measure and integration can refer to books by Cramer (1950), Halmos (1950), Loève (1955), Munroe (1953), Mcshane (1944), Pitt (1963), among others.

5.8. Exercise

1. *A r.v. X has an uniform distribution*

$$f(x) = \begin{cases} 0, & x < a \\ \dfrac{1}{b-a}, & a \leq x \leq b \\ 0, & x > b. \end{cases}$$

Find its d.f.

2. *A r.v. X has the p.d.f.*

$$f(x) = \lambda e^{-\lambda x}, x > 0$$

Find its d.f.

3. *A r.v. X has p.d.f.*

$$f(x) = \begin{cases} Kx^2 e^{-\frac{x}{2}}, & x > 0 \\ 0 & \text{otherwise} \end{cases}$$

where K is a suitable constant. Find its d.f.

4. *If X has a continuous distribution with p.d.f.*

$$f(x) = \begin{cases} \frac{2}{a}\left(1 - \frac{x}{a}\right), & 0 < x < a \\ 0, & \text{otherwise} \end{cases}$$

Find its d. f. Find $P\left(\frac{a}{2} < X < a\right)$. (This is called 'right triangular' distribution).

$\left[\text{Prob.} = \frac{1}{4}\right]$

5. *A r.v. X has the p.d.f.*

$$f(x) = Axe^{-h^2x^2}, x > 0$$

Find the factor A. Find its d.f. (This is called Rayleigh distribution).

$\left[A = 2h^2\right]$

6. *Is the function f(x) defined as*

$$f(x) = \begin{cases} 0, & x < 2 \\ \frac{1}{18}(3 + 2x), & 2 \le x \le 4 \\ 0, & x > 4 \end{cases}$$

a density function ? Find the probability that a r.v. having this density will fall in the interval [2,3].

$\left[\text{Yes. } \frac{4}{9}\right]$

7. *If f(x) has the density function* $f(x) = \frac{x}{2}, 0 \le x \le 2$, *find the probability that (a) both of two sample values exceed 1, (b)exactly two out of four sample values will exceed 1.*

$\left[\text{(a) } \frac{9}{16}, \text{(b) } \frac{27}{128}\right]$

8. *Does the function* $f(x) = \dfrac{x+1}{\theta(\theta+1)} e^{-x/\theta} (x > 0)$, *where* $\theta > 0$, *represent a p.d.f. ? Find the corresponding d.f.*

9. *Do the following functions define d.f.'s ? In case they are d.f.'s, determine the p.d.f.,* $f(x)$. *Draw the graph of* $F_X(x)$ *and* $f(x)$.

(a) $F_X(x) = 0, \quad x \le 1; \quad 1 - \dfrac{1}{x}, \quad 1 < x.$

(b) $F_X(x) = 1 - e^{-x}, \quad x \ge 0; \quad 0, x < 0$

(c) $F_X(x) = \dfrac{2}{\pi} \sin^{-1}(\sqrt{x}), 0 \le x \le 1$

(d) $F_X(x) = e^{3x}, \quad -\infty < x \le 0$

(e) $F_X(x) = \dfrac{x^3}{2} + \dfrac{1}{2}, \quad -1 \le x \le 1$

(f) $F_X(x) = \dfrac{1}{1 + \exp\left\{-\dfrac{x-\alpha}{\beta}\right\}}, \beta > 0$

(This is called Logistic distribution).

10. *A unbiased die is thrown. If the die shows 1 or 2, a number is selected at random from the interval* $[0, 1]$. *If the die shows 3,4,5 or 6, a number is selected at random from the interval* $[1, 2]$. *Show that the distribution function of the number selected is*

$$F_X(x) = \begin{cases} 0, & x < 0 \\ \dfrac{x}{3}, & 0 \le x < 1 \\ \dfrac{2x}{3} - \dfrac{1}{3}, & 1 \le x < 2 \\ 1, & 2 < x \end{cases}$$

11. *Two unbiased coins are tossed. The score is 1 for each head and 0 for each tail. Find the distribution function of the score.*

12. *Two unbiased dice are tossed. Find the distribution function of the total score.*

13. *Let X be a r.v. with d.f.*

$$F_X(x) = \begin{cases} 0, & x < 0 \\ \dfrac{1}{2}, & x = 0 \\ \dfrac{1}{2} + \dfrac{x}{2}, & 0 < x < 1 \\ 1, & 1 \le x \end{cases}$$

Show that $F_X(x)$ can be written as a mixture of two d.f.'s $F_1(x)$ and $F_2(x)$ when

$$F_1(x) = \begin{cases} 0, & x < 0 \\ 1, & x \geq 0 \end{cases} \quad \text{and} \quad F_2(x) = \begin{cases} 0, & x \leq 0 \\ x, & 0 < x < 1 \\ 1, & 1 \leq x \end{cases}$$

$$\left[F(x) = \frac{1}{2}(F_1(x) + F_2(x))\right]$$

14. *The percentage of a component (100X) in a certain medical compound may be considered as a r.v., where X has the p.d.f.*

$$F_X(x) = 20x^3(1 - x), \quad 0 < x < 1.$$

(a) Find the d.f. and sketch its graph. (b) Find $P\left(X \leq \frac{2}{3}\right)$. (c) When $x \leq \frac{1}{3}$ the compound sells for Rs. 10/- per litter. When $x > \frac{1}{3}$ it sells for Rs. 12/- per litre. If the production cost is Rs. 9.50 per litre find the probability distribution of net profit per litre.

$$\left[\text{(a) } 5x^4 - 4x^5, \text{ (c) Hints : } P_r \text{ (Profit= .50 Rs/litre)}=\frac{8}{243}\right]$$

15. *A r.v. X has a Laplace distribution $f(x) = ae^{-\lambda|x|}$, $\lambda > 0$, $-\infty < x < \infty$. Find the factor a. Find $F_X(x)$.* $\left[a = \frac{1}{2},\ F_X(x) = \frac{1}{2}e^{\lambda x}, x \leq 0;\ 1 - \frac{1}{2}e^{\lambda x},\ x > 0\right]$

16. *The r.v. X has a Cauchy distribution*

$$f(x) = \frac{a}{\pi(1 + x^2)}, \quad -\infty < x < \infty$$

Find the factor a. Find the d.f. Find $P(-1 < X < 1)$.

$$\left[\frac{1}{\pi},\ F_X(x) = \frac{1}{\pi} \text{ arc tan } x + \frac{1}{2},\ \text{Prob. } = \frac{1}{2}\right]$$

17. *A variable X has p.d.f.*

$$f(x) = \begin{cases} ae^{-a(x - x_o)}, & x > 10 \\ 0, & x < 10 \end{cases}$$

Find the factor a. Find its d.f.

18. *An angle θ is chosen at random from* $\left[-\frac{\pi}{2}, \frac{\pi}{2}\right]$ *and a line is drawn through the point (0, a) making an angle θ with the y-axis. Find the density function for the point x at which this line crosses the x-axis.* $\left[f(x) = \frac{a}{\pi(a^2 + x^2)}\right]$

19. *A r.v. X has a density $f_1(x)$ with probability p_1 and a density $f_2(x)$ with probability $p_2, p_1+p_2=1$. Find the p.d.f. and distribution function of X.*

20. *A r.v.. X has possible values $1.2.\cdots, j\cdots$ with probability $P(X=j)=\frac{1}{2^j}$, $j=1,2,\cdots$. Calculate (i) $P(X$ is odd) (ii) $P(X\geq 6)$, (iii) $P(X$ is divisible by 5).*

$$\left[\text{(i) } \frac{2}{3}\right]$$

21. *Suppose f and g are p.d.f. on the same interval, $\alpha\leq x\leq\beta$. (a) Show $f+g$ is not a p.d.f. on that interval, (b) Show that for every number $a, 0<a<1$, $af(x)+(1-a)g(x)$ is a p.d.f. on that interval.*

22. *If X_1 and X_2 are independently and identically distributed random variables, prove that*

(i) $P\{|X_1-X_2|>t\}\leq 2P\left\{|X_1|>\frac{t}{2}\right\}$

(ii) if $a\geq 0$ such that $P\{X_1\geq a\}\leq 1-p$ and $P\{X_1\leq -a\}\leq 1-p$, then $P\{|X_1-X_2|\geq\epsilon\}\geq pP\{|X_1|>a+\epsilon\}$

[(i) $\left\{|X_1|>\frac{t}{2}\right\}\cup\left\{|X_2|>\frac{t}{2}\right\}\supset\{|X_1-X_2|>t\}$

$$\therefore\quad P\{|X_1-X_2|>t\}\leq P\left\{|X_1|>\frac{t}{2}\right\}+P\left\{|X_2|>\frac{t}{2}\right\}=2P\left\{|X_1|>\frac{t}{2}\right\}$$

(ii) $\left\{X_1\geq a+\epsilon, X_2\leq a\right\}\cup\left\{X_1\leq -a-\epsilon, X_2\geq -a\right\}\subset\left\{|X_1-X_2|\geq\epsilon\right\}$

$$\therefore\quad P\left\{|X_1-X_2|\geq\epsilon)\right\}\geq P\left\{X_1>a+\epsilon, X_2\leq a\right\}+P\left\{X_1<-a-\epsilon, X_2\geq -a\right\}\geq pP\left\{X_1\geq a+\epsilon\right\}+pP\left\{X_1\leq -a-\epsilon\right\}=pP\left\{|X_1|>a+\epsilon\right\}$$

23. *For any real r.v. X with finite expectation, show that*

$$\sum_{n=1}^{\infty}P(|X|\geq n)\leq E(|X|)<1+\sum_{n=1}^{\infty}P(|X|\geq n).$$

[*Let $Y=|X|$ and suppose G is the d.f. of Y.*

$$E(Y)=\int_0^{\infty}ydG(y)=\sum_{i=0}^{\infty}\int_i^{(i+1)-}ydG(y)$$

$$\text{Now}\quad i\int_i^{(i+1)-}dG(y)\leq\int_i^{(i+1)-}ydG(y)\leq(i+1)\int_i^{(i+1)-}dG(y)$$

$$\text{i.e.}\quad iP[i\leq Y<i+1]\leq\int_i^{(i+1)-}ydG(y)\leq(i+1)P[i\leq Y<i+1].$$

Summing over all i

$$\sum_{i=0}^{\infty} iP[i \le Y < i+1] \le E(Y) \le \sum_{i=0}^{\infty} (i+1)P[i \le Y < i+1]$$

Now $$\sum_{i=0}^{\infty} iP[i \le Y < i+1] = \sum_{n=1}^{\infty} P(|X| \ge n).$$

$$\sum_{i=0}^{\infty} (i+1)P[i \le Y < i+1] = 1 + \sum_{n=1}^{\infty} P(|X| \ge n).\Bigg]$$

6 SOME CHARACTERISTICS OF A PROBABILITY DISTRIBUTION

6.1. Introduction

This chapter deals with some characteristics of probability distribution of a r.v., namely, its expectation, moments, measures of central tendency and dispersion and some probability inequalities. We study expectation in section 6.2, moments in sections 6.3-6.6, measures of central tendency in section 6.7, measures of dispersion in section 6.8, measures of skewness and kurtosis in section 6.9. Lastly some probability inequalities are discussed in section 6.10.

6.2. Expectation

Let X be a r.v. defined on a probability space $(\mathcal{S}, \mathcal{A}, \mathcal{P})$ and let F be its distribution function. The expectation of X is defined by

$$E(X) = \int_{-\infty}^{\infty} x dF(x) \tag{6.1a}$$

$$= \int_{S} X dP \tag{6.1b}$$

$$= \int X dP_X \tag{6.1c}$$

where the integral (6.1a) is the RS-integral of x w.r.t. $F(x)$ over R^1, the integral (6.1b) is the LS-integral of the r.v. X w.r.t. $(\mathcal{S}, \mathcal{A}, \mathcal{P})$ over $\mathcal{S}$ and the integral (6.1c) is the

LS-integral of X w.r.t. the induced probability space $(R^1, \mathcal{B}_1, P_X)$ over R^1. $E(X)$ is said to exist if and only if

$$\int_{-\infty}^{\alpha} |x| dF(x) < \infty. \tag{6.2}$$

Thus if X be a r.v. of discrete type with pmf $P(X = x_k) = p(x_k)$, $k = 1, 2, \cdots$ the series

$$E(X) = \sum_{k=1}^{\infty} x_k p(x_k) \tag{6.3}$$

is called the expectation of X provided the condition

$$\sum_{k=1}^{\infty} |x_k| p(x_k) < \infty \tag{6.4}$$

is satisfied. If X is a continuous r.v. with *pdf* $f(x)$, the integral

$$E(X) = \int_{-\infty}^{\infty} x f(x) dx \tag{6.5}$$

is called the expectation of X provided the condition

$$\int_{-\infty}^{\infty} |x| f(x) dx < \infty \tag{6.6}$$

is satisfied.

Remarks The expectation of a r.v. X may not always exist. Before concluding that $E(X)$ exists one must check if (6.4) or (6.6) is satisfied.

The notion of mathematical expectation is due to Huygens (1629-95), a Dutch mathematician.

❑ **EXAMPLE 6.1.** *A variate X has pdf*

$$f(x) = \frac{1}{b-a} \quad if \quad a \le x \le b$$

$$= 0 \qquad \text{otherwise}$$

$$E(X) = \int_a^b \frac{x dx}{b-a} = \frac{a+b}{2}$$

❑ **EXAMPLE 6.2.** *A variate X has pdf* $f(x) = \dfrac{1}{x\sqrt{2\pi}} e^{-\frac{1}{2}(\log x)^2}, x > 0$

$$E(X) = \int_0^{\infty} \frac{1}{\sqrt{2\pi}} e^{-\frac{1}{2}(\log x)^2} dx = \frac{1}{\sqrt{2\pi}} \int_{-\infty}^{\infty} e^{-\frac{t^2}{2}} e^t dt \quad (\text{putting} \quad t = \log x)$$

$$= \frac{\sqrt{e}}{\sqrt{2\pi}} \int_{-\infty}^{\infty} e^{-\frac{1}{2}(t-1)^2} dt = \sqrt{e} \quad \left[\because \quad \frac{1}{\sqrt{2\pi}} \int_{-\infty}^{\infty} e^{-y^2/2} dy = 1 \right]$$

❑ **EXAMPLE 6.3.** *The following example shows that (6.3) is finite though the condition (6.4) is not satisfied.*

Let X be a discrete r.v. with *pmf* $P(X = x) = \frac{6}{\pi^2 x^2}(0)$ for $x = 1, -2, +3, -4, \cdots$ (otherwise).

$$E(X) = \frac{6}{\pi^2}\sum_{x=1}^{\infty}\frac{(-1)^{x-1}}{x} < \infty$$

However, $E|X| = \frac{6}{\pi^2}\sum_{x=1}^{\infty}\frac{1}{x} = \infty$. Hence $E(X)$ does not exist.

❑ **EXAMPLE 6.4.** *The distribution* $f(x) = \frac{1}{\pi(1+x^2)}, -\infty \le x \le \infty$ *is known as the Cauchy distribution. Here* $E(X)$ *exists if*

$$E\{|X|\} = \lim_{\substack{a\to-\infty \\ b\to\infty}} \int_a^b \frac{|x|}{\pi(1+x^2)}dx < \infty$$

However

$$\int_0^b \frac{|x|}{\pi(1+x^2)}dx = \frac{1}{2\pi}\Big(\log(b^2+1)\Big) \to \infty$$

as $b \to \infty$. Hence $E(X)$ does not exist for this distribution.

We have seen in theorem 5.9, if X is a r.v. on $(\mathcal{S}, \mathcal{A}, \mathcal{P})$, so is any borel-measurable function $g(X)$. $E[g(X)]$, if it exists, is given by

$$E[g(X)] = \int_{-\infty}^{\infty} g(x)dF(x) \tag{6.7}$$

Thus if X is discrete with pmf $p(x_i)$,

$$E\Big[g(X)\Big] = \sum_{i=1}^{\infty} g(x_i)p(x_i); \tag{6.8}$$

if X is continuous with *pdf* $f(x)$,

$$E\Big[g(X)\Big] = \int_{-\infty}^{\infty} g(x)f(x)dx \tag{6.9}$$

provided the expectation exists.

Some results on expectation

(1) If the r.v. X is a constant c i.e. if $X(w) = c$ a.e. i.e. if $P\{X = c\} = 1$, $E(X) = c$.

Proof.

$$E(X) = \int_{-\infty}^{\infty} xdF(x) = \int_{-\infty}^{\infty} cdF(x) = c\int_{-\infty}^{\infty} dF(x) = c$$

(2) If $E(X)$ exists $E(aX+b)$ exists and is given by

$$E(aX+b) = aE(X)+b$$

where a, b are arbitrary constants.

Proof.

$$\int_{-\infty}^{\infty} |ax+b|dF(x) < \int_{-\infty}^{\infty} |ax|dF(x) + |b|\int_{-\infty}^{\infty} dF(x) < |a|\int_{-\infty}^{\infty} |x|dF(x) + |b| < \infty$$

since $E|X| < \infty$. Also

$$E(aX+b) = \int_{-\infty}^{\infty} (ax+b)dF(x) = a\int_{-\infty}^{\infty} xdF(x) + b\int_{-\infty}^{\infty} dF(x) = aE(X)+b.$$

(3) If $g(X) = g_1(X)+g_2(X)$ where $g_1(X)$, $g_2(X)$ are two borel measurable functions of r.v. X, then

$$E[g(X)] = E[g_1(X)] + E[g_2(X)]$$

provided the expectations exist.

Proof.
$$E[g(X)] = \int_{-\infty}^{\infty} \left[g_1(x)+g_2(x)\right] dF(x)$$

$$= \int_{\infty}^{\infty} g_1(x)dF(x) + \int_{-\infty}^{\infty} g_2(x)dF(x) = Eg_1(X)+Eg_2(X)$$

(4) If $g_1(X) \le g_2(X)$ for all possible values of X, then

(i) $Eg_1(X) \le Eg_2(X)$ (ii) $|Eg_1(X)| \le E|g_1(X)|$.

Proof. (i) $$Eg_1(X) = \int_{-\infty}^{\infty} g_1(x)dF(x) \le \int_{-\infty}^{\infty} g_2(x)dF(x) = Eg_2(X)$$

(ii) $$|Eg_1(X)| = \left|\int_{-\infty}^{\infty} g_1(x)dF(x)\right| \le \int_{-\infty}^{\infty} |g_1(x)|dF(x) = E|g_1(X)|$$

(5) ***Conditional expectation of X Given A :*** Let A be a measurable subset of the sample sapce $\mathcal{S}$. Consider the r.v. XI_A where I_A is the indicator function so that for a point w,

$$XI_A(w) = X(w) \quad \text{if} \quad w \in A$$
$$= 0 \quad \text{otherwise}$$

The conditional expected value of X given A is defined as

$$E(X|A) = \frac{E(XI_A)}{P(A)} \tag{6.10}$$

provided $E(XI_A)$ exists and $P(A) > 0$. Since $|XI_A| \le |X|$, $E(XI_A)$ exists if $E(X)$ exists. $\left(\text{Clearly } E(XI_A) = \sum_{w\in A} X(w)p(w)\right)$.

• **THEOREM 6.1.** *Let* $\{A_i, i = 1, \cdots, k\}$ *be a measurable partition of the sample sapce* $\mathcal{S}$ $\left(\text{i.e.} \bigcup_{i=1}^{k} A_i = \mathcal{S},\ A_iA_j = \phi,\ i \neq j = 1, \cdots, k\right)$ *such that* $P(A_i) > 0$ *for all* i *and let* $E(X)$ *exist. Then*

$$E(X) = \sum_{i=1}^{k} E(X|A_i)P(A_i)\Big(= E_A[E(X|A)]\Big) \tag{6.11}$$

$\Big(E_A(\cdot)$ denotes expectation with respect to A (i.e. for different subsets $A_i, i = 1, \cdots, k)\Big)$

Proof.

$$E(X) = \int X dP = \int_{\Sigma A_i} X dP = \sum_i \int X I_{A_i} dP$$

$$= \sum_i E\big(X I_{A_i}\big) = \sum_i E(X|A_i)P(A_i) \tag{6.12}$$

by (6.10).

Theorems 4.4 and 4.8 for expectation of sum and product of discrete random variables respectively hold good for continuous variables also. That is,

• **THEOREM 6.2.** *If* $X_1, \cdots X_n$ *are each one-dimensional r.v.s, then* $E(X_1 + \cdots X_n) = E(X_1) + \cdots + E(X_n)$.

• **THEOREM 6.3.** *If* $X_1, \cdots, X_n$ *are independent r.v.s,* $E(X_1 \cdots X_n) = E(X_1 \cdots E(X_n)$. *Proof of these theorems will be given in chapter 10.*

(6) *An inequality for convex functions :*

A continuous function $f(x)$ is said to be convex in an interval I, if for any two values x_1, x_2 in I,

$$\lambda f(x_1) + (1 - \lambda)f(x_2) \geq f(\lambda x_1 + (1 - \lambda)x_2)$$

for any $\lambda, 0 < \lambda < 1$ i.e. any point lying on the straight line joining any two points of $f(x)$, $x \in I$ lies above the point with the same value of abscissa on the curve (fig. 6.1). For such a function for every point x_o in I there exists a straight line which passes through $(x_o, f(x_o))$ and lies on or below the graph of $f(x)$. The following inequality holds in the case of a convex function .

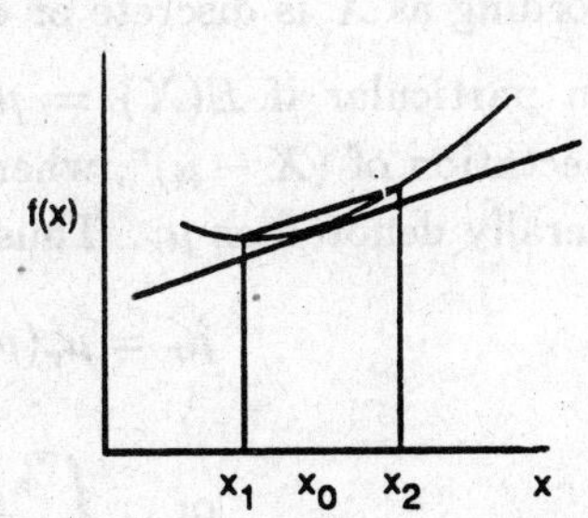

Fig. 6.1

Jensen's Inequality. Let X be a r.v. and $f(X)$ a convex function of X. Then

$$E[f(X)] \geq f[E(X)]$$

provided the expectations exist. The equality holds if and only if X has a degenerate distribution [i.e. $X(w) = k$, a constant, a.e.]

Proof. Since $f(X)$ is continuous and convex there exists a line $l(X) = a + bX$ which passes through the point $(E(X), f(E(X)))$ and lies below it i.e. $l(X) \leq f(X)$ for all values of X and $l[E(X)] = f(E(X))$. Now $l[E(X)] = a + bE(X) = E[l(X)]$. Hence $f[E(X)] = E[l(X)] \leq E[f(X)]$ i.e. $E[f(X)] \geq f[E(X)]$.

❑ **EXAMPLE 6.5.** *$f(x) = x^2$ is a convex function of x. Hence $E(X^2) \geq (E(X))^2$ i.e. $V(X) \geq 0$.*

❑ **EXAMPLE 6.6.** *Since $|X|^r$ is a convex function of X for $r > 1$, we have*

$$E|X|^r \geq \left[E|X|\right]^r.$$

6.3. Moments

Let X be a r.v. with *pmf* $p(x_k)$ or *pdf* $f(x)$. The r th raw (non-central) moment of the distribution of X, if it exists, is defined as

$$\mu'_r = E(X^r) = \sum_{i=1}^{\infty} x_i^r p(x_i) \tag{6.13}$$

$$\text{or} \quad \int_{-\infty}^{\infty} x^r f(x)dx \tag{6.14}$$

according as X is discrete or continuous. The series (6.13) or the integral (6.14) must be absolutely convergent for μ'_r to exist. The first moment μ'_1, if it exists, is the expectation of X. Clearly $\mu_o = 1$.

The rth moment of X about a point 'a', if it exists, is given by,

$$\mu'_r(a) = E(X - a)^r = \sum_{i=1}^{\infty}(x_i - a)^r p(x_i) \tag{6.15}$$

$$\text{or} \quad \int_{-\infty}^{\infty}(x - a)^r f(x)dx \tag{6.16}$$

according as X is discrete or continuous.

In particular if $E(X) = \mu'_1 = \mu$ (say) exists, then one may take $a = \mu$. The expectation of $(X - \mu)^r$, when it exists, is called the rth central moment of X and is generally denoted as μ_r. Thus

$$\mu_r = \mu'_r(\mu) = E(X - \mu)^r = \sum_{i=1}^{\infty}(x_i - \mu)^r p(x_i) \tag{6.17}$$

$$\text{or} \quad \int_{-\infty}^{\infty}(x - \mu)^r f(x)dx \tag{6.18}$$

according as x is discrete or continuous. Expanding $(X - \mu)^r$ in the integral (or summation), we get

$$\mu_r = \int_{-\infty}^{\infty}\left[x^r - \binom{r}{1}x^{r-1}\mu + \binom{r}{2}x^{r-2}\mu^2 - \cdots + (-1)^r\mu^r\right]f(x)dx$$

$$= \mu'_r - \binom{r}{1}\mu'_{r-1}\mu + \binom{r}{2}\mu'_{r-2}\mu^2 - \cdots + (-1)^r\mu^r \tag{6.19}$$

The relation (6.19) gives the expressions for the central moments in terms of raw moments. Thus,

$$\mu_1 = \mu_1' - \mu_0'\mu = 0 \qquad \mu_3 = \mu_3' - 3\mu\mu_2' + 2\mu^3$$

$$\mu_2(\text{or} \quad \sigma^2) = \mu_2' - \mu^2 \qquad \mu_4 = \mu_4' - 4\mu\mu_3' + 6\mu^2\mu_2' - 3\mu^4.$$

Similarly one may express μ_r' in terms of central moments by considering

$$\int_{-\infty}^{\infty} x^r\,dF(x) = \int_{-\infty}^{\infty} \Big\{(x-\mu)+\mu\Big\}^r dF(x)$$

when

$$\mu_r' = \mu_r + \binom{r}{1}\mu_{r-1}\mu + \binom{r}{2}\mu_{r-2}\mu^2 + \cdots + \mu^r \tag{6.20}$$

In particular,

$$\mu_1' = \mu \qquad \mu_3' = \mu_3 + 3\mu_2\mu + \mu^3$$

$$\mu_2' = \mu_2 + \mu^2 \qquad \mu_4' = \mu_4 + 4\mu_3\mu + 6\mu_2\mu^2 + \mu^4$$

In theorems 6.4 – 6.8 below we consider the problem of existence of moments.

- **THEOREM 6.4.** *If μ_r' exists for a r.v. X, then μ_s' exists, $0 < s < r$.*

Proof. We have for $|X| \geq 1$, $|X|^s \leq |X|^r$; for $|X| < 1$, $1 > |X|^s > |X|^r$. Now

$$E|X|^s = \int\limits_{|x|\leq 1} |x|^s\,dF(x) + \int\limits_{|x|>1} |x|^s\,dF(x)$$

$$\leq \int\limits_{|x|\leq 1} dF(x) + \int\limits_{|x|>1} |x|^r\,dF(x) < 1 + E|X|^r < \infty.$$

It can be shown similarly

- **THEOREM 6.5.** *If $\mu_r'(a)$ exists for a r.v. X, then $\mu_s'(a)$ exists where a is an arbitrary constant, $0 < s < r$.*
- **THEOREM 6.6.** *If μ_r exists for a r.v. X, then μ_s exists, $0 < s < r$.*
- **THEOREM 6.7.** *If μ_r' exists, $\mu_r'(a)$ exists and is given by*

$$\mu_r'(a) = \mu_r' - \binom{r}{1}\mu_{r-1}'a + \binom{r}{2}\mu_{r-2}'a^2 - \cdots + (-1)^r a^r \tag{6.21}$$

Proof. (6.21) follows as in (6.19) by exapnding $(x-a)^r$ in $\displaystyle\int_{-\infty}^{\infty}(x-a)^r\,dF(x)$.

Existence follows from the fact

$$\int_{-\infty}^{\infty} |x-a|^r\,dF(x) \leq \int_{-\infty}^{\infty} |x|^r\,dF(x) + \binom{r}{1}|a|\int_{-\infty}^{\infty}|x|^{r-1}dF(x)$$

$$+\binom{r}{2}|a|^2\int_{-\infty}^{\infty}|x|^{r-2}dF(x) + \cdots + \cdots |a|^r$$

and since μ_r' exists, μ_{r-1}', $\mu_{r-2}' \cdots$ exist.

• **THEOREM 6.8.** *If $\mu'_r(a)$ exists, μ'_r exists and is given by*

$$\mu'_r = \mu'_r(a) + \binom{r}{1} a\mu'_{r-1}(a) + \binom{r}{2} a^2\mu'_{r-2}(a) + \cdots + a^r \tag{6.22}$$

Proof. Follows as in (6.20) and theorem 6.7.

The quantity $E|X|^r = \nu_r$ (say), when it exists, is called the rth absolute moment of r.v. X.

❑ **EXAMPLE 6.7.** *Find the rth non-central moment of the rectangular distribution*

$$f(x) = \frac{1}{b-a}, a < x < b; \ = 0, \text{ otherwise.}$$

$$\mu'_r = \frac{1}{b-a}\int_a^b x^r\,dx = \frac{b^{r+1} - a^{r+1}}{(b-a)(r+1)}$$

Thus $\mu = \dfrac{b+a}{2}$, $\sigma^2 = \dfrac{(b-a)^2}{12}$.

❑ **EXAMPLE 6.8.** *A Continuous r.v. X has the distribution function $F(x)$ proportional to $\alpha x^\beta - \beta x^\alpha, \alpha > \beta \geq 1, 0 \leq X < 1$. Show that the rth moment of X about the origin is $\alpha\beta/(\alpha+r)(\beta+r)$.*

Let $F(x) = k(\alpha x^\beta - \beta x^\alpha)$ when k is a suitable constant.

$$f(x) = \frac{dF(x)}{dx} = k\alpha\beta\left(x^{\beta-1} - x^{\alpha-1}\right).$$

Since $\displaystyle\int_0^1 f(x)dx = 1, \quad k = \frac{1}{\alpha - \beta}.$

$$\mu'_r = \frac{(\alpha\beta)}{\alpha-\beta}\int_0^1 x^r\left(x^{\beta-1} - x^{\alpha-1}\right)dx = \frac{k\alpha\beta}{(\alpha+r)(\beta+r)}$$

❑ **EXAMPLE 6.9.** *Find the rth non-central moment of the distribution*

$$f(x) = \frac{1}{\Gamma(n)}\alpha^n e^{-\alpha x} x^{n-1}, x \geq 0, n > 0$$

(This is known as Gamma distribution. $\Gamma(n) = (n-1)!$ for integral n).

$$\mu'_r = \frac{\alpha^n}{\Gamma(n)}\int_0^\infty e^{-\alpha x} x^{n+r-1}dx = \frac{\Gamma(n+r)}{\alpha^r\,\Gamma(n)}$$

Hence $\mu'_1 = \mu = \dfrac{n}{\alpha}$, $\mu_2 = \mu'_2 - \mu^2 = \dfrac{n}{\alpha^2}$.

For $n = 1$, the distribution is known as exponential distribution when $\mu = \dfrac{1}{\alpha}$, $\sigma^2 = \dfrac{1}{\alpha^2}$.

❑ **EXAMPLE 6.10.** ***If*** $f(x)$ ***is an odd function of*** x ***of period*** $\frac{1}{2}$, ***show that*** $\int_0^\infty x^r x^{-\log x}$ $f(\log x)dx = 0$ ***for all integral values of*** r. ***Hence show that***

$$dF = x^{-\log x}\left[1 - \lambda \sin(4\pi \log x)\right]dx,\ 0 < x < \infty, 0 \le \lambda \le 1$$

have the same moments whatever be the value of λ.

Let $I = \int_0^\infty x^{r-\log x} f(\log x)dx$

Putting $\log_e x = y + \frac{1}{2}$,

$$I = \int_{-\infty}^{\infty} e^{(y+\frac{1}{2})(r-y-\frac{1}{2})} f\left(y+\frac{1}{2}\right) e^{y+\frac{1}{2}} dy$$

$$= e^{\frac{1}{4}(r+1)^2} \int_{\infty}^{\infty} e^{-(y-\frac{1}{2}r)^2} f(y)dy, \quad \left[\text{since } f\left(y+\frac{1}{2}\right) = f(y)\right] \qquad \text{(i)}$$

Again putting $\log_e x = -y + \frac{1}{2}$

$$I = -e^{\frac{1}{4}(r+1)^2} \int_{-\infty}^{\infty} e^{-(y+\frac{r}{2})^2} f(y)dy \quad \left[\text{since } f(-y) = -f(y)\right] \qquad \text{(ii)}$$

Adding (i) and (ii),

$$2I = e^{\frac{1}{4}(r+1)^2}\left[\int_{-\infty}^{\infty} e^{-(y-\frac{r}{2})^2} f(y)dy - \int_{-\infty}^{\infty} e^{-(y+\frac{r}{2})^2} f(y)dy\right] \qquad \text{(iii)}$$

Putting $y - \frac{r}{2} = u$ in the the first integral and $y + \frac{r}{2} = v$ in the second integral in (iii) and observing that $f\left(u+\frac{r}{2}\right) = f(u), f\left(v-\frac{r}{2}\right) = f(v)$,

$$2I = e^{\frac{1}{4}(r+1)^2}\left[\int_{-\infty}^{\infty} e^{-u^2} f(u)du - \int_{-\infty}^{\infty} e^{-v^2} f(v)dv\right] = 0$$

Hence $I = 0$. Now $\sin(4\pi x)$ is an odd function of x of period $\frac{1}{2}$.

Thus $\int_0^\infty x^{r-\log x} \sin(4\pi \log x) = 0$.

Hence the moments of the given distribution do not depend on the value of λ.

6.4. Some Inequalities Concerning Moments

First we consider *Cauchy-Schwartz inequality*. If X, Y are two random variables both defined on $(\mathcal{S}, \mathcal{A}, P)$ and with finite variances then

$$E(X^2)E(Y^2) \ge \left(E(XY)\right)^2 \qquad (6.23)$$

Proof. Since $|XY| \leq \frac{1}{2}(X^2 + Y^2)$, $E|XY| < \infty$, as $E(X^2), E(Y^2)$ exist and hence $E(XY)$ exists.

For any real number t, let $h(t) = E(tX - Y)^2$. Clearly $h(t)$ is real and ≥ 0. Hence the equation

$$h(t) = t^2 E(X^2) - 2tE(XY) + E(Y^2) = 0 \tag{6.24}$$

has either no (real) solution or only one solution. (6.24) has no solution if and only if $\left(E(XY)\right)^2 < E(X^2)E(Y^2)$. (6.24) has exactly one solution if and only if $\left(E(XY)^2\right) = E(X^2)\ E(Y^2)$. Hence the inequality (6.23). Equality in (6.23) holds if and only if there exists some (real) t for which $E(tX - Y)^2 = 0$ i.e. $P[tX - Y = 0 \text{ for some } t] = 1$ i.e. X is proportional to Y a.e.

Note : If in (6.23) we take $X - E(X), Y - E(Y)$ in place of X, Y respectively the inequality reduces to

$$V(X)V(Y) \geq \left[\text{Cov}\ (X, Y)\right]^2 \tag{6.25}$$

Note : (6.23) holds if X, Y are replaced respectively by some borel measurable functions $g_1(X), g_2(X)$ of r.v. X defined on $(S, \mathcal{A}, \mathcal{P})$ provided $E[g_1(X)]^2 < \infty$, $E[g_2(X)^2] < \infty$.

[Algebraic versions of $C - S$ inequality have been given in the appendix to this chapter]

• **THEOREM 6.9.** *If both μ'_{2r}, μ'_{2s} exist then*

(i) $\mu'_{2r}\mu'_{2s} \geq (\mu'_{r+s})^2$ (6.26a)

(ii) $\mu_{2r}\mu_{2s} \geq (\mu_{r+s})^2$ (6.26b)

In particular,

(iii) $\beta_2 = \dfrac{\mu_4}{\mu_2^2} \geq 1$, provided $\mu_2 > 0$ (6.26c)

Proof. If μ'_{2r}, μ'_{2s} exist μ_{2r}, μ_{2s} exist Also μ_{r+s}, μ_{r+s} exist (theorems 6.5 and 6.7).

Putting in (6.23) X^r, X^s in place of X and Y respectively the inequality (6.26a) follows. Putting in (6.23), $(X - E(X))^r$, $(X - E(X))^s$ in place of X, Y respectively (6.26b) follows. (6.26c) gives $\mu_4 \geq \mu_2^2$ which is (6.26b), when $r = 2, s = 0$.

Note : The inequality (6.26c) can also be proved as follows : Since

$$V(X^2) = E\left[X^2 - E(X^2)\right]^2 \geq 0 \tag{a}$$

$$E(X^4) - E\left(E(X^2)\right)^2 \geq 0 \quad \text{i.e.} \quad \mu'_4 \geq (\mu'_2)^2$$

Putting $X - E(X)$ in place of X in (a), (6.26c) follows—equality holds if and only if $\left(X - E(X)\right)^2 = k$ (a constant) for all values of X. This occurs if X takes only two values, x_1 and x_2 with equal probabilities when $\left(X - E(X)\right)^2 = \frac{1}{4}(x_1 - x_2)^2$.

• **THEOREM 6.10.** *For any r.v. X,*

$$\begin{vmatrix} \mu_{2a} & \mu_{a+b} & \mu_{a+c} \\ \mu_{a+b} & \mu_{2b} & \mu_{b+c} \\ \mu_{a+c} & \mu_{b+c} & \mu_{2c} \end{vmatrix} = \delta(a,b,c) \text{ (say)} \geq 0 \tag{6.27}$$

where a, b, c are non-negative integers. Hence, provided μ_4 exists,

$$\beta_2 - \beta_1 - 1 \geq 0 \tag{6.28}$$

$$\left(\text{where} \quad \beta_2 = \frac{\mu_4}{\mu_2^2}, \beta_1 = \frac{\mu_3^2}{\mu_2^3}\right)$$

Proof. We have

$$E\left[AX^a + BX^b + CX^c\right]^2 \quad \text{is always} \geq 0$$

i.e.

$$E\left[A^2X^{2a} + B^2X^{2b} + C^2X^{2c} + 2ABX^{a+b} + 2ACX^{a+c} + 2BCX^{b+c}\right] \geq 0 \tag{6.29}$$

for all values of A, B and C. If X is measured from $E(X)$, (6.29) ruduces to

$$A^2\mu_{2a} + B^2\mu_{2b} + C^2\mu_{2c} + 2AB\mu_{a+b} + 2AC\mu_{a+c} + 2BC\mu_{b+c} \geq 0 \tag{6.30}$$

for all values of A, B and C i.e. the left hand side of (6.30) is a positive semi-definite quadratic form in A, B, C.

A set of necessary and sufficient conditions for this quadratic form to be poistive semi-definite is that all the principal minors of δ are non-negative i.e.

$$\mu_{2a} \geq 0, \begin{vmatrix} \mu_{2a} & \mu_{a+b} \\ \mu_{a+b} & \mu_{2b} \end{vmatrix} \geq 0, \quad \delta \geq 0$$

For $a = 0$, $b = 1$, $c = 1$, (6.25) reduces to

$$\begin{vmatrix} \mu_0 & \mu_1 & \mu_2 \\ \mu_1 & \mu_2 & \mu_3 \\ \mu_2 & \mu_3 & \mu_4 \end{vmatrix} \geq 0 \quad \text{or} \quad \mu_2\mu_4 - \mu_3^2 - \mu_2^3 \geq 0 \quad \text{or} \quad \beta_2 - \beta_1 - 1 \geq 0$$

Since $\beta_1 \geq 0$, $\beta_2 \geq 1$.

COROLLARY 6.1. *For any r.v. X,*

$$\begin{vmatrix} \mu_0 & \mu_1 & \mu_2 & \mu_3 \\ \mu_1 & \mu_2 & \mu_3 & \mu_4 \\ \mu_2 & \mu_3 & \mu_4 & \mu_5 \\ \mu_3 & \mu_4 & \mu_5 & \mu_6 \end{vmatrix} \geq 0$$

• **THEOREM 6.11.** *For any r.v. X, $\nu_r^{\frac{1}{r}}$ is an increasing function of r i.e. $\nu_{r+1}^{\frac{1}{(r+1)}} \geq \nu_r^{\frac{1}{r}}$,* $r = 1, 2, \cdots$

Proof. We have

$$E\left[|X|^{\frac{r+1}{2}} u + |X|^{\frac{r-1}{2}} v\right]^2 \geq 0 \tag{6.31}$$

where u, v are arbitrary constants. Thus (6.29) means

$$u^2 \nu_{r+1} + v^2 \nu_{r-1} + 2uv\nu_r \geq 0$$

is a positive semi-definite quadratic form in u, v. *A* set of necessary and sufficient conditions for this is

$$\begin{vmatrix} \nu_{r+1} & \nu_r \\ \nu_r & \nu_{r-1} \end{vmatrix} \geq 0 \text{ i.e. } \nu_{r-1}\nu_{r+1} \geq \nu_r^2 \text{ or } \nu_{r-1}^r \nu_{r+1}^r \geq \nu_r^{2r} \tag{6.32}$$

Putting in (6.32), $r = 1.2, \cdots, r$ we have

$$\nu_0 \nu_2 \geq \nu_1^2 \tag{i}$$

$$\nu_1^2 \nu_3^2 \geq \nu_2^4 \tag{ii}$$

$$\nu_2^3 \nu_4^3 \geq \nu_3^6 \tag{iii}$$

$$\cdots\cdots\cdots\cdots \qquad \cdots$$

$$\nu_{r-1}^r \nu_{r+1}^r \geq \nu_r^{2r} \tag{r}$$

Multiplyting $(i), \cdots, (r)$,

$$\nu_{r+1}^r \geq \nu_r^{r+1} \text{ or } \nu_{r+1}^{\frac{1}{r+1}} \geq \nu_r^{\frac{1}{r}}$$

Thus $\dfrac{1}{r+1} \log \nu_{r+1} \geq \dfrac{1}{r} \log \nu_r$ or $\dfrac{1}{r} \log \nu_r$ is an increasing function of r for positive r.

• **THEOREM 6.12.** *For any r.v. X,*

$$\nu_{\alpha_1} \nu_{\alpha_2} \cdots \nu_{\alpha_k} \geq \left[\left(\nu_{\frac{\alpha_1 + \cdots + \alpha_k}{k}}\right)\right]^k \tag{6.33}$$

where $\alpha_1 \cdots, \alpha_k$ are positive integers.

Proof. Consider generalised Cauchy-Schwartz inequality : for positive r.v.s $Y_1 \cdots, Y_k$

$$E\left(Y_1^k\right) E\left(Y_2^k\right) \cdots E\left(Y_k^k\right) \geq \left\{E(Y_1 \cdots Y_k)\right\}^k \tag{6.34}$$

(also see Holder's inequality in section 6.A)

Putting in (6.32), $Y_1 = |X - \mu|^{\alpha_1/k}, \cdots, Y_k = |X - \mu|^{\alpha_k/k}$, the inequality (6.33) follows.

• **THEOREM 6.13.** *(Liapounoff's inequality). For any r.v. X,*

$$\nu_b^{a-c} \le \nu_c^{a-b} \nu_a^{b-c} \tag{6.35}$$

for $a \ge b \ge c \ge 0$.

Proof. Put in (6.33),

$$k = a - c; \quad \alpha_1 = \alpha_2 = \cdots = \alpha_{a-b} = c; \quad \alpha_{a-b+1} = \alpha_{a-b+2} = \cdots = \alpha_{a-c} = a.$$

Thus $\dfrac{\alpha_1 + \cdots + \alpha_k}{k} = b$ and (6.33) readily follows.

6.5. Moments of a Symmmetric Probability Distribution

A r.v. X is said to be symmetric about a point a if

$$P(X \ge a + x) = P(X \le a - x)$$

for all x. In terms of the distribution function of X this means, if

$$F_X(a - x) = 1 - F_x(a + x) + P\{X = a + x\}$$

holds for all $x \in R^1$, we say that the d.f. of r.v. X is symmetric with a as the centre of symmetry. If $a = 0$, then for every x

$$F_X(-x) = 1 - F_X(x) + P\{X = x\}$$

If X is a r.v. of continuous type, X is symmetric with centre a if and only if the *pdf* $f(x)$ of X satisfies

$$f(a - x) = f(a + x) \quad \text{for all} \quad x \in R^1.$$

If $a = 0$, we say simply that X is symmetric (or F is symmetric).

It follows immediately that if X is symmetric about a, $E(X)$ exists and is a.

• **THEOREM 6.14.** *If X is symmetric about a, $E(X) = a$ and μ_{2r+1} (if it exists)=0, $r = 0, 1, 2, \cdots$*

Suppose X is a continuous r.v. with *pdf* $f(x)$.

$$E(X) = \int_{-\infty}^{\infty} xf(x)dx = \int_{-\infty}^{a} xf(x)dx + \int_{a}^{\infty} xf(x)dx = I_1 + I_2(\text{say}).$$

In I_1 put $y = a - x$; $I_1 = \displaystyle\int_0^\infty (a - y)f(a - y)dy$

In I_2 put $y = x - a$; $I_2 = \displaystyle\int_0^\infty (a + y)f(a + y)dy$

Hence $E(X) = a$.

Again $\mu_{2r+1} = \int_{-\infty}^{\infty} (x-a)^{2r+1} f(x)dx = \int_{-\infty}^{a} (x-a)^{2r+1} f(x)dx + \int_{a}^{\infty} (x-a)^{2r+1}$ $f(x)dx = I_3 + I_4$ (say). Now $I_3 = -\int_{0}^{\infty} z^{2r+1} f(a-z)dz$ (putting $z = a - x$); $I_4 = \int_{0}^{\infty} z^{2r+1} f(a+z)dz$ (putting $z = x - a$). Hence $\mu_{2r+1} = 0$. In case X is discrete, the result can similarly be proved.

6.6. Factorial Moments

Let X be a discrete r.v. with pmf $P(X = k) = p_k$, $k = 0, 1, 2, \cdots$. Then $E\Big[(X)_r\Big]$, if it exists, is called the rth factorial moment of X about origin and will be denoted as $\mu_{(r)}$.

$$\mu_{(r)} = E\Big[X(X-1)\cdots(X-r+1)\Big] = \sum_{x=r}^{\infty} x(x-1)\cdots(x-r+1)p_x$$

It readily follows that $\mu_{(r)}$ exists if and only if μ'_r exists. It is easy to check

$$\mu_{(1)} = \mu'_1 \qquad \mu_{(3)} = \mu'_3 - 3\mu'_2 + 2\mu'_1$$
$$\mu_{(2)} = \mu'_2 - \mu'_1 \qquad \mu_{(4)} = \mu'_4 - 6\mu'_3 + 11\mu'_2 - 6\mu'_1$$

Again since $x^2 = (x)_2 + x$, $x^3 = (x)_3 + 3(x)_2 + x$, $x^4 = (x)_4 + 6(x)_3 + 7(x)_2 + x, \cdots$ it follows that

$$\mu'_1 = \mu_{(1)} \qquad \mu'_3 = \mu_{(3)} + 3\mu_{(2)} + \mu_{(1)}$$
$$\mu'_2 = \mu_{(2)} + \mu_{(1)} \qquad \mu'_4 = \mu_{(4)} + 6\mu_{(3)} + 7\mu_{(2)} + \mu_{(1)}$$

❑ **EXAMPLE 6.11.** *Consider the poisson distribution*

$$P(X = x) = e^{-\lambda}\frac{\lambda^x}{x!}, \quad x = 0, 1, 2, \cdots, \quad \lambda > 0$$

Hence

$$\mu_{(r)} = \sum_{0}^{\infty} x_{(r)} e^{-\lambda}\frac{\lambda^x}{x!} = \sum_{x=r}^{\infty} \frac{e^{-\lambda}\lambda^x}{(x-r)!} = \sum_{y=0}^{\infty} e^{-\lambda}\frac{\lambda^{y+r}}{y!} = \lambda^r.$$

Thus $\mu = \lambda$, $\sigma^2 = \lambda^2 + \lambda - \lambda^2 = \lambda$.

❑ **EXAMPLE 6.12.** *Consider the discrete uniform distribution* $P(X = x) = \frac{1}{N}$, $x = 1, 2, \cdots, N$.

Hence

$$\mu_{(r)} = \frac{1}{N}\sum_{x=1}^{N}(x)_r$$

Now

$$\sum_{j=a}^{b}(j)_n = \frac{(b+1)_{(n+1)} - (a)_{(n+1)}}{n+1}$$

Hence

$$\mu_{(r)} = \frac{(N+1)_{(r+1)}}{N(r+1)}, r = 1, \cdots, N-1; \quad \mu_{(N)} = (N-1)!$$

Thus

$$\mu = \frac{N+1}{2}, \quad \mu_{(2)} = \frac{N^2-1}{3}, \quad \sigma^2 = \frac{(N+1)(N-1)}{12}.$$

6.7. Different Measures of Central Tendency

The r.v. X takes values in the borel sets in $\mathcal{B}_1$ according to some probability law which is called the probability distribution of X. However, the different values of X will have a tendency to cluster around some central value. This value may be taken as a representative value of the distribution of X and measures the central tendency (location parameter) of the distribution. Different measures of central tendency are given below.

(a) *Mean* This is the most commonly used measure of central tendency of the probability distribution of a r.v. X. This is also called the expectation of X and if it exists, is given by

$$E(X) = \int_{-\infty}^{\infty} x dF(x)$$

(b) *Median* If μ_e is a point which divides the whole probability mass of the distribution into two equal parts, each containing probability mass equal to $\frac{1}{2}$, μ_e is called the median of the distribution.

Consider the curve of probability distribution function $F(x)$ of a continuous r.v. X. The abscissa of the point of intersection of this curve with the straight line $y = \frac{1}{2}$ is the unique median μ_e of this distribution. Thus μ_e satisfies

$$F(\mu_e - 0) \leq \frac{1}{2} \leq F(\mu_e) \tag{6.36}$$

For a discrete distribution it may so happen that $F(x) = \frac{1}{2}$ for all x in the interval $a \leq x \leq b$. Then median is taken to be $\frac{1}{2}(a+b)$.

The mean does not exist always, whereas a median always exists (an example is Cauchy distribution discussed in section 9.5). Even in case where mean does exist, the median is sometimes preferred as a measure of central tendency, since the value of the mean may sometimes be largely affected by the presence of some stray probability masses at either end.

(c) *Mode* Mode is that item or value in the population which occurs most frequently. It corresponds to the maximum ordinate of the frequency curve $y = f(x)$. If the pmf or *pdf* of X has a single relative maximum (unique mode) m_o, then m_o may be used as a measure of the central tendency of the distribution of X. Clearly for a discrete r.v., m_o, if it exists, will be the value of X which has the highest probability. For a continuous r.v. with *pdf* $f(x)$, if $f(x)$ is twice differentiable, m_o can be found out from the conditions $f'(m_o) = 0, f''(m_o) < 0$. In case $f(x)$ is not differentiable twice, m_o has to be found out from the condition $f(m_o) > f(x)$ for all $x \neq m_o$. For a multimodal distribution mode cannot be used as a measure of central tendency.

For a non-negative r.v. X one may sometimes use the geometric mean (G.M.), G as a measure of central tendency. This is given by

$$\log G = \int_{-\infty}^{\infty} (\log x) dF(x)$$

Sometimes harmonic mean (H. M.), H is used as a masure. This is given by

$$H = \frac{1}{\int_{-\infty}^{\infty} \left(\frac{1}{x}\right) dF(x)}.$$

In a perfectly symmetrical distribution the mean, median and the mode coincide. For a moderately asymmetric distribution occuring in practice, it is found empirically

$$\text{Mean} - \text{Mode} = 3\ (\text{ Mean} - \text{Median }).$$

❑ **EXAMPLE 6.13.** *Find the arithmatic mean (A.M.), geometric mean (G.M.) and the harmonic mean (H.M.) of the distribution*

$$dF(x) = \frac{1}{B(p,q)}(1-x)^{p-1}x^{q-1}dx, \quad 0 \le x \le 1; \quad p, q > 0$$

where

$$B(p,q) = \int_0^1 (1-x)^{p-1}x^{q-1}dx = \frac{\Gamma(p)\Gamma(q)}{\Gamma(p+q)}$$

and $\Gamma(r) = (r-1)!$ for integral r.

The A.M. is given by $E(X) = \frac{p}{p+q}$.

The G.M., G is given by

$$\log G = E(\log X) = \int_0^1 \log x (1-x)^{p-1} x^{q-1} dx$$

Since

$$\int_0^1 (1-x)^{p-1} x^{q-1} dx = B(p,q)$$

differentiating both sides w.r.t. q (the differentiation under the sign of integration being permissible as the integral converges uniformly)

$$\int_0^1 (1-x)^{p-1} x^{q-1} \log x dx = \frac{\partial}{\partial q} B(p,q)$$

Hence

$$\log G = \frac{1}{B(p,q)} \frac{\partial}{\partial q} B(p,q) = \frac{\partial \log B(p,q)}{\partial q}$$

$$= \frac{\partial}{\partial q} \log \left[\frac{\Gamma(p)\,\Gamma(q)}{\Gamma(p+q)} \right] = \frac{\partial}{\partial q} [\log \Gamma(q) - \log \Gamma(p+q)]$$

The H.M. is given by

$$\frac{1}{H} = E\left(\frac{1}{X}\right) = \frac{1}{B(p,q)} \int_0^1 (1-x)^{p-1} x^{q-2} dx$$

$$= \frac{B(p,q-1)}{B(p,q)} = \frac{\Gamma(q-1)\,\Gamma(p+q)}{\Gamma(p+q-1)\,\Gamma(q)} = \frac{p+q-1}{q-1}$$

Hence

$$H = \frac{q-1}{p+q-1}$$

❑ **EXAMPLE 6.14.** *Find the position of the mode and median of the curve*

$$f(x) = \frac{\alpha\beta x^{\alpha-1}}{(1+\beta x^\alpha)^2}, \quad \beta > 0, \alpha > 1, x \geq 0$$

Let x_m, μ_e be the modal value and the median of the distribution respectively.

$$\frac{df(x)}{dx} = \frac{\alpha\beta x^{\alpha-2}(1+\beta x^\alpha)\{\alpha - 1 - \beta x^\alpha(1+\alpha)\}}{(1+\beta x^\alpha)^4}$$

Hence $\dfrac{df(x)}{dx} = 0 \Rightarrow x_m = \left[\dfrac{\alpha-1}{\beta(1+\alpha)}\right]^{1/\alpha}$. It can be checked $\left[\dfrac{d^2 f(x)}{dx^2}\right]_{x_m} < 0$

Again, $\displaystyle\int_0^{\mu_e} \frac{\alpha\beta x^{\alpha-1}}{(1+\beta x^2)^2} dx = \frac{1}{2}$

Putting $1+\beta x^2 = t$, $\displaystyle\int_1^{1+\beta\mu_e^\alpha} \frac{dt}{t^2} = \frac{1}{2}$ when $\mu_e = \left(\dfrac{1}{\beta}\right)^{1/\alpha}$

❑ **EXAMPLE 6.15.** *For the distribution in example 6.8, find the median when* $\alpha = 2\beta$.

When $\alpha = 2\beta$, $f(x) = 2\beta\left(x^{\frac{\alpha}{2}-1} - x^{\alpha-1}\right)$. If $\hat{x}$ is the median of the distribution,

$$\int_0^{\hat{x}} f(x)dx = \frac{1}{2} \quad \text{i.e.} \quad 2\hat{x}^{\alpha} - 4\hat{x}^{\alpha/2} + 1 = 0 \quad \text{when} \quad \hat{x} = \left(1 - \frac{1}{\sqrt{2}}\right)^{2/\alpha}.$$

6.8. Measures of Dispersion

Measures of spread or dispersion (sometimes called measures of concentration) are parameters which give an idea of the scatter (or cluster) of the probability mass of the distribution about the location parameter. These are given below.

(a) *Range* If X has a distribution such that

$$\begin{array}{ll} F(x) = 0 & \text{if } x < a \\ 0 < F(x) < 1 & \text{if } a \le x \le b \\ F(x) = 1 & \text{if } x \ge b \end{array}$$

then the difference $b-a$, called the range of X, may be taken as a measure of dispersion. This measure does not take into account how the probability mass is distributed over $[a, b]$ and is thus very unsuitable as a good measure of the dispersion of distribution of X.

(b) *Mean deviation* It is the expected value of the absolute deviation of X about A, $E|X - A| = MD_A$ (say), where A is an arbitrary number. The absolute value of $(X - A)$ is taken because dispersion should be indicated by the magnitude of deviation and not by its sign. Its value is large when there is a great seattering and small when there is a great concentration about A. Mean deviation is generally taken about the mean.

• **THEOREM 6.15.** *Mean deviation is least when taken about the median i.e.*

$$E|X - A| \ge E|X - \mu_e|$$

where A is any number.

Proof. Suppose : $A < \mu_e$. Let us write $\mu_e = m$ (say) for simplicity.

$$E|X - A| = \int_{-\infty}^{\infty} |x - A| dF(x) = \int_{-\infty}^{A} (A - x)dF(x) + \int_{A}^{\infty} (x - A)dF(x)$$

$$= \int_{-\infty}^{A} (A - m + m - x)dF(x) + \int_{A}^{\infty} (x - m + m - A)dF(x)$$

$$= (A - m)F(A) + (m - A)(1 - F(A)) + \int_{-\infty}^{A} (m - x)dF(x) + \int_{A}^{\infty} (x - m)dF(x)$$

$$= (m - A)(1 - 2F(A)) + \int_{-\infty}^{m} (m - x)dF(x) - \int_{A}^{m} (m - x)dF(x)$$

$$+ \int_{A}^{m} (x - m)dF(x) + \int_{m}^{\infty} (x - m)dF(x)$$

$$= (m - A)(1 - 2F(A)) + E|X - m| + \int_A^m \Big[(x - m) - (m - x)\Big] dF(x)$$

$$= E|X - m| + (m - A)(1 - 2F(A)) + 2\int_A^m (x - m) dF(X)$$

$$= E|X - m| + (m - A)(1 - 2F(A)) + 2\int_A^m \Big[(x - A) + (A - m)\Big] dF(x)$$

$$= E|X - m| + (m - A)(1 - 2F(A)) - 2(m - A)\left[\int_{-\infty}^m dF(x) - \int_{-\infty}^A dF(x)\right]$$

$$+2\int_A^m (x - A) dF(x)$$

$$= E|X - m| + (m - A)(1 - 2F(m)) + 2\int_A^m (x - A) dF(x).$$

Since $F(m) = \frac{1}{2}$, the middle term vanishes. Since $A < m$ the integral is also positive. Hence the result.

The result for the case $\mu_e < A$ can be proved similarly.

(c) Variance, Mean Square Error, Standard Deviation :

For measuring the deviation of X around A one may also consider $E(X - A)^2$ which like $E|X - A|$ is free from sign and measures the deviations of X around A. This is called the mean square error of X around $A\left[MSE_A(X)\right]$. When A is $E(X) = \mu$, this is called the variance of X, often denoted as $V(X)$. Clearly this is the μ_2 of the distribution of X.

$$V(X) = E(X - \mu)^2$$

Clearly

$$E(X - A)^2 = E(X - \mu)^2 + (\mu - A)^2.$$

so that

$$MSE_A(X) = V(X) + (\mu - A)^2$$

Another measure is the positive square root of $V(X)$, called the standard deviation (s.d.) of X, often denoted as σ_X $\left(\sigma_X = +\sqrt{V(X)}\right)$. The positive square root of $MSE_A(X)$ is called the root mean square deviation of X about A.

• **THEOREM 6.16.** *$V(X)$ exists if and only if $E(X)$ and $E(X^2)$ exists.*

Proof is easy.

• **THEOREM 6.17.** *For any r.v. X, $V(X) \le E(X - A)^2$, $A \ne \mu$.*

• **THEOREM 6.18.** *For any r.v. X, $E|X - \mu| \le \sqrt{E(X - \mu)^2}$ where $\mu = E(X)$.*

Proof. In Cauchy-Schwartz inequality put $|X - \mu|$ in place of X and 1 in place of Y. Taking positive square root of both sides of the inequality, the result follows.

Conditional variance of X given A. Let A be a measurable subset of S. The conditional variance of X given A is

$$V(X|A) = E\left[\left\{X - E(X|A)\right\}^2 |A\right] = \frac{E\left\{(X - E(X|A))^2 I_A\right\}}{P(A)}$$

• **THEOREM 6.19.** *Let $\{A_i\}$ be a measurable partition of the sample sapce S such that $P(A_i) > 0$ for each $i = 1, \cdots k$ and let $V(X)$ exist. Then*

$$V(X) = \sum_{i=1}^{k} P(A_i)V(X|A_i) + \sum_{i=1}^{k} P(A_i)\left[E(X|A_i) - E(X)\right]^2 \tag{6.37}$$

$$= E_A\left[V(X|A)\right] + V_A\left[E(X|A)\right]$$

where $E(X|A_i)$, $V(X|A_i)$ are respectively the conditional expected value of X given A_i (defined in section 6.1) and conditional variance of X given A_i.

Proof. $V(X) = E(X - E(X))^2 = \sum_i P(A_i)E\left\{\left(X - E(X)\right)^2 \Big| A_i\right\}$ **(by theorem 6.1)**

Now

$$P(A_i)E\left\{\left(X - E(X)\right)^2 \Big| A_i\right\} = E\left[\left\{\left(X - E(X)\right)^2\right\} I_{A_i}\right] \quad \text{(by 6.10)}$$

$$= \int_{A_i} \left(X - E(X)\right)^2 dP = \int_{A_i} \left\{X - E(X|A_i) + E(X|A_i) - E(X)\right\}^2 dP$$

$$= \int_{A_i} \left(X - E(X|A_i)\right)^2 dP + \int_{A_i} \left(E(X|A_i) - E(X)\right)^2 dP$$

$$= P(A_i)V(X|A_i) + \left(E(X|A_i) - E(X)\right)^2 P(A_i).$$

Hence the theorem.

Note that we can also write the right hand side of (6.37) as $E_A\left[V(X|A)\right] + V_A\left[E(X|A)\right]$ where $V_A(\cdot)$ denotes variance of $(\cdot)$ w.r.t. A (i.e. for different subsets of A), because, by theorem 6.1, $E(X) = E_A E(X|A)$.

(d) *Quartile deviation* Consider a quantity $0 < p < 1$. The value $x = \zeta_p$ (say) of r.v. X which satisfies

$$F(\zeta_p - 0) \leq p \leq F(\zeta_p) \tag{6.38}$$

is said to be a quantile of X of order p. Comparing with (6.36), median is quantile of order $\frac{1}{2}$.

If X is a discrete r.v., ζ_p may be defined as the smallest value of X (with non-zero probability) satisfying $F(x) \geq p$.

The quantiles $\zeta_{1/4}, \zeta_{2/4}, \zeta_{3/4}$ are called the first quartile (Q_1) second quartile (Q_2 ; also median), third quartile (Q_3) respectively. The quantiles $\zeta_{.1}, \zeta_{.2}, \cdots, \zeta_{.9}$ are called the first, second, $\cdots$, nineth decile respectively. The quantiles $\zeta_{.01}, \zeta_{.02}, \cdots$ are called the first, second, $\cdots$ percentiles respectively.

The quantity

$$Q = \frac{Q_3 - Q_1}{2}$$

called the quartile deviation (or semi-interquartile range) is also taken as a measure of dispersion. The higher the dispersion in values of X the higher the value of Q.

(e) *Gini's Mean difference* The Italian Statistician Corrado Gini suggested the quantity Δ_1 based on the expected, absolute difference between pair of units as a measure of dispersion.

$$\Delta_1 = \int_{-\infty}^{\infty}\int_{-\infty}^{\infty} |x-y| f(x)f(y)dxdy \tag{6.39}$$

if X is a continuous r.v. with pdf $f(x)$, where x, y are two arbitrary values of X.

$$\Delta_1 = \sum_{i=1}^{\infty}\sum_{j=1}^{\infty} |x_i - x_j| p(x_i)p(x_j) \tag{6.40}$$

if X is discrete with pmf $P(X = x_k) = p(x_k)$. Clearly Δ_1 does not depend on any measure of central tendency.

Another measure Δ_2 based on the square of difference of pairs of observations is obtained by replacing $|x-y|$ by $(x-y)^2$ and $|x_i - x_j|$ by $(x_i - x_j)^2$ in (6.39) and (6.40) respectively.

Lemmea 6.1 $\Delta_2 = 2\sigma_x^2$

Proof. Let X be continuous with $E(X) = \mu$.

$$\Delta_2 = \int_{-\infty}^{\infty}\int_{-\infty}^{\infty} \left[(x-\mu) - (y-\mu)\right]^2 f(x)f(y)dxdy$$

$$= \int_{-\infty}^{\infty} (x-\mu)^2 f(x)dx + \int_{-\infty}^{\infty} (y-\mu)^2 f(y)dy + 2\int_{-\infty}^{\infty} (x-\mu)f(x)dx$$

$$\int_{-\infty}^{\infty} (y-\mu)f(y)dy = 2\sigma_x^2$$

Similarly when X is discrete.

COROLLARY 6.2.

$$\frac{1}{N}\sum_{i=1}^{N}(x_i - \bar{x})^2 = \frac{1}{2N^2}\sum_{i=1}^{N}\sum_{j=1}^{N}(x_i - x_j)^2, \bar{x} = \frac{1}{N}\sum_{i=1}^{N} x_i \tag{6.41}$$

Proof. If X takes N values x_i ($i = 1, \cdots, N$) each with probability $\frac{1}{N}$ the left hand side of (6.41) is σ_x^2 and the right hand side is $\Delta_2/2$.

Lemma 6.2. If X is a continuous r.v. with distribution function $F(x)$,

$$\Delta_1 = 2\int_{-\infty}^{\infty} F(x)\left\{1 - F(x)\right\}dx \tag{6.42}$$

Proof.

$$\Delta_1 = \int_{-\infty}^{\infty}\int_{-\infty}^{\infty}|x-y|dF(x)dF(y) = 2\int_{-\infty}^{\infty}\int_{y=-\infty}^{x}(x-y)dF(x)dF(y)$$

$$= 2\int_{-\infty}^{\infty} dF(x)\int_{-\infty}^{x}(x-y)dF(y)$$

Now

$$\int_{-\infty}^{x}(x-y)dF(y) = \int_{-\infty}^{x}F(y)dy$$ (on integration by parts taking $(x-y)$ as the first function)

Hence

$$\Delta_1 = 2\int_{-\infty}^{\infty} dF(x)\int_{-\infty}^{x}F(y)dy = 2\int_{-\infty}^{\infty} dF(x)\phi(x) \quad \text{where} \quad \phi(x) = \int_{\infty}^{x}F(y)dy$$

$$= 2F(x)\phi(x)\Big]_{-\infty}^{\infty} - 2\int_{-\infty}^{\infty}F(x)F(x)dx \quad \text{(integration by parts)}$$

$$= 2\int_{-\infty}^{\infty}F(x)\Big[1 - F(x)\Big]dx$$

since

$$F(x)\phi(x)\Big]_{-\infty}^{\infty} = \int_{-\infty}^{\infty}F(x)dx.$$

(f) ***Coefficient of Variation*** If $E(X)$ and σ_X exists coefficent of variation of a r.v. X is defined as $\dfrac{\sigma_X}{E(X)} = C_X$ (say) and is taken as a measure of dispersion. Unlike the previous measures it has no unit and is often expressed in percentages ($100C_X\%$).

(g) Gini's coefficent of concentration defined as $G = \dfrac{\Delta_1}{2E(X)}$ is another measure often used by the scientists in measuring social inequality in terms of concentration of money, resources, etc.

❑ **EXAMPLE 6.16.** *X is a continuous r.v. with pdf $f(x) = Ce^{-x/\sigma}$, $x \geq 0$, $\sigma > 0$. Find its mean, median, variance and quartile deviation.*

Since $\displaystyle\int_0^{\infty} f(x)dx = 1$, $C = \dfrac{1}{\sigma}$

$$\mu = \frac{1}{\sigma}\int_0^{\infty} xe^{-x/\sigma}dx = -xe^{-x/\sigma}\Big]_0^{\infty} + \int_0^{\infty} e^{-x/\sigma}dx = \sigma.$$

If m be the median,

$$\int_0^m f(x)dx = \frac{1}{2} \quad \text{i.e.} \quad \frac{1}{\sigma}\int_0^m e^{-x/\sigma}dx = \frac{1}{2}$$

This gives $e^{m/\sigma} = 2$ or $m = \sigma \log_e 2$.

Again $\int_0^{\zeta_{1/4}} \frac{1}{\sigma}e^{-x/\sigma}dx = \frac{1}{4}$

This gives $\zeta_{1/4} = \sigma(\log 4 - \log 3)$.

Similarly $\int_0^{\zeta_{3/4}} \frac{1}{\sigma}e^{-x/\sigma}dx = \frac{3}{4}$ implies $\zeta_{3/4} = \sigma \log 4$.

Hence quartile deviation is $\frac{1}{2}\left(\zeta_{3/4} - \zeta_{1/4}\right) = \frac{\sigma \log 3}{2}$, $\mu_2' = \frac{1}{\sigma}\int_0^\infty x^2 e^{-x/\sigma}dx = 2\sigma^2$.

Thus $\mu_2 = \sigma^2$.

6.9. Measures of Skewness and Kurtosis

By skewness we mean the asymmetry of the distribution. A probability density curve (*pm*-histogram) is positively skewed if it has more probability mass to the right of the mode (fig. 6.2); is negatively skewed if more mass is distributed to the left of the mode (fig. 6.3). A measure of skewness should be unit-free and should take positive (negative) values for a positively (negatively) skewed distribution. For a symmetrical distribution $\mu_3 = 0$; μ_3 is positive (negative) for a positively (negatively) skewed distribution. To make it dimensionless we divide it by σ^3. Thus

$$\sqrt{\beta_1} = \frac{\mu_3}{\sigma^3} = \frac{\mu_3}{\mu_2^{3/2}} \tag{6.43}$$

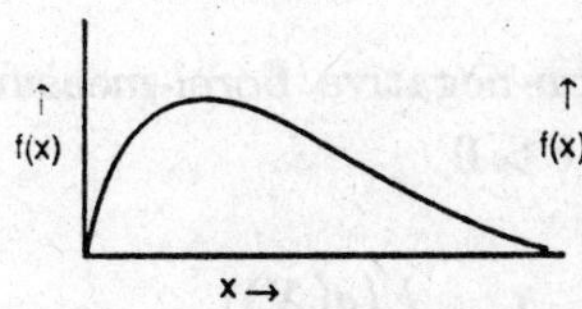

Fig. 6.2

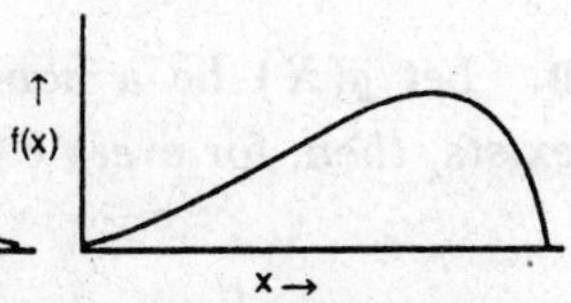

Fig. 6.3

or β_1 is often taken as a measure of skewness (if μ_3 exists). For a symmetrical distribution $\beta_1 = 0$. The value of β_1 gives a measure of departure from symmetry i.e. of skewness. It can take any value between $-\infty$ and ∞.

Other measures are :

(i) $\dfrac{E(X) - m_o}{\sigma}$ (6.44)

(ii) $\dfrac{3(E(X) - \mu_e)}{\sigma}$ (since mean−mode $\simeq$ 3(mean−median) empirically) (6.45)

(iii) For a symmetrical distribution, $Q_3 - \mu_e = \mu_e - Q_1$. Hence $(Q_3 - \mu_e) - (\mu_e - Q_1)$ can be taken as a measure of skewness. To make it dimensionless we divide it by $Q_3 - Q_1$ to obtain

$$\frac{Q_3 - 2Q_2 + Q_1}{Q_3 - Q_1}. \tag{6.46}$$

as another measure.

For a symmetrical distribution each measure has value zero. For a positively skewed distribution, $\mu_3 > 0$, $E(X) > \mu_e > m_o$; $Q_3 - Q_2 > Q_2 - Q_1$ and each measure takes positive value. For a negatively skewed distribution the inequalities are reversed and each measure takes a negative value.

Since $|E(X) - \mu_e| \le E|X - \mu_e| \le E|X - \mu| \le \sigma$ (6.45) varies between -3 and 3. (6.44) and (6.46) being approximately equal have the same limits. For a extremely positively (negatively) skewed distribution, $Q_2 - Q_1 \to 0$ $(Q_3 - Q_2 \to 0)$ and hence (6.46) $\to 1(-1)$.

Measures of kurtosis :

By kurtosis we mean the flatness or peakedness of the curve of probability distribution. The quantity

$$\beta_2 - 3 = \frac{\mu_4}{\mu_2^2} - 3 = \gamma \quad \text{(say)}$$

is taken as a measure of kurtosis.

For a normal distribution (section 9.6), $\beta_2 = 3$ and $\gamma = 0$. If for a distribution $\beta_2 > 3$ the peak of the distribution is sharper than that of a normal curve it is called a Lepto-kurtic distribution.

If $\beta_2 < 3$ i.e. $\gamma < 0$ the distribution is said to be Platykurtic. Here the peak is flatter than that of the normal curve.

6.10. Some Probability Inequalities

• **THEOREM 6.20.** *Let $g(X)$ be a non-negative borel-measureable function of a r.v. X. If $Eg(X)$ exists, then, for every $\epsilon > 0$*

$$P\{g(X) \ge \epsilon\} \le \frac{E(g(X))}{\epsilon}.$$

Proof.

$$Eg(X) = \int_A g(x)dF(x) + \int_{A^c} g(x)dF(x)$$

where $A = \{x : g(x) \ge \epsilon\}$.

$$Eg(X) \ge \int_A g(x)dF(x) \ge \epsilon \int_A dF(x) = \epsilon P\{g(X) \ge \epsilon\}.$$

COROLLARY 6.3. *If $g(X) = |X|^r$ and $\epsilon = t^r$, $t > 0$, $r > 0$ and ν_r exists,*

$$P\{|X| \geq t\} \leq \frac{E|X|^r}{t^r}.$$

For $r = 1$, we have

$$P\{|X| \geq t\} \leq \frac{E|X|}{t} \tag{6.47}$$

which is known as Markov's inequality.

Taking $g(X) = (X - \mu)^2$, $t = \epsilon^2\sigma^2$, we have

$$P\{|X - \mu| \geq \epsilon\sigma\} \leq \frac{1}{\epsilon^2}$$

which is known as Chebychev's inequality, where $E(X) = \mu$, $V(X) = \sigma^2$.

• **THEOREM 6.21.** *(Chebychev's inequality). If X be a r.v. with finite variance $\sigma^2(> 0)$ and $E(X) = \mu$, then*

$$P\Big\{|X - \mu| \geq \epsilon\Big\} \leq \frac{\sigma^2}{\epsilon^2} \tag{6.48}$$

for any $\epsilon > 0$.

Chebyshev's inequality gives an upper bound for the probability that X deviates from its mean by more than c units. No assumption on the distribution of X is required except that it has a finite variance. Many probability inequalities have been developed starting from Chebychev's inequality and many sharper bounds for $P\Big(|X - \mu| \geq \epsilon\Big)$ than the one given by this inequality exist. However, in general, the bound given by the Chebychev's inequality can hardly be improved upon.

❑ **EXAMPLE 6.17.** *Let $X_1, \cdots, X_n$ be independent and identically distributed r.v.s with mean zero and unit variance. Then thier sum $X = X_1 + \cdots + X_n$ is also a r.v. with $E(X) = 0$, $V(X) = n$. Hence from (6.48),*

$$P\Big\{|X| \geq \epsilon\Big\} \leq \frac{n}{\epsilon^2}.$$

When $\epsilon = \sqrt{n}$,

$$P\Big\{|X| \geq \sqrt{n}\Big\} \leq 1.$$

❑ **EXAMPLE 6.18.** *Suppose $X_1, \cdots, X_{n+1}$ are the results of $(n + 1)$ Bernoulli trials with $P(X_i = 1) = p$, $P(X_i = 0) = 1 - p$, $i = 1, \cdots, n + 1$. Suppose also that Y_n is equal to be number of $i(i = 1, \cdots, n)$ such that $X_i = X_{i+1} = 1$. Using Chebychev's inequality find an upper bound to*

$$P\left\{\left|\frac{Y_n}{n} - p^2\right| > \epsilon\right\}, \quad \epsilon > 0.$$

Let $Z_i = 1$ if $X_i = 1$, $X_{i+1} = 1$ and 0 otherwise. Then $Y_n = Z_1 + \cdots + Z_n$. $E(Z_i) = p^2$. $V(Z_i) = p^2(1-p^2)$. $\text{Cov}(Z_i, Z_{i+1}) = E(Z_i Z_{i+1}) - p^4 = p^3(1-p)$, because $E(Z_i Z_{i+1}) = P(Z_i Z_{i+1} = 1) = P(Z_i = 1)P(Z_{i+1} = 1|Z_i = 1) = p^2\left(\frac{p^3}{p^2}\right) = p^3$. Hence $E(Y_n) = np^2$, $V(Y_n) = p^2\left[n(1+3p)(1-p) + 2p(p-1)\right]$. By (6.48),

$$P\left(\left|\frac{Y_n}{n} - p^2\right| > \epsilon\right) < \frac{V(Y_n)}{n^2\epsilon^2} = \frac{p^2(1+3p)(1-p)}{n\epsilon^2} + \frac{2p^3(p-1)}{n^2\epsilon^2} < \frac{p^2(1+3p)(1-p)}{n\epsilon^2}.$$

• **THEOREM 6.22.** *(One-sided Chebychev inequality). If X be a random variable with finite variance $\sigma^2(>0)$ and $E(X) = \mu$ then*

$$P\left[X \geq \mu + t\sigma\right] \leq E\left(\frac{X - \mu - a\sigma}{t\sigma - a\sigma}\right)^2 \tag{6.49}$$

for any $t > 0$ and any a.

Proof. Define a r.v. Y such that

$$Y = 1 \quad \text{if} \quad X \geq \mu + t\sigma$$
$$= 0 \quad \text{otherise}$$

Hence

$$E(Y) = P(X \geq \mu + t\sigma)$$

Now $\left(\frac{X - \mu - a\sigma}{t\sigma - a\sigma}\right)^2 \geq Y$ for all values of X a.e. Taking expectation of both sides (6.49) follows.

It can be seen that equality in (6.49) holds when X takes only two values $\mu + t\sigma$ and $\mu + a\sigma$ with probabilities p and $1 - p$ respectively. The conditions $E(X) = \mu$ and $V(X) = \sigma^2$ give, in this case,

$$a = -\frac{1}{t}, p = \frac{1}{1+t^2}$$

in which case

$$E\left(\frac{X - \mu - a\sigma}{t\sigma - a\sigma}\right)^2 = p.$$

The inequality (6.49) can thus be put in the alternative form :

$$P[X \geq \mu + t\sigma] \leq \frac{1}{1+t^2} \tag{6.50}$$

COROLLARY 6.4.

$$P[X \leq \mu - t\sigma] = P[-X \geq -\mu + t\sigma] \leq \frac{1}{1+t^2} \tag{6.51}$$

COROLLARY 6.5. *The distribution function $F(x)$ of a r.v. X with mean μ and variance σ^2 satisfies the inequalities :*

(i) $F(x) \leq \dfrac{\sigma^2}{\sigma^2 + (x-\mu)^2}$ for $x \leq \mu$

(ii) $F(x) \geq \dfrac{(x-\mu)^2}{\sigma^2 + (x-\mu)^2}$ for $x \geq \mu$.

Proof.

(i) Put $x = \mu - t\sigma$ in (6.51)

(ii) Put $x = \mu + t\sigma$ in (6.50)

6.11. Exercise

1. *Show that for the distribution* $P(X = k) = \dfrac{1}{2^k}$, $k = 1, 2, \cdots$ *$E(X)$ does not exist.*

2. *For the continuous distribution*

$$dF = \frac{3}{4}x(2-x)dx, \quad 0 \leq x \leq 2$$

find the first four moments, mean deviation about mean, skewness and kurtosis.

$$\left[\mu = 1,\ \mu_2 = \frac{1}{5},\ \mu_3 = 0,\ \mu_4 = \frac{3}{35},\ \text{m.d.} = \frac{3}{8},\ \beta_1 = 0,\ \beta_2 = -\frac{6}{7}\right]$$

3. *Find the mean, variance, mean deviation about mean of the r.v. X with pdf*

(a) $f(x) = \dfrac{1}{2}e^{-|x|}(-\infty < x < \infty)$

(b) $f(x) = \begin{cases} \dfrac{1}{2b} & \text{if } |x-a| \leq b \\ 0 & \text{otherwise} \end{cases}$

$$\left[(a)\mu = 0,\ \mu_2 = 2;\ 1 \quad (b)\ \mu = a,\ \mu_2 = \frac{b^2}{3};\ \frac{b}{2}\right].$$

4. *Show that for the triangular distribution*

$$dF = \frac{1}{\alpha}\left[1 - \frac{|x-\beta|}{\alpha}\right]dx, \quad |x-\beta| > \alpha$$

the mean is β and variance is $\dfrac{\alpha^2}{6}$.

5. *X has a continuous distribution with*

$$dF = k(x - x^2)dx, \quad 0 \leq x \leq 1,$$

k being a constant. Find the arithmatic mean, harmonic mean, mode and median.

$$\left[k = 6,\ \text{a.m.} = \frac{1}{2},\ \text{h.m.} = \frac{1}{3},\ \text{mode} = \frac{1}{2},\ \text{median} = \frac{1}{2}\right].$$

6. *Show that for the rectangular distribution*

$$dF = dx, \quad 0 \leq x \leq 1$$

$\mu_1' = \frac{1}{2}$, $\mu_2 = \frac{1}{12}$, *mean deviation* $= \frac{1}{4}$, $\Delta_1 = \frac{1}{3}$. *Show also that* $E\left(\frac{1}{X}\right)$ *does not exist.*

7. *X has pdf*

$$f(x) = k(1-x), \quad 0 \leq x \leq 1.$$

Find k and determine the arithmatic mean, median and variance.

$$\left[k = 2, \text{ mean} = \frac{1}{3}, \text{ median} = 1 - \frac{1}{\sqrt{2}}, \text{ variance} = \frac{1}{18}\right].$$

8. *Obtain (quartile deviation)/ (standard deviation) for the distribution with pdf*

$$f(x) = \begin{cases} 2(b+x)/b(a+b), & -b \leq x \leq 0 \\ 2(a-x)/a(a+b), & 0 \leq x \leq a \end{cases}$$

9. *Show that the geometric mean G of the distribution*

$$dF = 6(2-x)(x-1)dx, \quad 1 \leq x \leq 2$$

is given by $6\log(16G) = 19$.

10. *A continuous r.v. X has the pdf proportional to*

$$(\alpha + |x|)e^{-\beta|x|}, \quad -\infty < x < \infty$$

where α, β *are positive constants. Find* β_2*, the coefficent of kurtosis of X and show that* $\frac{1}{3} < \beta_2 < 3$*, irrespective of the values of* α *and* β.

11. *A continuous r.v. X has the distribution function*

$$F(x) = 1 - e^{\alpha \tan x}, \quad 0 \leq x \leq \frac{\pi}{2}, \quad 0 < \alpha < 1$$

Find the pdf of X and show that the mode of the distribution is $m_o = \frac{1}{2}(\pi - \sin^{-1}\alpha)$.

12. *For a r.v. X with pdf*

$$f(x) = \begin{cases} \frac{x}{2}, & 0 \leq x \leq 1 \\ \frac{1}{2}, & 1 < x \leq 2 \\ \frac{1}{2}(3-x), & 2 < x \leq 3 \end{cases}$$

show that moments of all orders exist. Find the mean and variance of X.

13. *For a continuous r.v. X defined in the range $a \le x \le b$ and having pdf proportional to $g(x)$, prove that m, the median of distribution satisfies the relation,*

$$2\int_a^{m} g(x)dx = \int_a^b g(x)dx$$

14. *For the Poisson r.v. with pmf*

$$P(X = x) = e^{-\lambda}\frac{\lambda^x}{x!}, \quad x = 0, 1, 2 \cdots$$

show that $E(X) = \lambda$, $E(X^2) = \lambda + \lambda^2$, $E(X^3) = \lambda + 3\lambda^2 + \lambda^3$, $E(X^4) = \lambda + 7\lambda^2 + 6\lambda^3 + \lambda^4$, $\mu_2 = \mu_3 = \lambda$, $\mu_4 = \lambda + 3\lambda^2$.

15. *For a continuous r.v. X the distribution function is*

$$F(x) = 1 - e^{-\beta x^2}, \quad \beta > 0, \quad 0 \le x \le \infty$$

Find the median of the distribution. Also if m, m_o, σ denote respectively the mean, mode and the s.d. of the distribution, show that

$$2m_o^2 - m^2 = \sigma^2 \quad \text{and} \quad m_o = m\sqrt{\frac{2}{\pi}}$$

$$\left[\text{median} = \sqrt{\frac{\log_e 2}{\beta}},\ m = \frac{1}{2}\sqrt{\frac{\pi}{\beta}},\ \sigma^2 = \frac{4-\pi}{4\beta},\ m_o = \frac{1}{\sqrt{2\beta}}\right]$$

16. *If X is a continuous r.v. with pdf $f(x)$ and if m is the median of the distribution of X and c is a real constant, show that*

$$E|X - c| = E|X - m| + 2\int_m^c (c - x)f(x)dx$$

17. *Let X be a r.v. with pdf*

$$f(x) = \frac{ak^a}{x^{a+1}}, \quad a > 0, \quad x \ge k > 0$$

(This is called Pareto distribution). Show that the d.f. is

$$F(x) = 1 - \left(\frac{k}{x}\right)^a;$$

$E|X|^r < \infty$ *for* $r < a$ *and* $\mu_r' = \dfrac{ak^r}{a-r}$; $\mu = \dfrac{ak}{a-1}$ *(if $a > 1$) and* $V(X) = \dfrac{ak^2}{(a-1)(a-2)}$ *(if $a > 2$), the mean deviation is* $\dfrac{2k}{a-1}(1 - a^{-1})^{a-1}$; *the geometric mean is* $G = ke^{1/a}$; *the harmonic mean is* $H = k\left(1 + \dfrac{1}{a}\right)$; *median is* $m = k2^{1/a}$.

18. *If X has pdf proportional to $x/(1+x)^3$ $(0 \le x \le \infty)$, show that the median of the distribution is $(1 + \sqrt{2})$; prove that mode is $\frac{1}{2}$ and hence deduce the skewness of the distribution.*

19. *Show that if a distribution*

$$dF = \frac{1}{\sqrt{2\pi}} e^{-x^2/2} dx, \quad -\infty \le x \le \infty$$

is transformed by $x = a\log(y-b)+c$, *the distribution of* y *has*

$$\frac{\text{mean - mode}}{\text{mean - median}} = -\exp\left(\frac{1}{2a^2}\right).$$

20. *Show that for the Cauchy pdf*

$$f(x) = \frac{1}{\pi(1+x^2)}, \quad -\infty \le x \le \infty$$

$E(X)$ *does not exist and the median of* X *is zero.*

21. *For the Poisson variate at exercise 14, find* $E(kXe^{-kX})$, *where* k *is a constant.*

$$\left[E(\cdot) = \sum_{0}^{\infty} (kxe^{-kx})\left(e^{-\lambda}\frac{\lambda^x}{x!}\right) = \lambda k \exp\left\{\lambda(e^{-k}-1)-k\right\}\right].$$

22. *For the distribution*

$$dF = k\left(1-\frac{x^2}{a^2}\right)^{-h} dx, \quad 0<h<1, \quad |x|<a,$$

k, *a constant, show that*

$$\mu'_{r+1} = \frac{ra^2}{r+2-2h}\mu'_{r-1}.$$

[Integrate by parts].

23. *For the distribution*

$$dF = k\left(1+\frac{x}{\alpha}\right)^{m-1} e^{-\frac{mx}{\alpha}} dx, \quad -\alpha \le x \le \infty$$

find the constant k *and the first four moments. Show that* $8\beta_2 = 4 + 9\beta_1$.

$$\left[k = m^m e^{-m}/a\overline{|(m)}, \mu'_1 = \alpha, \mu_2 = \frac{\alpha^2}{2m}, \mu_3 = \frac{\alpha^3}{3m^2}, \mu_4 = \frac{\alpha^4(m+2)}{8m^2}\right]$$

24. *A continuous r.v.* X *has pdf*

$$f(x) = \frac{ak(x-x_0)^{a-1}}{\left\{1+k(x-x_0)^a\right\}^2}, \quad k>0, \quad a>1, \quad x_0 \le x \le \infty$$

Find the median and modal value.

$$\left[\text{median} = x_0 + \left(\frac{1}{k}\right)^{1/a} \cdot \text{mode} = x_0 + \left\{\frac{a-1}{k(a+1)}\right\}^{1/2}\right]$$

25. *For the curve*

$$y = y_0 x^r (1-x)^s \quad 0 \le x \le 1$$

find skewness S_k = *(mean - median)/s.d. and* $\sqrt{\beta_1}$.

$$\left[S_k = \frac{s-r}{s+r}\sqrt{\frac{r+s+3}{(r+1)(s+1)}},\ \sqrt{\beta_1} = \frac{2(s-r)}{r+s+4}\sqrt{\frac{r+s+3}{(r+1)(s+1)}} \right]$$

26. *For the distribution*

$$dF = \frac{x}{\sigma^2} e^{-\frac{x^2}{2\sigma^2}} dx, \quad x > 0$$

find $Q_3 - Q_1$ *and* $(Q_3 - Q_1)$/s.d.

$$\left[\sigma\sqrt{2}\left[\sqrt{\log 4} - \sqrt{\log\frac{4}{3}}\right];\ \left(\sqrt{\log 4} - \sqrt{\log\frac{4}{3}}\right) \Big/ \sqrt{1-\frac{\pi}{2}} \right]$$

27. *The distribution function of a r.v. X is*

$$F(x) = \begin{cases} 0 \quad if, & x < -1 \\ a + b \text{ arc } \sin x \quad \text{if} & -1 \le x \le 1 \\ 1 \quad if & x > 1 \end{cases}$$

Find $E(X), V(X)$.

$$\left[E(X) = 0, \quad V(X) = \frac{1}{2} \right]$$

28. *A r.v. X has pdf*

$$f(x) = k \cdot e^{-b(x-a)}, \quad a \le x \le \infty$$

k, a, b *being constants. Show that* $k = b = \dfrac{1}{\sigma}$, $a = m - \sigma$ *where* m, σ *respectively are the mean and standard deviation of the distribution. Find* β_1, β_2 *of the distribution.*

$$[\beta_1 = 4, \quad \beta_2 = 9]$$

29. *Prove that if* $y = g(x)$ *be a monotone function and* μ_e *be the median of X, then* $g(\mu_e)$ *is median of* $Y = g(X)$.

30. *A target is made of 3 concentric circles of radii* $\dfrac{1}{\sqrt{3}}$, *1 and* $\sqrt{3}$ *feet respectively. Shots within the inner circle count 4 points, shots within the next ring outside this circle 3 points, shots within the next ring 2 points and shots outside count zero. Supposing the probability of a hit at a distance r from the centre is given by* $p(r) = \dfrac{2}{\pi(1+r)^2}$, *find the expectation of the score after 5 shots and the probability that the score exceeds 15.*

$$\left[\text{Probability} = \frac{2}{\pi}\tan^{-1} r,\ \text{Expected score} = \frac{65}{6} \right]$$

31. ***If for a r.v. X, μ_{2r+2} exists then show that***

$$(\mu_{2r+1})^2 \geq \mu_{2r}\mu_{2r+2}$$

[In Cauchy-Schwartz inequality $E(Y^2)E(Z^2) \geq \{E(YZ)\}^2$, put $Y = (X-\mu)^r$, $Z = (X-\mu)^{r+1}$ where $\mu = E(X)$]

32. ***Suppose X is a non-negative r.v. whose mean (μ) exists. Prove that***

(a) $\lim_{t\to\infty} t[1-F(t)] = 0$

and hence in case X is absolutely continuous

(b) $\mu = \int_0^\infty [1-F(x)]dx$

[We prove the results when X is absolutely continuous with *pdf* f.

$$\mu = \int_0^\infty xf(x)dx = \lim_{t\to\infty}\int_0^t xf(x)dx \qquad \text{(i)}$$

On integration by parts,

$$\int_0^t xf(x)dx = tF(t) - \int_0^t F(x)dx = -t[1-F(t)] + \int_0^t [1-F(x)]dx \qquad \text{(ii)}$$

but

$$t[1-F(t)] = t\int_t^\infty f(x)dx < \int_t^\infty xf(x)dx \qquad \text{(iii)}$$

Since $E|X| < \infty$, taking limit of both sides of (iii), $\lim_{t\to\infty} t[1-F(t)] = 0$ which proves (a). Hence from (i) and (ii)

$$\mu = \lim_{t\to\infty}\int_0^t [1-F(x)]dx = \int_0^\infty [1-F(x)]dx \qquad \text{(iv)}$$

which proves (b)].

33. ***Show that if the mean exists, $\Delta_1 = 4\int_{-\infty}^{\infty} x\left[F(x)-\frac{1}{2}\right]dF$.***

$$\Big[\Delta_1 = 2\int_{-\infty}^{\infty}\int_{-\infty}^{\infty}(x-y)dF(x)dF(y) = 2\int_{-\infty}^{\infty}dF(x)\int_{-\infty}^{x}(x-y)dF(y)$$

$$= 2\int_{-\infty}^{\infty}dF(x)\int_{-\infty}^{x}F(y)dy$$

[on integration by parts taking $(x-y)$ as the first factor].

$$= 2\int_{-\infty}^{\infty}dF(x)\left\{\Big[yF(y)\Big]_{-\infty}^{x} - \int_{-\infty}^{x} ydF(y)\right\}$$

$$= 2\int_{-\infty}^{\infty}xF(x)[dF(x)] - 2\int_{-\infty}^{\infty}dF(x)\int_{-\infty}^{x}ydF(y) \qquad \text{(i)}$$

The least term in

$$\text{(i)} \quad = 2\int_{-\infty}^{\infty} ydF(y)\int_{y}^{\infty} dF(x) = 2\int_{-\infty}^{\infty} ydF(y)(1-F(y))$$

Hence

$$\text{(i)} \quad = 2\int_{-\infty}^{\infty} xF(x)dF(x) - 2\int_{-\infty}^{\infty} x(1-F(x))dF(x) = 4\int_{-\infty}^{\infty} x\left[F(x)-\frac{1}{2}\right]dF(x)\Bigg]$$

34. *Show that if m be the mode of X then $a+bm_o(b \neq 0)$ is the mode of $Y = a+bX$.*

35. *Let S_n be the total score obtained in n tosses in a symmetric die. Using Chebychev's inequality, make an upper estimation of*

$$P\left\{\left|\frac{Sn}{n} - 3.5\right| > \epsilon\right\}, \quad \epsilon > 0 \qquad \left[\frac{35}{12n\epsilon^2}\right]$$

36. *Let X have a Poisson distribution with parameter λ. Prove that*

(i) $P\left(X \leq \frac{\lambda}{2}\right) \leq \frac{4}{\lambda}$

(ii) $P(X \geq 2\lambda) \leq \frac{1}{\lambda}$

37. *For the geometric distribution $p_n = 2^{-n}$, $n = 1, 2, \cdots$ show Chebychev's inequality gives $P[|X-2| > 2] < \frac{1}{2}$ while the actual probability is $\frac{1}{16}$.*

38. *If $X_1, \cdots, X_k$ are independently and identically distributed r.v.s having zero means and unit variance then prove that*

$$P\left(\sum_{1}^{k} X_i{}^2 \geq k\lambda\right) \leq \frac{1}{\lambda} \quad \text{for} \quad \lambda > 0$$

[Apply Markov inequality]

39. *Show that for any distribution* $\int_{-\infty}^{\infty}\left(1-\frac{x^2}{t^2}\right)dF(x) \leq \int_{-t}^{t} dF(x)$ *and hence deduce that* $P[|X - E(X)] > k\sigma] \leq \frac{1}{k^2}$ *when* $k > 0$ *and* $\sigma^2 = V(X)$.

$$\left[\int_{-\infty}^{\infty}\left(1-\frac{x^2}{t^2}\right)dF(x) = \int_{-\infty}^{-t}\left(1-\frac{x^2}{t^2}\right)dF(x) + \int_{-t}^{t}\left(1-\frac{x^2}{t^2}\right)dF(x)\right.$$

$$+\int_{t}^{\infty}\left(1-\frac{x^2}{t^2}\right)dF(x)$$

$$\leq \int_{-t}^{t}\left(1-\frac{x^2}{t^2}\right)dF(x) \leq \int_{-t}^{t} dF(x)$$

Now $\quad P(|X| > k\sigma) = 1 - \int_{-k\sigma}^{k\sigma} dF(x) \le 1 - \int_{-\infty}^{\infty}\left(1 - \frac{x^2}{k^2\sigma^2}\right)dF(x)$

$= \int_{-\infty}^{\infty} \frac{x^2}{k^2\sigma^2} dF(x) = \frac{1}{k^2}$ when the origin is at the mean.]

40. *(a) Let X be a non-negative r.v. Considering the discontinuous Heaveside function $H(X - a)$ where*

$$H(Y) = 1 \quad (Y > 0), = 0 \quad (Y \le 0)$$

'a' being a positive constant, prove Markov inequality $P(X > a) < \frac{E(X)}{a}$.

(b) Let X be a r.v. with finite $E(X^2)$. Considering the inequality

$$H\left(|X - b| - a\right) \le \left(\frac{X - b}{a}\right)^2$$

where $a(> 0), b$ are arbitray constants, show that

$$P(|X - b| > a) \le \frac{E(X - b)^2}{a^2}.$$

[(a) $H(X - a) = 1(0)$ if $\frac{X}{a} > 1$ (othersiwe). Hence $H(X - a) \le \frac{X}{a}(X \ge 0)$. Thus $E[H(X - a)] = P(X > a) \le \frac{E(X)}{a}$.

(b) $H(|X - b| - a) = 1(0)$ if $\frac{|X - b|}{a} > 1$ (otherwise). Hence $H(|X - b| - a) \le \frac{(X - b)^2}{a^2}$. The result follows by taking expectation.]

41. *Show that if $M(X)$ is an increasing positive function of $|X|$, then $P[|X| > a] \le \frac{E[M(X)]}{M(a)}$, provided $E[M(X)]$ exists.*

[Define a r.v. Y such that

$$Y = 1(0) \quad \text{if} \quad |X| > a \quad \text{(otherwise)}. \quad \text{Hence} \quad E(Y) = P(|X| > a).$$

Now $M(X) \ge YM(a)$. The result follows by taking expectation.]

42. *Let X be such that $E(e^{aX})$ exists, $a > 0$. Show that for any t,*

$$P(X > t) < \frac{E(e^{eX})}{e^{at}}$$

[Proceed as in 41; e^{ax} is a non-negative function of x]

43. *Let g be a non-negative non-decreasing function. Prove that if* $E\left[(g(|X-\mu|)\right]$ *where* $\mu = E(X)$ *exits, then*

$$P(|X-\mu|>t) < \frac{E\left[g(|X-\mu|)\right]}{g(t)} \qquad \left[\textit{Proceed as in 41}\right]$$

Appendix 6.A

Some Inequalities

1. Cauchy-Schwartz inequality. Let $(x_1, \cdots, x_n)$, $(y_1, \cdots, y_n)$ be two sets of real numbers. Then

$$\sum_1^n x_i^2 \sum_1^n y_i^2 \geq \left(\sum_1^n x_i y_i\right)^2 \tag{6.A.1}$$

equality holds when and only when $a_i \infty b_i$, $i = 1, \cdots, n$.

Proof. Consider the quadratic form in two real variables a, b

$$\sum_{i=1}^n (ax_i + by_i)^2 = a^2 \sum_{i=1}^n x_i^2 + 2ab \sum_{i=1}^n x_i y_i + b^2 \sum_{i=1}^n y_i^2$$

which is always ≥ 0 for all values of a, b and hence has a non-negative determinant Δ. The equality is attained when $\Delta = 0$, which means there exists real numbers a, b such that $ax_i + by_i = 0$ for all $i = 1, \cdots, n$.

(6.A.1) also applies to infinite sums i.e. $\sum x_i^2 \sum y_i^2 \geq (\sum x_i y_i)^2$ and also to complex numbers i.e. $\sum |x_i|^2 \sum |y_i|^2 \geq \sum |x_i y_i|^2$ where x_i, y_i are arbitrary complex numbers and $|t|$ denotes modulus of complex number t.

Inequality (6.A.1) also applies to integrals. Let $f(x)$ and $g(x)$ be two arbitrary real functions of x. Then

$$\int f^2 dx \int g^2 dx \geq \left\{\int fg dx\right\}^2 \tag{6.A.2}$$

Holder's inequality

Let $(x_1, \cdots, x_n)$, $(y_1, \cdots, y_n)$ be two sets of non-negative real numbers. Then

$$\left(\sum_{i=1}^n x_i^p\right)^{\frac{1}{p}} \left(\sum_{i=1}^n y_i^q\right)^{\frac{1}{q}} \geq \sum_{i=1}^n x_i y_i \tag{6.A.3}$$

where $\dfrac{1}{p} + \dfrac{1}{q} = 1$, $p > 1$. The equality holds if and only if $y_i \infty x_i^{p-1}$.

(6.A.3.) also applies to integrals. Let $f(x), g(x)$ be two arbitrary real non-negative functions of x. Then

$$\left(\int f^p dx\right)^{\frac{1}{p}} \left(\int g^q dx\right)^{\frac{1}{q}} \geq \int fg dx$$

provided $\dfrac{1}{p} + \dfrac{1}{q} = 1$, $p > 1$.

More generally, we have, for sets of non-negative real numbers $(x_1, \cdots, x_n)$, $(y_1, \cdots, y_n)$, $(z_1, \cdots, z_n), \cdots$

$$\left(\sum x_i^p\right)^{\frac{1}{p}}\left(\sum y_i^q\right)^{\frac{1}{q}}\left(\sum z_i^q\right)^{\frac{1}{r}}\cdots \geq \sum x_i y_i z_i \cdots$$

if $\frac{1}{p}+\frac{1}{q}+\frac{1}{r}+\cdots=1$.

Minkowski's inequality

For two sets of non-negative real numbers $(x_1, \cdots, x_n)$, $(y_i, \cdots, y_n)$ and $p \geq 1$,

$$\left(\sum x_i^p\right)^{\frac{1}{p}}+\left(\sum y_i^p\right)^{\frac{1}{p}} \geq \left[\sum (x_i+y_i)^p\right]^{\frac{1}{p}}.$$

7 GENERATING FUNCTIONS

7.1. Introduction

We consider different generating functions in this chapter. Section 7.2 deals with probability generating function of a random variable. We consider moment generating function in section 7.3, factorial moment generating function in section 7.4, cumulants and cumulant generating function in section 7.5 and characteristic function in section 7.6.

7.2. Probability Generating Function

Let X be an integer-valued r.v. with $P(X = k) = p_k$, $k = 0, 1, 2, \cdots$ and $\sum_{k=0}^{\infty} p_k = 1$.

Definition 7.1: The function defined by

$$P_X(t) = E(t^X) = \sum_{k=0}^{\infty} p_k t^k, \quad |t| < 1 \tag{7.1}$$

is called the probability generating function (pgf) of X. Since $P_X(1) = 1$, series (7.1) is uniformly and absolutely convergent in $|t| \leq 1$ and $P_X(t)$ is a continuous function of t.

Every *pgf* determines a unique probability distribution i.e. a unique set of probabilities $\{p_k\}$. Coefficient of t^k in the expansion of $P_X(t)$ gives $P(X = k)$. The *pgf* is used for discrete variables only.

❑ **EXAMPLE 7.1.** *X is uniformly distributed over* $0, 1, \cdots, n$ *with* $P(X = k) = \frac{1}{n+1}$, $k = 0, 1, \cdots, n$. $P_X(t) = \frac{1}{n+1}\sum_{k=0}^{n} t^k = \frac{1}{(1-t)(n+1)}$ $[1 - t^{n+1}]$ $|t| < 1$.

❑ **EXAMPLE 7.2.** *X has a geometric distribution,* $P(X = k) = pq^k$, $k = 0, 1, \cdots$,

$$0 < p < 1, q = 1 - p, P_X(t) = \sum_0^{\infty} t^k pq^k = \frac{p}{1 - tq}, |t| < 1.$$

❑ **EXAMPLE 7.3.** *X has a poisson distribution,* $P(X = k) = e^{-\lambda}\frac{\lambda^k}{k!}$, $k = 0, 1, \cdots$

$$P_X(t) = \sum_0^{\infty} (\lambda t)^k \frac{e^{-\lambda}}{k!} = e^{-\lambda(1-t)}, |t| < 1.$$

❑ **EXAMPLE 7.4.** *X has a binomial distribution,* $P(X = k) = \binom{n}{k} p^k (1-p)^{n-k}$,

$$0 < p < 1.\ P_X(t) = \sum_0^{n} t^k \binom{n}{k} p^k (1-p)^{n-k} = (1 + p(t-1))^n$$

❑ **EXAMPLE 7.5.** *Negative binomial distribution with parameters* n, p : $P(X = x) = \binom{-n}{x} p^n (p-1)^x$, $x = 0, 1, 2, \cdots$

$$P_X(t) = \sum_0^{\infty} t^x \binom{-n}{x} p^n (p-1)^x = p^n \sum_0^{\infty} \binom{-n}{x} \Big[t(p-1)\Big]^x$$
$$= p^n \Big[1 + t(p-1)\Big]^{-n}, \quad \text{provided } \Big|t(p-1)\Big| < 1.$$

Note that when $n = 1$, the *pgf* becomes identical with the *pgf* of a geometric distribution with parameter p (example 7.2). Hence if $X_1, \cdots, X_n$ are independently and identically geometrically distributed r.v.s with parameter p, their sum $X_1 + \cdots + X_n$ follows a negative binomial distribution with parameters n, p (follows by theorem 7.2).

• **THEOREM 7.1.** *Let X be an integer-valued r.v. Consider* $Y = aX + b$, a, b *being arbitrary constants. The pgf of Y is*

$$P_Y(t) = E\left(t^{aX+b}\right) = t^b P_{aX}(t) \tag{7.2}$$

• **THEOREM 7.2.** *Let* $X_1, \cdots X_n$ *be n independent r.v.s with pgfs* $P_1(t), \cdots, P_n(t)$ *respectively. Consider the sum* $X_1 + \cdots + X_n = S_n$. *The pgf of* S_n *is*

$$P_{S_n}(t) = P_1(t) \cdots P_n(t) \tag{7.3}$$

Proof. $P_{S_n}(t) = E\left(t^{S_n}\right) = E(t^{X_1}) \cdots E\left(t^{X_n}\right) = P_1(t) \cdots P_n(t)$, $|t| < 1$, since each of $t^{X_1}, \cdots, t^{X_n}$ are independent r.v.s.

❑ **EXAMPLE 7.6.** *Independent Bernoullian trials with probability p of a success are performed, $0 < p < 1$. Let X_k denote the number of failures following the $(k-1)$th success and preceding the kth success. Then $S_r = X_1 + \cdots + X_r$ is the total number of failures preceding the rth success. Here $X_1, \cdots, X_r$ are independently and indentically distributed r.v.s. each X_i having the geometric distribution mentioned in example 7.2. Hence $P_{S_r}(t) = \left\{P_X(t)\right\}^r = \left(\dfrac{p}{1-tq}\right)^r$, $|t| < 1$. Coefficient of t^k is $\binom{-r}{k} p^r(-q)^k$, $k = 0, 1, \cdots$ which is the probability of k failures preceding the rth success. (vide section 8.7)*

We now consider the pgf of a sum of indicator r.v.s.

• **THEOREM 7.3.** *Suppose $I = I_1 + \cdots + I_n$ where each I_i is an indicator r.v. Let A_i be the event for which I_i is an indicator r.v. i.e. $P(I_i = 1) = P(A_i)$. The pgf of I is*

$$P_I(t) = 1 + (t-1)B_1 + (t-1)^2 B_2 + \cdots + (t-1)^n B_n, \quad |t| < 1 \tag{7.4}$$

where

$$B_k = \sum_{\substack{i_1 < \cdots < i_k \\ =1}}^{n} \cdots \sum P(A_{i_1} \cdots A_{i_k})$$

Proof. Consider the random function

$$Y = [1 + (t-1)I_1] \cdots [1 + (t-1)I_n] \tag{7.5}$$

Suppose for a sample point w, $I(w) = k$. For this w, $Y(w) = t^k$. Hence $P[Y = t^k] = P(I = k)$, $k = 0, 1, \cdots n$. Thus

$$\begin{aligned} E(Y) &= t^{\circ} P(Y = t^{\circ}) + tP(Y = t) + \cdots + t^n P(Y = t^n) \\ &= t^{\circ} P(I = 0) + tP(I = 1) + \cdots t^n P(I = n) = P_I(t) \end{aligned} \tag{7.6}$$

Againg, from (7.5),

$$Y = 1 + (t-1)\sum_i I_i + (t-1)^2 \sum_{i<j}\sum I_i I_j + \cdots + (t-1)^n I_1 \cdots I_n$$

$$E(Y) = 1 + (t-1)B_1 + (t-1)^2 B_2 + \cdots + (t-1)^n B_n \tag{7.7}$$

Combining (7.6) and (7.7) the result follows.

❑ **EXAMPLE 7.7.** *A coin is tossed n times. Find the pgf of number of heads. Let $X_i = 1(0)$ if the ith toss results in a head (tail); then $X = X_1 + \cdots + X_n$ = total number of heads. Here*

$$B_k = \text{Probability of getting } k \text{ heads} = \binom{n}{k}\frac{1}{2^k}$$

$$P_X(t) = \sum_{k=0}^{n} \binom{n}{k} \frac{(t-1)^k}{2^k} = \left(\frac{1+t}{2}\right)^n, \quad |t| < 1$$

❑ **EXAMPLE 7.8.** *Let X be a r.v. with* $P(X = k) = p_k$, $k = 1, 2, \cdots$. *Let* $P(X > k) = q_k$, $k = 0, 1, \cdots$ *and* $Q(t) = \sum_{j=0}^{\infty} q_j t^j$. *Show that the pgf of tail probabilities* q_k *is*

$$Q(t) = \frac{1 - P(t)}{1 - t}, \quad |t| < 1 \tag{7.8}$$

where $P(t)$ is the *pgf* of X. Show also that

$$\lim_{t \uparrow 1} \frac{1 - P(t)}{1 - t} = E(X) \tag{7.9}$$

when $t \uparrow 1$ means t approaches 1 from left.

Here $Q(t) = 1 + q_1 t + q_2 t^2 + \cdots = 1t^{\circ} + (1 - p_1)t + (1 - p_1 - p_2)t^2 + \cdots$

$$= (1 + t + t^2 + \cdots) - p_1 t(1 + t + t^2 + \cdots) - p_2 t^2(1 + t + t^2 + \cdots) - \cdots$$

$$= \frac{1 - P(t)}{1 - t}$$

Hence $\sum_{j=0}^{\infty} t^j P(X > j) = \frac{1 - P(t)}{1 - t}$. Taking limit of both sides as $t \uparrow 1$, (7.9) follows from lemma 4.1.

If a distribution is specified by its *pgf* its factorial moments (about origin) can be obtained as follows :

$$\frac{\partial^r P_X(t)}{\partial t^r} = \sum_x x(x-1)\cdots(x-r+1)t^{x-r} p_x$$

Hence $$\left.\frac{\partial^r P_x(t)}{\partial t^r}\right]_{t=1} = \mu_{(r)}$$

❑ **EXAMPLE 7.9.** *In example 7.2,* $P'(t) = \frac{pq}{(1 - tq)^2}$, $P''(t) = \frac{2pq^2}{(1 - tq)^3}$. Hence

$$E(X) = \frac{q}{p}, \; E(X^2) = \frac{q}{p} + \frac{2pq^2}{p^3} \text{ when } V(X) = \frac{q}{p^2}.$$

In example 7.3, $P'(t) = \lambda e^{-\lambda(1-t)}$, $P''(t) = \lambda^2 e^{-\lambda(1-t)}$. Hence $E(X) = \lambda$, $E(X(X - 1)) = \lambda^2$, when $V(X) = \lambda$.

• **THEOREM 7.4.** *Suppose X is non-negative and integer-valued and that moments of all orders exist. Then*

$$P_X(t) = \sum_{k=0}^{\infty} \left\{E(X)_k\right\} \frac{(t-1)^k}{k!}, \quad |t| \le 1 \tag{7.10}$$

Proof. We have $P_X(t) = p_{\circ} + p_1 t + p_2 t^2 + \cdots$

Now $t^i = (t-1+1)^i = \sum_{k=0}^{i} \binom{i}{k}(t-1)^k$

Hence $P_X(t) = \sum_{i=0}^{\infty} p_i \sum_{k=0}^{i} \binom{i}{k}(t-1)^k = \sum_{k=0}^{\infty}(t-1)^k \sum_{i=k}^{\infty} \binom{i}{k} p_i$

The rearrangement in the order of summation is valid because (7.10) is absolutely convergent for $|t| \le 1$. Since $\sum_{i=k}^{\infty} \binom{i}{k} p_i = \frac{1}{k!} E\{(X)_k\}$, the result follows.

7.3. Moment Generating Function

Definition 7.2. Let X be a r.v. defined on $(\mathcal{S}, \mathcal{A}, P)$. The function

$$M_X(t) = E(e^{tX})$$

defined for all real values of t is known as the moment generating function (mgf.) of X, if the expectation of e^{tX} exists for some $|t| < T$. For a discrete r.v. with *pmf* $P(X = x_k) = p_k$, $k = 0, 1, 2, \cdots$

$$M_X(t) = \sum_{0}^{\infty} e^{tx_k} p_k$$

In the case of a continuous probability distribution, specified by a *pdf* f, the *mgf* is

$$M_X(t) = \int_{-\infty}^{\infty} e^{tx} f(x)dx.$$

Since for every fixed t, e^{tx} is a positive function of x, it follows that $M_X(t)$ is either finite or infinite. We say that the probability law possesses a *mgf* if there exists a positive number T such that $M_X(t)$ is finite for $|t| \le T$. In this case all moments of distribution of X exist and are given by the successive derivatives of $M_X(t)$ at $t = 0$.

If a *mgf* exists for some t in $[-T, T]$, then one may form its successive derivatives by successively differentiating under the integral or summation sign. Thus

$$M'(t) = \frac{d}{dt} M(t) = E\left(\frac{\partial}{\partial t} e^{tX}\right) = E\left(X e^{tX}\right)$$

$$M''(t) = \frac{d^2}{dt^2} M(t) = E\left(\frac{\partial}{\partial t} X e^{tX}\right) = E\left(X^2 e^{tX}\right)$$

$$\cdots \quad \cdots \quad \cdots \quad \cdots$$

Letting $t = 0$ we have

$$M^{(k)}(0) = E(X^k), \quad \text{for all positive integral } k.$$

Thus $M'(0) = E(X)$, $M''(0) = E(X^2)$.

If the mgf $M(t)$ is finite for some $|t| \le T$ (for some $T > 0$) one can express $M(t)$ (uniquely) in a Maclaurin series expansion :

$$M(t) = \sum_{j=0}^{\infty} \frac{M^{(j)}(0)}{j!} t^j = \sum_{j=0}^{\infty} \frac{\mu_j'}{j!} t^j \tag{7.11}$$

so that μ_j' is the coefficient of $\frac{t^j}{j!}$ in the expresion (7.11).

We have seen there exists probability laws without finite moments (e.g. Cauchy distribution). Consequently there also exist probability laws which do not possess mgfs. The following results hold for a mgf :

(i) • **THEOREM 7.5.** *A mgf, if it exists, uniquely determines the distribution function of a r.v. X and conversely if the mgf exists it is unique.*

Thus if X, Y be two r.v.s with mgf $M_X(t)$, $M_Y(t)$ respectively, such that $M_X(t) = M_Y(t)$ for all values of t, then X, Y have the same probability distribution.

(ii) Even if the moments of all orders exist for a r.v. X it does not follow that mgf exists.

(iii) If not all the moments of a r.v. X exist one can not find the distribution function of X from the moments, even from the moments that do exist.

Since there exist probability laws for which mgf does not exist, utility of mgf in determining the distribution function is somewhat limited. It is more convenient to use the characteristic function (discussed in section 7.6), $\varphi_X(t) = E(e^{itX})$, where $i = \sqrt{-1}$, the imaginary quantity and t is any real number, which always exists and uniquely determines the distribution of X. $\varphi_X(t)$ can also be used, like mgf, to determine those moments which do exist.

❑ **EXAMPLE 7.10.** *Consider again binomial distribution,* $P(X = k) = \binom{n}{k} p^k q^{n-k}$, $0 < p < 1$, $q = 1 - p$.

$$M(t) = \sum_{k=0}^{n} e^{tk} \binom{n}{k} p^k q^{n-k} = \left(q + pe^t\right)^n \text{ exists for } -\infty < t < \infty$$

$$M'(t) = npe^t \left(pe^t + q\right)^{n-1}$$

$$M''(t) = npe^t \left(pe^t + q\right)^{n-1} + n(n-1)p^2 e^{2t} \left(pe^t + q\right)^{n-2}$$

Thus

$$M'(0) = E(X) = np$$

$$M''(0) = E(X^2) = npq + n^2 p^2$$

$$V(X) = npq$$

❑ **EXAMPLE 7.11.** *Poisson distribution with parameter λ :*

$$P(X=k)=e^{-\lambda}\frac{\lambda^k}{k!},\quad k=0,1,2,\cdots\quad M(t)=\sum_{O}^{\infty}e^{-\lambda}\frac{\lambda^k}{k!}e^{tk}=e^{\lambda(e^t-1)}$$

$$M'(t)=e^{\lambda(e^t-1)}\lambda e^t,\quad M''(t)=\lambda^2e^{\lambda(e^t-1)}e^{2t}+\lambda e^te^{\lambda(e^t-1)}$$

Thus $E(X)=M'(0)=\lambda$, $E(X^2)=M''(0)=\lambda^2+\lambda$, $V(X)=\lambda$.

For truncated Poisson distribution, truncated at $X=1$,

$$P(X=k)=\frac{e^{-\lambda}\lambda^k}{k!(1-e^{-\lambda})},\quad k=1,2,\cdots$$

$$M_X(t)=\left(e^{\lambda e^t}-1\right)\Big/(e^{\lambda}-1).$$

❑ **EXAMPLE 7.12.** *Geometric distribution with parameter $p, 0<p<1$: $P(X=k)=pq^k$, $q=1-p$, $k=0,1,2,\cdots$ $M(t)=\dfrac{p}{1-qe^t}$ exists for t such that $qe^t<1$. It follows $E(X)=\dfrac{q}{p}$, $V(X)=\dfrac{q}{p^2}$.*

❑ **EXAMPLE 7.13.**

$$f(x)=\frac{6}{\pi^2x^2},\quad x=1,2,\cdots=0\quad \textit{otherwise}$$

$$M_X(t)=\frac{1}{\pi^2}\sum_{o}^{\infty}\frac{6e^{tx}}{\pi^2x^2}$$ *is infinite for every $t>0$. The mgf, therefore, does not exist.*

The mgf of $Z=\dfrac{X-a}{h}$ $(a,h$ are arbitrary constents) is

$$M_z(t)=Ee^{tZ}=e^{-\frac{at}{h}}M_X\left(\frac{t}{h}\right).$$

The momdent generating function of X about its mean μ, $M_{X-\mu}(t)$ gives the central moments of X, μ_r. $M_{X-\mu}(t)=\sum_{j=0}^{\infty}\frac{\mu_j}{j!}t^j$, provided the mgf exists for some $|t|\le T$.

• **THEOREM 7.6.** *The mgf of the sum $S_n=X_1+\cdots+X_n$ of n independent r.v.s $X_1,\cdots,X_n$ is*

$$M_{S_n}(t)=M_{X_1}(t)\cdots M_{X_n}(t)$$

where $M_{X_i}(t)$ is the mgf of X_i, provided all the mgfs exist.

Proof.

$$MS_n(t)=E\left(e^{t\Sigma_1^nX_i}\right)=E\left(\prod_{i=1}^{n}e^{tX_i}\right)$$

$$=\prod_{i=1}^{n}E(e^{tX_i}),$$

because of independence.

❑ **EXAMPLE 7.14.** *X is normally distributed with mean μ and s.d. σ.*

$$f(x) = \frac{1}{\sigma\sqrt{2\pi}} e^{-\frac{(x-\mu)^2}{2\sigma^2}}, \quad -\infty \le x \le \infty$$

The mgf of X about μ is

$$M_{X-\mu}(t) = E\left[e^{t(X-\mu)}\right] = \frac{1}{\sigma\sqrt{2\pi}} \int_{-\infty}^{\infty} e^{t(x-\mu)} e^{-\frac{(x-\mu)^2}{2\sigma^2}}\, dx$$

$$= \frac{e^{t^2\sigma^2/2}}{\sigma\sqrt{2\pi}} \int_{-\infty}^{\infty} e^{-\frac{1}{2\sigma^2}\left\{(x-\mu)-\sigma^2 t\right\}^2} dx = e^{t^2\sigma^2/2} = \sum_{j=0}^{\infty} \frac{(t^2\sigma^2)^j}{2^j j!}$$

Hence $\quad E(X-\mu) = 0 \quad$ or $\quad E(X) = \mu$

$$E(X-\mu)^{2r+1} = 0, \quad E(X-\mu)^{2r} = \mu_{2r} = \frac{\sigma^{2r}(2r)!}{2^r \cdot r!}$$

$$= (2r-1)(2r-3)\cdots 3.1\cdots\sigma^{2r} \quad \text{for } r = 1, 2, \cdots$$

The mgf of X is $e^{\mu t + \frac{t^2\sigma^2}{2}}$

❑ **EXAMPLE 7.15.** *Consider example 7.14. Let $Y = \alpha X + \beta$ when α, β are constants. The mgf of Y is*

$$M_Y(t) = e^{\beta t} M_X(\alpha t) = e^{(\beta+\alpha\mu)t + (\alpha\sigma)^2 \frac{t^2}{2}}$$

Thus Y follows a normal distribution with mean $\beta + \alpha\mu$ and variance $\sigma^2\alpha^2$ (by theorem 7.5).

❑ **EXAMPLE 7.16.** *Consider n independent Poisson variables $X_1, \cdots, X_n$ (vide example 7.11). By theorem 7.6, the mgf of $S_n = X_1 + \cdots + X_n$ is $\exp\left\{\sum_i^n \lambda_i(e^t - 1)\right\}$. Thus by theorem 7.5, S_n has a Poisson distribution with parameter $\sum_{i=1}^n \lambda_i$ (This is called the reporductive property of Poisson distribution. The Poisson distribution is discussed in details in section 8.5).*

7.4. Factorial Moment Generating Function

Defintion 7.3. The factorial moment generating function ($fmgf$) of a discrete probability distribution of X about the origin is defined as

$$\phi_X(t) = E\left[(1+t)^X\right] \tag{7.12}$$

$$= E\left[e^{X \log_e(1+t)}\right] = M_X\left[\log_e(1+t)\right].$$

Since $(1+t)^x = \sum_{r=0}^{\infty} \frac{(x)_r}{r!} t^r$, we get $\mu_{(r)}$ = coefficient of $\frac{t^r}{r!}$ in the expansion of $\phi_X(t)$ in ascending powers of t. Also

$$\frac{d^r}{dt^r}\phi_X(t)\Big]_{t=0} = \mu_{(r)}.$$

❑ **EXAMPLE 7.17.** *In the familar matching problem with n cards, the probability distribution of number of matches X is*

$$P[X=m] = \frac{1}{m!}\sum_{k=0}^{n-m}(-1)^k\frac{1}{k!}, \quad m = 0, 1, \cdots, n \quad \text{(vide example 2.27)}$$

The mgf of X is

$$M_X(t) = E(e^{tX}) = \frac{1}{m!}\sum_{k=0}^{n-m}(-1)^k\frac{e^{tX}}{k!} = \sum_{r=0}^{n}\frac{1}{r!}(e^t-1)^r$$

Hence the $fmgf$ is

$$\phi_X(t) = \sum_{r=0}^{n}\frac{t^r}{r!}$$

7.5. Cumulant Generating Function

Defintion 7.4. If the logarithm of the mgf of the distribution of a r.v. X can be expanded in a convergent series in powers of t viz.

$$K_X(t) = \log_e M_X(t) = \sum_{r=1}^{\infty} k_r(X)\frac{t^r}{r!}$$

then the coefficients $k_1(X), k_2(X), \cdots$ are called the first cumulant, second cumulant, $\cdots$ etc of the distribution of X and $K_X(t)$ is called the cumulant generating function (cgf) of the distribution of X.

We have

$$k_r(X) = \frac{d^r}{dt^r}\log_e M_X(t)\Big]_{t=0}$$

If the mgf exists, so does the cgf. Evidently $k_r(X)$ is a function of the moments of X. The first few cumulants are as follows :

$$k_1 = \mu_1' = \mu$$

$$k_2 = \mu_2' - (\mu_1')^2 = \sigma^2$$

$$k_3 = \mu_3' - 3\mu_2'\mu_1' + 2(\mu_1')^3 = \mu_3$$

$$\cdots\cdots\cdots\cdots\cdots\cdots\cdots\cdots\cdots$$

and conversely,

$$\mu_1' = k_1$$
$$\mu_2' = k_2 + k_1^2$$
$$\mu_3' = k_3 + 3k_1k_2 + k_1^3$$
$$\cdots\cdots\cdots\cdots\cdots\cdots$$

Now, for any constants a and h, the mgf of $Z = \dfrac{X-a}{h}$ is

$$M_Z(t) = E\left[e^{t\left(\frac{(X-a)}{h}\right)}\right] = e^{-\frac{at}{h}} M_X\left(\frac{t}{h}\right).$$

Hence

$$K_Z(t) = \log_e M_{\frac{X-a}{h}}(t) = -\frac{at}{h} + \log_e M_X\left(\frac{t}{h}\right)$$
$$= -\frac{at}{h} + \frac{t}{h}k_1(X) + \frac{t^2}{2!h^2}k_2(X) + \frac{t^3}{3!h^3}k_3(X) + \cdots$$
$$= (-a + k_1(X))\frac{t}{h} + \frac{t^2}{2!h^2}k_2(X) + \frac{t^3}{3!h^3}k_3(X) + \cdots \qquad (7.12)'$$

Thus

$$k_1(Z) = \frac{1}{h}[-a + k_1(X)], \quad k_r(Z) = \frac{k_r(X)}{h^r}, \quad r \geq 2$$

Thus only the first cumulant of Z is affected by the change of origin. All the cumulants of Z are affected by the change of scale. For this reason, the cumulants are sometimes called semi-invariants. The cumulants were originally defined and studied by Thiele (1903).

Putting $a = \mu$ and $h = 1$ in (7.12)$'$ we get the central cumulants (cumulants about the mean μ) which are the same as the cumulants about zero except for the first central cumulant which becomes zero.

• **THEOREM 7.7.** *If $X_1, \cdots, X_n$ are independent r.v.s and if the mgf $M_{X_i}(t)$ exists $(i = 1, \cdots, n)$ the cgf of $S_n = X_1 + \cdots + X_n$ is*

$$K_{S_n}(t) = \sum_{i=1}^{n} K_{X_i}(t)$$

Proof. $K_{S_n}(t) = \log_e\{M_{S_n}(t)\} = \log_e\left\{\prod_{i=1}^{n} M_{X_i}(t)\right\}$. Hence the proof.

It follows that

$$k_r\left(\sum_{i=1}^{n} X_i\right) = \sum_{i=1}^{n} k_r(X_i) \quad \text{for all } r.$$

Thus the cumulant of the sum is the sum of the cumulants. This makes the name 'cumulant' appropriate.

❑ **EXAMPLE 7.18.** *Poisson distribution* : $P(X = k) = e^{-\lambda}\frac{\lambda^k}{k!}$. *The cgf is* $\lambda(e^t - 1)$. *Hence* $k_r = \lambda$, $r = 1, 2, \cdots$

Normal distribution (vide example 7.14). The cgf about mean μ is $\frac{t^2\sigma^2}{2}$. Hence $k_1 = \mu$, $k_2 = \sigma^2$, $k_r = 0$, $r \geq 2$.

7.6. Characteristic Function

Definition 7.5. The characteristic function (c.f.) of a r.v. X is defined as

$$\psi_X(t) = E(e^{itX}) \tag{7.13}$$

where t is any real number and $i = \sqrt{-1}$. Hence for a discrete r.v. with pmf p_k ($k = 0, 1, 2, \cdots$)

$$\psi_X(t) = \sum_{k=0}^{\infty} e^{itx_k} p_k \tag{7.13a}$$

and for a continuous r.v. with pdf $f(x)$

$$\psi_X(t) = \int_{-\infty}^{\infty} e^{itx} f(x) dx \tag{7.13b}$$

Since $e^{itx} = \cos tx + i \sin tx$ and $\cos tx$ and $\sin tx$ are both integrable over R^1 for any real t, $\psi(t)$ is a complex number whose real and imaginery parts are finite for every value of t.

The r th non-central moment μ_r' of X, when it exists, can be determined from the c.f. by the formula

$$\left.\frac{d^r}{i^r dt^r}\right]_{t=0} = \mu_r'$$

The properties of c.f. are :

(i) $\psi(0) = 1$.

(ii) $|\psi(t)| \leq 1$ for all values of t, because $|\psi(t)| = \left|E(e^{itX})\right| \leq E\left|e^{itX}\right| = E(1) = 1$.

(iii) If $\psi(t)$ is the c.f. of X, $\overline{\psi(t)}$, complex conjugate of $\psi(t)$, is the c.f. of $(-X)$.

Proof. $E(e^{-itX}) = E(\cos tX - i \sin tX) = E(\cos tX) - iE\sin(tX) = \overline{\psi(t)}$.

(iv) $\psi(t)$ is uniformly continuous on R^1.

Proof.

$$\psi(t) = \int_{-\infty}^{\infty} e^{itx} dF(x).$$

$$\left|\psi(t + h) - \psi(t)\right| = \int_{-\infty}^{\infty} e^{itx}(e^{ihx} - 1) dF(x)$$

$$\leq \int_{-\infty}^{\infty} |e^{ihx} - 1| dF(x) = 2\int_{-\infty}^{\infty} \left|\sin\frac{hx}{2}\right| dF(x)$$

Hence
$$|\psi(t+h)-\psi(t)| \le 2\int_{-\infty}^{-A}\left|\sin\frac{hx}{2}\right|dF(x)+2\int_{-A}^{B}\left|\sin\frac{hx}{2}\right|dF(x)$$
$$+2\int_{B}^{\infty}\left|\sin\frac{hx}{2}\right|dF(x) \quad \text{for} \quad A>0, B>0 \qquad (7.14)$$

The right hand side of (7.14) is independent of t. By choosing A and B sufficiently large and h sufficiently small the first and the third term on the right hand side of (7.14) can be made sufficiently small when the proof follows.

We state below the following properties without proof.

(v) If μ_r' exists for the distribution of X, then $\psi^{(j)}(t)$ exists and is continuous for $1 \le j \le r$ and
$$\psi(t)=\sum_{j=0}^{r}\frac{(it)^j}{j!}\mu_j'+o(t^r)$$
where $\lim_{t\to 0}\frac{o(t^r)}{t^r}=0$.

(vi) Suppose the c.f. $\psi(t)$ has a derivative of order $2r$ at zero, $\psi^{(2r)}(0)$ (2r is an even integer). Then the $(2r)$th moment μ_{2r}' exists and there is a $2r$th derivative for all real t given by
$$\psi^{(2r)}(t)=(-1)^r\int_{-\infty}^{\infty}e^{itx}x^{2r}dF(x).$$

(vii) The necessary and sufficient conditions for a bounded and continuous function to be the characteristic function of a distribution are that $\psi(0)=1$ and the function
$$\Psi(x,A)=\int_{O}^{A}\int_{O}^{A}\varphi(t-u)e^{i(t-u)x}dt\,du$$
is real and non-negative for all x and $A>0$.

(viii) • **THEOREM 7.8.** *The Uniqueness theorem. Two distribution functions $F_1(x)$ and $F_2(x)$ are identical if and only if their characteristic functions $\psi_1(t)$ and $\psi_2(t)$ are identical.*

The uniqueness theorem states that, given a function $\psi(t)$, if $\psi(t)$ is the c.f. of a r.v. X, we can determine its d.f. $F(x)$ from it. The process of determining the distribution function from the c.f. is given by the following inversion theorem.

(ix) • **THEOREM 7.9.** *Inversion theorem. Let $\psi(t)$ be the characteristic function of the distribution $F(x)$. Then for any two continuity points a and $a+h$ $(h>0)$ of $F(x)$,*
$$F(a+h)-F(a)=\lim_{T\to\infty}\frac{1}{2\pi}\int_{-T}^{T}\frac{1-e^{-ith}}{it}e^{-ita}\psi(t)dt \qquad (7.15)$$

Chosing a $= x-\delta$ and $h=2\delta$, (7.15) can be written as :
$$\frac{F(x+\delta)-F(x-\delta)}{2\delta}=\lim_{T\to\infty}\frac{1}{2\pi}\int_{-T}^{T}\frac{\sin t\delta}{t\delta}e^{-itx}\psi(t)dt \qquad (7.16)$$
provided $x+\delta$, $x-\delta$ are continuity points of $F(x)$.

(x) If c.f. $\psi(t)$ is aboslutely integrable over R^1 (i.e. $\int_{-\infty}^{\infty} |\psi(t)|dt < \infty$) then the derivative of $F(x)$ exists, which is bounded and continuous on R^1 and is given by

$$f(x) = F'(x) = \frac{1}{2\pi}\int_{-\infty}^{\infty} e^{-itx}\psi(t)dt \tag{7.17}$$

for every x in R^1.

$f(x)$ is continuous for all values of x on R^1. (7.17) enables one to determine the density function $f(x)$ by means of the c.f. when the latter is absolutely integrable.

(xi) $\psi(t)$ is real if and only if $F(x)$ is symmetric.

Proof.

$$\psi_X(t) = \int_{-\infty}^{\infty} (\cos tx + i\sin tx)dF(x)$$

If X is symmetric, the contribution of sine term vanishes, because

$$\int_{-\infty}^{0-} \sin tx dF_X(x) + \int_{0+}^{\infty} \sin tx dF_X(x) + (\sin 0)(P(X=0))$$

$$= \int_{0+}^{\infty} -\sin tx dF_X(x) + \int_{0+}^{\infty} \sin tx dF_X(x) = 0$$

Conversely, if $\psi_X(t)$ is real, consider the r.v. $Y = -X$.

$$\psi_Y(t) = \int_{-\infty}^{\infty} e^{-itx} dF_X(x) = \psi_X(-t) = \overline{\psi_X(t)}.$$

Since $\psi_X(t)$ is real, $\psi_X(t) = \overline{\psi_X(t)}$. Thus X and Y have the same c.f.

Again, $P(X \le -x) = P(-X \ge x) = P(Y \ge x)$ i.e. $F_X(-x) = 1 - P(Y < x) = 1 - F_Y(x-0) = 1 - F_X(x-0)$ (by theorem 7.8). Hence F_X is symmetric about zero.

For proof of these different theorems, the reader may refer to Lukacs (1960).

❑ **EXAMPLE 7.19.** *For the normal distribution*

$$f(x) = \frac{1}{\sqrt{2\pi}\sigma} e^{-\frac{1}{2}\frac{(x-\mu)^2}{\sigma^2}} \qquad -\infty < x < \infty$$

$$\psi(t) = \int_{-\infty}^{\infty} \frac{e^{itx}}{\sqrt{2\pi}\sigma} e^{-\frac{1}{2}\frac{(x-\mu)^2}{\sigma^2}} dx. \qquad \text{Let } \frac{x-\mu}{\sigma} = y$$

$$\psi(t) = \int_{-\infty}^{\infty} \frac{1}{\sqrt{2\pi}} e^{it(\mu+\sigma y)} e^{-\frac{y^2}{2}} dy = \frac{e^{it\mu}}{\sqrt{2\pi}} \int_{\infty}^{\infty} \cdot e^{-\frac{1}{2}(y^2 - 2it\sigma y)} dy$$

$$= e^{it\mu - \frac{t^2\sigma^2}{2}} \int_{\infty}^{\infty} \frac{1}{\sqrt{2\pi}} \cdot e^{-\frac{1}{2}(y - it\sigma)^2} dy = e^{it\mu - \frac{t^2\sigma^2}{2}},$$

the integral being unity, since it is the integral of a normal distribution. Hence,

$$\psi'(t) = e^{it\mu - \frac{t^2\sigma^2}{2}} \quad (i\mu - t\sigma^2)$$

$$\mu''(t) = e^{it\mu - \frac{t^2\sigma^2}{2}} \left[(i\mu - t\sigma^2)^2 - \sigma^2\right]$$

Thus $\psi'(0) = i\mu, \quad \psi''(0) = i^2(\mu^2 + \sigma^2)$.

❑ **EXAMPLE 7.20.** *For the gamma distribution*

$$f(x) = \frac{\lambda^n}{\lceil(n)} e^{-\lambda x} x^{n-1}, \quad 0 < x < \infty, \quad \lambda > 0$$

$$\psi(t) = \int_0^\infty \frac{\lambda^n}{\lceil(n)} e^{-(\lambda - it)x} x^{n-1} dx$$

$$= \frac{\lambda^n}{(\lambda - it)^n} \int_0^\infty \frac{(\lambda - it)^n}{\lceil(n)} e^{-(\lambda - it)x} x^{n-1} dx$$

$$= \left(\frac{\lambda}{\lambda - it}\right)^n = \left(1 - \frac{it}{\lambda}\right)^{-n}$$

For the exponential distribution, which is a gamma distribution with parameter $n = 1$,

$$\psi(t) = \left(1 - \frac{it}{\lambda}\right)^{-1}, \quad \psi'(t) = (-n)\left(1 - \frac{it}{\lambda}\right)^{-n-1}\left(-\frac{i}{\lambda}\right).$$

$$\psi''(t) = (-n)(-n-1)\left(1 - \frac{it}{\lambda}\right)^{-n-2}\left(\frac{i^2}{\lambda^2}\right)$$

$$\psi'(0) = \frac{in}{\lambda}, \quad \psi''(0) = \frac{n(n+1)i^2}{\lambda^2}.$$

❑ **EXAMPLE 7.21.** *Laplace distribution* $f(x) = \frac{1}{2}e^{-|x|}, \quad -\infty \le x \le \infty.$

$$\psi(t) = \frac{1}{2}\int_{-\infty}^\infty e^{itx} e^{-|x|} dx = \frac{1}{2}\int_{-\infty}^\infty (\cos tx + i \sin tx) e^{-|x|} dx = \int_0^\infty \cos tx e^{-x} dx$$

Now $\int_0^\infty e^{-x} \cos tx dx = 1 - t\int_0^\infty e^{-x} \sin tx dx$ (on integration by parts, taking $\cos tx$ as first fimction)

Similarly $\int_0^\infty e^{-x} \sin tx dx = t\int_0^\infty e^{-x} \cos tx dx.$

Hence $\psi(t) = \dfrac{1}{1 + t^2}.$

Note that $\psi(t)$ is absolutely integrable and by (7.17)

$$\int_{-\infty}^\infty \frac{e^{-itx}}{1 + t^2} dt = \pi e^{-|x|} \tag{a}$$

❑ **EXAMPLE 7.22.** *For the Cauchy distribution*

$$f(x) = \frac{1}{\pi(1 + x^2)}, \quad -\infty < x < \infty, \quad \psi(t) = \int_{-\infty}^\infty \frac{e^{itx}}{\pi(1 + x^2)}$$

To evaluate this consider the result (a) in example 7.19. Replacing t by $-t$ and interchanging t and x

$$\int_{-\infty}^{\infty} \frac{e^{itx}}{1+x^2} dx = \pi e^{-|t|}$$

Hence $\psi(t) = e^{-|t|}$.

$\psi(t)$ is not differentiable at the point $t = 0$. This shows that for the cauchy distribution no moment exists.

• **THEOREM 7.10.** *If a and b are real constants, then c.f. of $Y = aX + b$ is*

$$\psi_Y(t) = e^{itb}\psi_X(at)$$

In particular

$$\Psi_{\frac{X-\mu}{\sigma}} = e^{-it\mu/\sigma}\psi_X\left(\frac{t}{\sigma}\right).$$

❑ **EXAMPLE 7.23.** *Consider the normal distribution in example 7.19. The c.f. of*

$$Y = \frac{X-\mu}{\sigma} \text{ is } e^{-\frac{t^2}{2}} = \sum_{j=0}^{\infty} \frac{1}{j!}\left(-\frac{t^2}{2}\right)^j$$

Since the coefficient of $\dfrac{(it)^j}{j!}$ gives the jth moment, μ_j of distribution of Y is

$$\frac{j!}{2^{j/2}\left(\frac{j}{2}\right)!} \quad \text{if } j \text{ is even,}$$

$$0 \quad \text{if } j \text{ is odd.}$$

7.7. Exercises

1. *Find the pgf of the r.v. X when*

(a) $P(X = k) = \left[e^{-\mu}/(1-e^{-\mu})\right]\frac{\mu^k}{k!}, \quad k = 1, 2, \cdots, \quad \lambda > 0$

(b) $P(X = k) = pq^k/(1-q^{N+1}), \quad k = 0, 1, \cdots, N; \quad 0 < p < 1, q = 1-p.$

2. *Let X be the number of tosses required to obtain a head in tossing a unbiased coin. Find pgf of X.*

$\left[\frac{1}{2-t}, |t| < 1.\right.$

3. *In the familar matching problem with 2 decks of n cards each let X be the number of matches. Show that pgf of X is* $\displaystyle\sum_{k=0}^{n} \frac{(t-1)^k}{k!}$

4. (a) *If $P(t)$ is a pgf show that* $\dfrac{1}{2-P(t)}$ *is a pgf.*

(b) Show that $P(t) = \dfrac{2}{1+t}$ *is not a pgf even though* $P(1) = 1$.

[(a) $$Q(t) = \frac{1}{2 - P(t)} = \frac{1}{2}\left(1 - \frac{P(t)}{2}\right)^{-1}$$
$$= \frac{1}{2}\left[1 + \frac{1}{2}P(t) + \left(\frac{P(t)}{2}\right)^2 + \left(\frac{P(t)}{2}\right)^3 + \cdots\right]$$

$Q(1) = 1$, since $P(1) = 1$; also coefficient of $(P(t))^i$ is non-negative, $i = 1, 2, \cdots$

(b) Coefficient of t^i, $i = 1, 2, \cdots$ are not all non-negative]

5. ***Balls are distributed among m cells. Let Z be the number of balls that must be distributed to occupy all m cells. Find the pgf of Z.***

[$Z = m + X_1 + X_2 + \cdots + X_{m-1}$ where X_i is the number of balls following the occupancy of i cells for the first time and preceding the occupancy of $(i+1)$ cells for the first time.

X_1 follows a gemotric distribution with probability of success $p = \dfrac{m-1}{m}$. The *pgf* of X_1 is $\dfrac{(m-1)/m}{1 - \frac{t}{m}}$ (by example 7.2).

Similarly X_r follows a geometric distribution with $p = \dfrac{m-r}{m}$, $r = 2, 3, \cdots, m-1$.

Also $X_1, \cdots, X_{m-1}$ are independent. Hence *pgf* of Z is $t^m \prod_{r=1}^{m-1} \dfrac{(m-r)}{(m-tr)}$]

6. ***n balls are randomly distributed among m cells. Find the pgf of X, the number of unoccupied cells.***

[Let $X_i = 1(0)$ if the ith cell is empty (occupied); $X = \sum_1^m X_i$ is the number of unoccupied cells. By the notation of theorem 7.3, $B_i = \binom{m}{i}\left(\dfrac{m-i}{m}\right)^n$. Hence by (7.4)

$$P_X(t) = \sum_{i=0}^{m} \binom{m}{i}\left(\frac{m-i}{m}\right)^n (t-1)^i$$

$$\text{Hence} \quad P(X = k) = \sum_{i=0}^{m-k} (-1)^i \binom{k+i}{k}\binom{m}{k+i}\left(\frac{m-k-i}{m}\right)^n\Bigg]$$

7. ***For any integer-valued r.v. X, show that***

$$\sum_{n=0}^{\infty} t^n P(X \le n) = (1-t)^{-1} P(t), \text{ where } P(t) \text{ is the pgf of } X.$$

8. *Let X, Y be independent r.v.s each assuming the values $1, 2, \cdots, a$ with probability $\frac{1}{a}$. Find the pgf of $X - Y$ and hence determine the probability that X exceeds Y.*

[$P_{X-Y}(t) = E\left(t^{X-Y}\right) = E(t^X)E(t^{-Y})$, $(X, -Y)$ are independent), $E(t^X) = \frac{t(1-t^a)}{a(1-t)}$, $E(t^{-Y}) = \frac{1-t^a}{at^a(1-t)}$. Hence $P_{X-Y}(t) = \frac{1}{a^2t^{a-1}}\left(\frac{1-t^a}{1-t}\right)^2$

$$P(X > Y) = \sum_{r=1}^{a-1} \text{ coefficient of } t^r \text{ in } P_{X-Y}(t)$$

$$= \sum_{r=1}^{a-1} \text{ coefficient of } t^{r+a-1} \text{ in } \frac{1}{a^2}\left(\frac{1-t^a}{1-t}\right)^2 = \frac{1}{a^2}\sum_{r=1}^{a-1}(a-r) = \frac{1}{2} - \frac{1}{2a}.$$

9. *Find the pgf of the total score obtained when n balanced dice are rolled.*

[*Let X_i denote the score on the ith die ($i = 1, \cdots, n$), X_i has a uniform distribution over $1, \cdots, 6$ and its pgf is $P_i(t) = \frac{t(1-t^6)}{6(1-t)}$. Hence pgf of $S_n = \sum_{i=1}^{n} X_i$ is*

$$P(t) = \left[\frac{t(1-t^6)}{6(1-t)}\right]^n = \left(\frac{t}{6}\right)^n (1-t^6)^n(1-t)^{-n}$$

$$= \left(\frac{t}{6}\right)^n \left\{\sum_{i=0}^{n}\binom{n}{i}(-1)^i t^{6i}\right\}\left\{\sum_{0}^{\infty}\binom{-n}{j}(-1)^j t^j\right\}$$

$$= \left(\frac{t}{6}\right)^n \sum_{i=0}^{n}\sum_{j=0}^{\infty}\binom{n}{i}\binom{-n}{j}(-1)^{i+j}t^{6i+j}$$

$$= \left(\frac{t}{6}\right)^n \sum_{k=0}^{\infty} t^k \sum_{i}\binom{n}{i}\binom{-n}{k-6i}(-1)^{i+k}$$

where $k = j + 6i$.

Prob $(S_n = n + k)$ = *coefficient of* t^{n+k} *in* $P(t)$

$$= \frac{1}{6^n}\sum_{i}\binom{n}{i}\binom{-n}{k-6i}(-1)^{i+k} \qquad \text{(a)}$$

where the ith term in the summand in (a) is zero whenever $i > n$ or $6i > k$.]

10. *For any integer-valued r.v. X, show that*

$$\sum_{n=0}^{\infty} t^n P(X \le n) = (1-t)^{-1}P(t)$$

where $P(t)$ is the pgf of X.

11. *Let X, Y be independent variables with pgf $U(t)$, $V(t)$, respectively. Show that $P(X - Y) = j$ is the coefficient of t^j in $U(t)V\left(\frac{1}{t}\right)$, when $j = 0, \pm 1, \pm 2, \cdots$*

12. A r.v. X has *pmf*

$$P(X = x) = \frac{1}{x!}\left[1 - \frac{1}{1!} + \frac{1}{2!} \cdots \pm \frac{1}{(n-x)!}\right], \quad x = 0, 1, \cdots, n.$$

Show that *pgf* is $P(t) = \sum_{i=0}^{n} \frac{(t-1)^i}{i!}$. Hence find the mean, variance, factorial moments of the distribution of X.

13. Let $X_1, \cdots, X_n$ be *iid* r.v. with $P(X_i = k) = \frac{1}{a}$, $k = 0, 1, \cdots, a-1$; $i = 1, \cdots, n$. Let $X = X_1 + \cdots + X_n$. Show that *pgf* of X is $P_X(t) = \left\{\frac{1 - t^a}{a(1-t)}\right\}^n$ and hence $P\{X = j\}$

$$= \frac{1}{a^n}\sum_{u=0}^{\infty}(-1)^{u+j+au}\binom{n}{u}\binom{-n}{j - au}.$$

The probability $P\{X \le j\}$ has the *pgf* $\frac{P(t)}{1-t}$ and hence

$$P\{X \le j\} = \frac{1}{a^n}\sum_u (-1)^u \binom{n}{u}\binom{j - au}{n}.$$

If, further, $a \to \infty$, $j \to \infty$, such that $\frac{j}{a} \to x$, then

$$P\{X \le j\} \to \frac{1}{n!}\sum_u (-1)^u \binom{n}{u}(x - u)^n.$$

14. *Suppose X has pgf $P(t)$ and that $E(X^3)$ exists.*

Show that (i) $E(X^3) = P^{(3)}(1) + 3P^{(2)}(1) + P^{(1)}(1)$ (ii) $E(X^2) = P^{(2)}(1) + P^{(1)}(1)$, when $P^{(i)}(1)$ denotes the ith derivative of $P(t)$ at $t = 1$.

15. *Show that the function $E\left(\frac{1}{1 - tX}\right)$ can be used to generate the moments of the r.v. X.*

16. *Find the mgf, expectation and variance of X having pdf $f(x) = \frac{1}{2}e^{-\frac{x}{2}}$, $x > 0$*

$$\left[mgf = \frac{1}{1 - 2t}\right]$$

17. *Show that*

$$e^{itx} = 1 + (e^{it} - 1)x_{(1)} + (e^{it} - 1)^2\frac{x_{(2)}}{2!} + \cdots + (e^{it} - 1)^r\frac{x_{(r)}}{r!} + \cdots$$

where $x_{(r)} = x(x-1)\cdots(x - r + 1)$. *Hence show that* $\mu_{(r)} = \Delta^r\phi(t)\Big]_{t=0}$ where $\Delta = \frac{d}{d(e^t)}$, $\phi(t) = E(e^{itX})$.

[Let $e^{itx} = A_o + A_1 x_{(1)} + A_2 \frac{x_{(2)}}{2!} + \cdots + A_r \frac{x_{(r)}}{r!} + \cdots$ an identity in $\frac{x_{(i)}}{i!}$.

Putting $x = 0, 1, 2, \cdots$ and solving for $A_o, A_1, A_2, \cdots$, $A_o = 1$, $A_1 = e^{it} - 1$, $A_2 = (e^{it} - 1)^2, \cdots$. Hence

$$E(e^{itX}) = A_o + A_1 E(X_{(1)}) + \frac{A_2}{2!} E(X_{(2)}) + \cdots + \frac{A_r}{r!} E(X_{(r)}) + \cdots$$

$$= A_o + A_1 \mu_{(1)} + \frac{A_2}{2!} \mu_{(2)} + \cdots + \frac{A_r}{r!} \mu_{(r)} + \cdots$$

Hence $\Delta^r \phi(t)\Big]_{t=0} = \mu_{(r)}$.

18. *Define a factorial mgf and show that if the factorial mgf of a function $f(x)$ is $G(t)$, then the factorial mgf of $(-\nabla)^r f(x)$ is $t^r G(t)$ where $\nabla f(x) = f(x) - f(x-1)$.*

Factorial mgf, $G(t) = E\Big[(1+t)^X\Big] = \sum_x (1+t)^x f(x)$.

$$fmgf \text{ of } f(X-1) - f(X) = \sum_x (1+t)^x \{f(x-1) - f(x)\}$$

$$= (1+t) \sum_x (1+t)^{x-1} f(x-1) - \sum_x (1+t)^x f(x) = tG(t)$$

Now $\nabla^2 f(x) = \nabla(\nabla f(x)) = f(x) - 2f(x-1) + f(x-2)$.

$fmgf$ of $\nabla^2 f(x) = t^2 G(t)$.

Thus the result follows by induction.]

19. *With usual notations, prove that*

$$\frac{\partial \mu'_r}{\partial k_j} = \binom{r}{j} \mu'_{r-j}.$$

[We have,

$$\exp\left\{k_1 t + k_2 \frac{t^2}{2!} + \cdots + k_r \frac{t^r}{r!} + \cdots\right\} = 1 + \mu'_1 t + \mu'_2 \frac{t^2}{2!} + \cdots$$

Differentiating wrt k_j,

$$\frac{t^j}{j!} \exp\left\{k_1 t + k_2 \frac{t^2}{2!} + \cdots + k_r \frac{t^r}{r!} + \cdots\right\} = \sum_{r=0}^{\infty} \frac{\partial \mu'_r}{\partial k_j} \frac{t^r}{r!}$$

$$\text{or} \quad \frac{t^j}{j!} \left\{1 + \mu'_1 t + \mu'_2 \frac{t^2}{2!} + \cdots\right\} = \sum_{r=0}^{\infty} \frac{\partial \mu'_r}{\partial k_j} \frac{t^r}{r!}$$

The result follows equating the coefficient of t^r from both sides.]

20. *For the logistic distribution in exercise 9(f) of section 5.7, find the mgf. Hence show that $E(X) = \alpha$, $V(X) = \beta^2 \pi^2/3$.*

[Substituting $Y = \dfrac{X-\alpha}{\beta}$, the mgf of Y is $M_Y(t) = B(1-t, 1+t)$ where $B(p,q)$ is the Beta function with parameters p, q. Beta function has been discussed in section 9.4]

21. *For the power series distribution*

$$P\{X = j\} = a_j \frac{\theta^j}{f(\theta)}, \quad j = 0, 1, 2, \cdots$$

$\theta > 0$, *where* $a_j \geq 0$, $f(\theta) = \sum_{j=0}^{\infty} a_j \theta^j$, *find the* pgf *and* mgf

Show that, for this distribution

$$\mu'_{r+1} = \theta \frac{d\mu'_r}{d\theta} + \mu'_r \mu'_1$$

$$\theta \frac{d\mu_r}{d\theta} = \mu_{r+1} - r\theta \frac{d\mu'_1}{d\theta} \mu_{r-1}$$

$$\mu'_1 = \theta \frac{d \log f(\theta)}{d\theta}, \quad \mu_2 = \theta^2 \frac{d^2 \log f(\theta)}{d\theta^2} + \mu'_1$$

$$= \theta^2 \frac{f''(\theta)}{f(\theta)} - \theta^2 \left(\frac{f'(\theta)}{f(\theta)}\right)^2 + \theta \frac{f'(\theta)}{f(\theta)}.$$

22. *Show that for a symmetric distribution all the cumulants of odd order except the first vanish.*

[*If the distribution of* X *is symmetric about* a, *we trnasform from* X *to* $X - a$. *The* mgf *about* a *is*

$$M_{X-a}(t) = E\left(e^{t(X-a)}\right) = \left[1 + \mu_2 \frac{t^2}{2!} + \mu_4 \frac{t^4}{4!} + \cdots + \mu_{2r} \frac{t^{2r}}{2r!} + \cdots\right]$$

since all the existent moments of odd order about a *vanish.*

$$K_{X-a}(t) = \log\left[1 + \mu_2 \frac{t^2}{2!} + \cdots\right]$$

Hence $k_1(X) = a + k_1(X - a) = a$, $k_{2r-1}(X) = k_{2r-1}(X - a) =$ *coefficient of* $\dfrac{t^{2r-1}}{(2r-1)!} = 0$]

23. *A variate* X *has moments* $\mu'_r = r!$. *Find its characteristic function and deduce its distribution.*

[$\phi(t) = \frac{1}{1-it}; f(x) = e^{-x}, \quad 0 \leq x \leq \infty$]

24. *Show that the distribution for which the characteristic function is* $e^{-|t|}$ *has pdf* $f(x) = \dfrac{1}{\pi(1+x^2)}, \quad -\infty \leq x \leq \infty.$

8 SOME DISCRETE DISTRIBUTION ON R^1

8.1. Introduction

We consider in this chapter some special univariate discrete distributions and study their properties. The Discrete Uniform distribution is considered in section 8.2, Bernoulli distribution in 8.3, Binomial distribution in §.4 and Hypergeometric distribution in 8.5. Sections 8.6 through 8.8 consider respectively Poissön distribution, Geometric distribution and Negative Binomial distribution. Along with each distribution its truncated form has been considered. We conclude this chapter by a brief discussion on the Power-Series distribution of which all the above distributions come as special cases.

An inportant discrete distribution not considered here is multinomial distribution which being a multivariate distribution, would be taken up in chapter 10.

8.2. The Discrete Uniform Distribution

A r.v. X is said to have a discrete uniform distribution over the range $[1, N]$ if its *pmf* is

$$P(X = x) = \begin{cases} \dfrac{1}{N} & \text{for } x = 1, 2, \cdots, N \\ 0 & \text{otherwise.} \end{cases} \tag{8.1}$$

N is the parameter of the distribution and lies in the set of all positive integers. (8.1) is also called a discrete rectangular distribution.

$$E(X) = \frac{1}{N}\sum_{1}^{N} i = \frac{N+1}{2}, \quad E(X^2) = \frac{1}{N}\sum_{1}^{N} i^2 = \frac{(N+1)(2N+1)}{6}$$

$$V(X) = E(X^2) - \left(E(X)\right)^2 = \frac{(N+1)(N-1)}{12}$$

The pgf is

$$P_X(t) = E(t^X) = \frac{1}{N}\sum_{1}^{N} t^x = \frac{t(1-t^N)}{N(1-t)}, \quad t \neq 1 (= 1 \text{ if } t = 1).$$

The mgf is

$$M_X(t) = E(e^{tX}) = \frac{1}{N}\sum_{1}^{N} e^{ti} = \frac{e^t(1-e^{Nt})}{N(1-e^t)}.$$

When X is distributed uniformly over the range $[0, N]$, its pmf is

$$P[X = x] = \begin{cases} \dfrac{1}{N+1}, & x = 0, 1, \cdots, N \\ 0 & \text{otherwise} \end{cases}$$

Here the pgf is

$$P_X(t) = \frac{1}{N+1} \cdot \frac{1-t^{N+1}}{1-t}, t \neq 1 (= 1, \text{ if } t = 1).$$

The mgf is

$$M_X(t) = \frac{1}{N+1} \cdot \frac{e^{(N+1)t} - 1}{e^t - 1}.$$

❑ **EXAMPLE 8.1.** *Let $X_i (i = 1, \cdots, n)$ be mutually independent r.v.'s each having rectangular distribution $P(X_i = x) = \frac{1}{N}$, $x = 1, \cdots, N$, $i = 1, \cdots, n$. Find the distribution of M_n and N_n where $M_n = \min_{i=1,\cdots,n} X_i$, and $N_n = \max_{i=1,\cdots,n} X_i$.*

$$P(M_n = r) = P(X_i \geq r, \quad i = 1, \cdots, n) - P(X_i \geq r+1, \quad i = 1, \cdots, n)$$
$$= \left(\frac{N-r+1}{N}\right)^n - \left(\frac{N-r}{N}\right)^n = \frac{(N-r+1)^n - (N-r)^n}{N^n}$$

Similarly,

$$P(N_n = r) = P(X_i \leq r, \quad i = 1, \cdots, n) - P(X_i \leq r-1, \quad i = 1, \cdots, n)$$
$$= \left(\frac{r}{N}\right)^n - \left(\frac{r-1}{N}\right)^n = \frac{r^n - (r-1)^n}{N^n}.$$

8.3. The Bernoulli Distribution

A r.v. X is said to have a Bernoulli distribution with parameter p if its pmf is given by

$$P(X = x) = \begin{cases} p^x(1-p)^{1-x} & \text{for } x = 0, 1 \\ 0 & \text{otherwise} \end{cases} \tag{8.2}$$

The parameter p satisfies $0 \leq p \leq 1$. Often $(1-p)$ is denoted as q.

$$E(X) = 1 \cdot p + 0 \cdot q = p, \quad E(X^2) = p, \quad V(X) = pq$$

The pgf is

$$P_X(t) = pt + q$$

The mgf is

$$M_X(t) = q + pe^t$$

A random experiment whose outcomes are of two types, success S and failure F, occuring with probabilities p and q respectively is called a Bernoulli trial. If for this experiment, a r.v. X is defined such that it takes value 1 when S occurs and 0 if F occurs, then X follows a Bernoulli distribution.

❑ **EXAMPLE 8.2.** *For an arbitrary probability space $(\mathcal{S}, \mathcal{A}, P)$ and for an event $A \in \mathcal{A}$ consider the indicator r.v. $I_A(w) = 1(0)$ if $w \in A(\notin A)$. Then $P(I_A = 1) = P(w \in A) = P(A)$. I_A follows a Bernoulli distribution with parameter $p = P(A)$.*

8.4. The Binomial Distribution

A r.v. X is said to have a Binomial distribution with parameters n and p if its pmf is given by

$$P(X = x) = b(x; n, p) = \begin{cases} \binom{n}{x} p^x q^{n-x} & \text{for } x = 0, 1, \cdots, n \\ 0 & \text{otherwise,} \end{cases} \tag{8.3}$$

n ranges over the set of positive integers and p satisfies $0 \leq p \leq 1$. The probabilities in (8.3) are the terms of binomial expansion of $(q+p)^n$. For this reason the distribution is called a binomial distribution.

For $n = 1$, (8.3) reduces to the Bernoulli distribution. For this reason, sometimes Bernoulli distribution is called a point binomial distribution.

A binomial distribution arises in the following circumstances. Consider n independent Bernoulli trials each having probability of success p. Let X be the r.v. denoting the number of successes in these n trials. The sample space for such an experiment is given by

$$\mathcal{S} = \{(t_1, \cdots, t_n); t_i = S \text{ or } F, i = 1, \cdots, n\},$$

t_i denoting the result of the ith trial. Since the trials are independent, the probability of any specified outcome, say, $(S, S, F, F, \cdots, S)$ is $p.p.q.q.\cdots p$. Hence $P(X = x) = P$ [there are x successes and $(n-x)$ failures] $= \binom{n}{x} p^x q^{n-x}$, since any outcome with x successes and $(n-x)$ failures has probability $p^x q^{n-x}$ and there are $\binom{n}{x}$ such mutually exclusive outcomes. Hence X follows a binomial distribution.

❑ **EXAMPLE 8.3.** *An urn contains a white balls and b black balls. From this n balls are sampled with replacement. Find the probability that the sample contains x white balls.*

Since sampling is with replacement probability that a white ball is selected at any particular trial is $\frac{a}{a+b}$. We have therefore a series of n independent bernoulli trials with probability of success $\frac{a}{a+b}$ at each trial. Hence X, the number of white balls in the sample follows a binomial distribution with parameters n and $\left(\frac{a}{a+b}\right)$, when

$$P(X = x) = \binom{n}{x}\left(\frac{a}{a+b}\right)^x\left(\frac{b}{a+b}\right)^{n-x}, \quad x = 0, 1, \cdots, n$$

Considering (8.3),

$$\sum_{x=0}^{n} P(X = x) = \sum_{x=0}^{n}\binom{n}{x} p^x q^{n-x} = (p+q)^n = 1.$$

The rth factorial moment of X is

$$\mu_{(r)} = E[(X)_r] = \sum_{x=0}^{n}(x)_r\binom{n}{x} p^x q^{n-x} = (n)_r p^r \sum_{x'=0}^{n-r}\binom{n-r}{x'} p^{x'} q^{n-r-x'}$$

where $x' = x - r$

$$= (n)_r p^r \quad (p+q)^{n-r} = (n)_r p^r.$$

Thus

$$\mu_1' = np$$

$$\mu_2' = \mu_{(2)} + \mu_{(1)} = np + n(n-1)p^2$$

$$\mu_3' = \mu_{(3)} + 3\mu_{(2)} + \mu_{(1)} = np + 3n(n-1)p^2 + n(n-1)(n-2)p^3$$

$$\mu_4' = \mu_{(4)} + 6\mu_{(3)} + 7\mu_{(2)} + \mu_{(1)}$$
$$= np + 7n(n-1)p^2 + 6n(n-1)(n-2)p^3 + n(n-1)(n-2)(n-3)p^4.$$

The central moments are

$$\mu_2 = npq \quad (< \mu_1'), \quad \mu_3 = npq(q-p), \quad \mu_4 = 3(npq)^2 + npq(1-6pq),$$

$$\sqrt{\beta_1} = \frac{q-p}{\sqrt{npq}}, \quad \beta_2 = 3 + \frac{1-6pq}{npq}, \quad \beta_2 - \beta_1 - 3 = -\frac{2}{n}.$$

Thus, the skewness of the distribution is positive (negative) if $p < (>)\frac{1}{2}$. The distribution is symmetrical if and only if $p = \frac{1}{2}$. For a fixed value of p the point (β_1, β_2) falls on the straight line

$$\frac{\beta_2 - 3}{\beta_1} = 1 - \frac{2pq}{(q-p)^2}.$$

As $n \to \infty$, p remaining finite, the point approaches the point (0, 3) as a limit i.e. the distribution tends to be symmetrical and mesokurtic ($\beta_2 = 3$). Figures 8.1(a) and 8.1(b) represent binomial distributions with parameters (10, .35) and (10, .45) respectively.

The following recurrence relation holds for the moments.

Lemma 8.1. For the binomial distribution,

$$\mu_{r+1} = pq\left(nr\mu_{r-1} + \frac{d\mu_r}{dp}\right) \tag{8.4}$$

Proof. $\mu_r = \sum_{j=0}^{n} \binom{n}{j} p^j q^{n-j}(j-np)^r$

$$\frac{d\mu_r}{dp} = \sum \binom{n}{j} jp^{j-1}q^{n-j}(j-np)^r - \sum (n-j)\binom{n}{j} p^j q^{n-j-1}(j-np)^r$$
$$- nr\sum \binom{n}{j} p^j q^{n-j}(j-np)^{r-1}$$

$$= \sum \binom{n}{j} p^{j-1}q^{n-j-1}(j-np)^r(jq - np + jp) - nr\mu_{r-1}$$

Hence

$$pq\frac{d\mu_r}{dp} = \sum \binom{n}{j} p^j q^{n-j}(j-np)^{r+1} - npqr\mu_{r-1}$$

when $\mu_{r+1} = pq\left(\frac{d\mu_r}{dp} + nr\mu_{r-1}\right)$.

A similar relation holds for the non-central moments :

$$\mu'_{r+1} = pq\left(\frac{n}{q}\mu'_r + \frac{d\mu'_r}{dp}\right) \tag{8.5}$$

The mean deviation about mean is

$$E[|X - np|] = n\binom{n-1}{[np]} p^{[np]+1} q^{n-[np]} \tag{8.6}$$

where $[\theta]$ denotes the highest integer contained in θ.

The ratio

$$\frac{b(k+1;n,p)}{b(k;n,p)} = \frac{n-k}{k+1}\cdot\frac{p}{q}, \quad k = 0, 1, \cdots, n-1$$

shows that $b(k+1;n,p) > (<) b(k;n,p)$ if $k < (>)(n+1)p-1$. Hence as k increases the quantities $P(X=k)$ increases to a maximum value at the integer k^* satisfying

$$(n+1)p-1 < k^* < (n+1)p,$$

provided $(n+1)p$ is not an integer, and thereafter they decrease. k^* is the mode of the distribution.

If $(n+1)p$ is an integer,

$$P[X=(n+1)p-1] = P[X=(n+1)p]$$

and this common value is the maximum among the values of $P[X=k]$. Here mode is at $X=(n+1)p-1$ and $X=(n+1)p$.

The *pgf* is

$$P_X(t) = E(t^X) = (q+pt)^n, \quad |t| \leq 1 \qquad \text{(vide example 7.4)}$$

The *mgf* about origin is

$$M_X(t) = E(e^{tX}) = (q+pe^t)^n, \quad |t| < \infty \qquad \text{(vide example 7.10)}.$$

The *mgf* about mean is

$$M_{X-\mu_1'}(t) = (qe^{-pt} + pe^{tq})^n.$$

The *cgf* about mean is

$$\begin{aligned} K_{X-\mu_1'}(t) &= n\log(q+pe^t)^n \\ &= n\log\left[1 + pq\frac{t^2}{2!} + pq(q-p)\frac{t^3}{3!} + pq(q^2-pq+p^2)\frac{t^4}{4!} + \cdots\right] \end{aligned}$$

Equating the coefficients of $\frac{t^2}{2!}, \frac{t^3}{3!}, \cdots$, on either side we obtain, after some simplifications,

$$k_2 = npq, \quad k_3 = npq(q-p), \quad k_4 = npq(1-6pq)$$

The factorial *mgf*

$$= E[(1+t)^X] = \sum_0^n (1+t)^x \binom{n}{x} p^x q^{n-x} = (1+pt)^n.$$

The rth factorial moment is

$$\mu_{(r)} = (n)_r p^r$$

• **THEOREM 8.1.** *Let X_i $(i = 1, \cdots, k)$ be independent Binomial variables with parameters (n_i, p). Then $S_k = \sum_{i=1}^{k} X_i$ has a Binomial distribution with parameters (n, p) where $n = \sum_1^k n_i$.*

Proof. The mgf of S_k

$$= P_{S_k}(t) = \prod_{i=1}^{k} P_{X_i}(t) = \prod_{i=1}^{k} (q + pe^t)^{n_i} = (q + pe^t)^n$$

which is the mgf of a binomial r.v. with parameters (n, p). The result follows from the uniqueness theorem of mgf (Theorem 7.5).

For large n, the calculation of values of the cumulative probabilities $F_X(k) = P(X \le k)$ is facilitated by the use of K. Pearson's Tables of the Incomplete Beta Function.

$$\sum_{x=0}^{k} b(x,; n, p) = \sum_{x=0}^{k} \binom{n}{x} p^x q^{n-x}$$

$$= \frac{n!}{k!(n-k-1)!} \sum_{x=0}^{k} \binom{k}{x} \frac{(n-k-1)!(k-x)!}{(n-x)!} p^x q^{n-x}$$

$$= \frac{1}{B(n-k, k+1)} \sum_{x=0}^{k} \binom{k}{x} B(n-k, k-x+1) p^x q^{n-x},$$

since $B(a, b) = \dfrac{(a-1)!(b-1)!}{(a+b-1)!}$ for integral values of a and b [The Beta function $B(a, b)$ has been discussed in section 9.4].

$$= \frac{1}{B(n-k, k+1)} \sum_{x=0}^{k} \binom{k}{x} \left[\int_0^1 y^{n-k-1} (1-y)^{k-x} dy \right] p^x q^{n-x}$$

$$= \frac{1}{B(n-k, k+1)} \int_0^1 y^{n-k-1} q^{n-k} \left[\sum_{x=0}^{k} \binom{k}{x} p^x \{q(1-y)\}^{k-x} \right] dy$$

(by changing the order of summation and integration)

$$= \frac{1}{B(n-k, k+1)} \int_0^1 (qy)^{n-k-1} (1-qy)^k d(qy)$$

$$= \frac{\int_0^q z^{n-k-1} (1-z)^k dz}{B(n-k, k+1)} \quad \text{where } z = qy$$

$$= I_q(n-k, k+1) = 1 - I_p(k+1, n-k).$$

The function $I_x(p, q)$ has been tabulated by K. Pearson in his Tables of Incomplete Beta function.

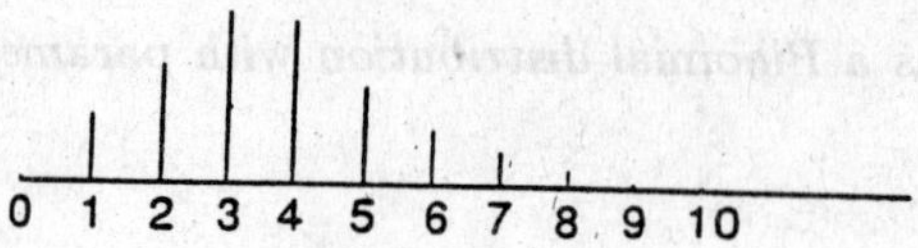

n = 10, p = .35
Fig. 8.1(a)

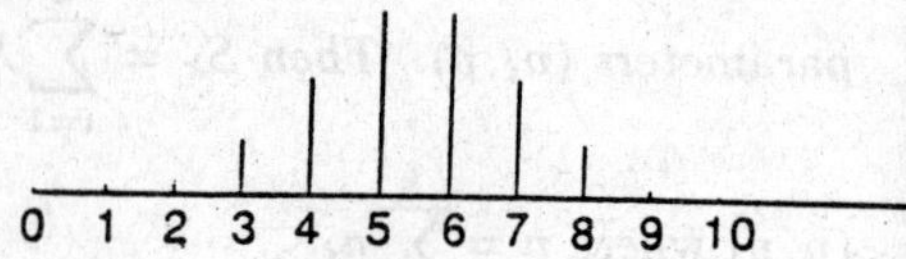

n = 10, p = .55
Fig. 8.1(b)

Binomial Probabilities

Consider two independent variables X_1, X_2 distributed as $b(x; n_1, p)$ and $b(x; n_2, p)$ respectively. The conditional probability $X_1 = k$ given that $X_1 + X_2 = x$ is

$$P[X_1 = k | X_1 + X_2 = x] = \frac{P(X_1 = k, X_2 = x - k)}{P(X_1 + X_2 = x)}$$

$$= \frac{\binom{n_1}{k} p^k q^{n_1 - k} \binom{n_2}{x-k} p^{x-k} q^{n_2 - x + k}}{\binom{n_1 + n_2}{x} p^x q^{n_1 + n_2 - x}}$$

$$= \frac{\binom{n_1}{k}\binom{n_2}{x-k}}{\binom{n_1+n_2}{x}}, \tag{8.7}$$

$\max(0, x - n_2) \le k \le \min(n_1, x)$. (8.7) is known as a hypergeometric distribution (discussed in section 8.4).

The distribution of the difference $(X_1 - X_2)$ is

$$P(X_1 - X_2 = k) = \sum_{x_1} \binom{n_1}{x_1}\binom{n_2}{x_1 - k} p^{2x_1 - k} q^{n_1 + n_2 - (2x_1 - k)} \tag{8.8}$$

where the summation is over the limits, $\max(0, k) \le x_1 \le \min(n_1, n_2 + k)$. When $p = q = \frac{1}{2}$, (8.8) reduces to

$$P(X_1 - X_2 = k) = \frac{\binom{n_1 + n_2}{n_2 + k}}{2^{n_1 + n_2}} (-n_2 \le k \le n_1) \tag{8.9}$$

The distribution of the standardised binomial variate

$$X' = \frac{X - np}{\sqrt{npq}}$$

tends to the unit normal distribution as $n \to \infty$. This has been discussed in section 9.6.

If $n \to \infty$ and $p \to 0$ in such a way that $np = \lambda$ remains constant, then

$$b(x; n, p) \to e^{-\lambda}\frac{\lambda^x}{x!}$$

This approximation has been discussed in lemma 8.3.

8.4.1. The Truncated Binomial Distribution

A singly truncated binomial distribution is formed if the binomial distribution is truncated at only one end i.e. if only the values $0, 1, \cdots, K_1 - 1 (K_1 \geq 1)$ or the values $n - K_2 + 1, \cdots, n$ $(K_2 \geq 1)$ are omitted from the values of a binomial variable. The distribution obtained by omitting the value 0 only , is

$$P(X = r) = \frac{\binom{n}{r} p^r q^{n-r}}{1 - q^n}, \quad r = 1, \cdots, n \tag{8.10}$$

The jth raw moment of the truncated distribution (8.10) is obtained by dividing the corresponding moment of the complete distribution (8.3) by $1 - q^n$. In particular, its mean and variance are $\dfrac{np}{1 - q^n}$ and $\dfrac{npq}{1 - q^n} - \dfrac{n^2 p^2 q^n}{(1 - q^n)^2}$ respectively.

A binomial distribution truncated on the left at $x = K$ is

$$\frac{\binom{n}{x} p^x q^{n-x}}{\sum_K^n \binom{n}{x} p^x q^{n-x}} = \frac{1}{H_K(z)}\binom{n}{x} z^x, \qquad x = K, K+1, \cdots, n \tag{8.11}$$

where $z = \dfrac{p}{q}$, $H_k(z) = \sum_K^n \binom{n}{x} z^x$.

Mean of the distribution is

$$\mu_1' = \frac{1}{H_K(z)} \sum_K^n x \binom{n}{x} z^x = \frac{z H_K'(z)}{H_K(z)} \tag{8.12}$$

For the distribution (8.11) the recurrence relation (8.4) holds i.e.

$$\mu_{r+1} = pq\left(nr\mu_{r-1} + \frac{d\mu_r}{dp} \right). \tag{8.13}$$

Proof. $\mu_r = \dfrac{1}{H_K(z)} \sum_K^n (x - \mu_1')^r \binom{n}{x} z^x$

$$z\frac{d\mu_r}{dz} = \frac{d}{dz}\left\{ \sum_K^n (x - \mu_1')^r \binom{n}{x} z^x \right\} \frac{z}{H_K(z)} - \left\{ \sum_K^n (x - \mu_1')^r \binom{n}{x} z^{x+1} \right\} \frac{H_K'(z)}{(H_K(z))^2}$$

$$= \frac{1}{H_K(z)} \sum_K^n \binom{n}{x} (x - \mu_1')^r x z^x - zr\frac{d\mu_1'}{dz} \frac{1}{H_K(z)} \sum_K^n \binom{n}{x} z^x (x - \mu_1')^{r-1}$$

$$- \frac{z H_K'(z)}{H_K(z)} \cdot \frac{1}{H_K(z)} \sum_K^n (x - \mu_1')^r \binom{n}{x} z^x \tag{8.14}$$

The first and the last term together on the rhs of (8.14) give

$$\frac{1}{H_K(z)}\sum_{K}^{n}\binom{n}{z}z^x(x-\mu_1')^{r+1}=\mu_r+1,\quad \text{using (8.12)}$$

Hence (8.14) reduces to

$$\mu_{r+1}=z\left[\frac{d\mu_r}{dz}+r\frac{d\mu_1'}{dz}\mu_{r-1}\right]$$

(8.13) follows, using $\dfrac{dz}{dp}=\dfrac{1}{q^2}$.

The relation (8.5) holds similarly for the distribution (8.11).

A doubly truncated binomial variable is formed from (8.3) by omitting the values of x such that $0 \le x < K_1$, and $x - K_2 < x < n$ (with $0 \le K_1 \le n - K_2 \le n$). For the resulting distribution

$$P(X=j)=\frac{\binom{n}{j}p^j q^{n-j}}{\sum_{r=K_1}^{n-K_2}\binom{n}{r}p^r q^{n-r}},\quad j=K_1,K_1+1,\cdots,n-K_2.$$

8.5. The Hypergenometric Distribution

A r.v. X is said to have a hypergeometric distribution with parameters N, M and n if its *pmf* is given by

$$P(X=x)=H(x;N,M,n)(\text{ say})=\begin{cases}\dfrac{\binom{M}{x}\binom{N-M}{n-x}}{\binom{N}{n}} & \text{for } \max(0,n-N+M)\\ & \le x \text{ (integer)} \le \min(M,n)\\ 0 & \text{otherwise}\end{cases}\tag{8.15}$$

where N is a positive integer, M is a positive integer not exceeding N and n is a positive integer that is at most N.

The formula in (8.15) is valid for all integer values of x such that, for each symbol of the form $\binom{a}{b}$ in (8.15), $a \ge b$. This means that

$$0 \le x \le M \text{ and } 0 \le (n-x) \le N-M;$$

that is,

$$\max(0,n-N+M) \le x \le \min(M,n).$$

Note that the same distribution is obtained if n and M are interchanged.

The quantities in the rhs of (8.15) are successive terms in the expansion of

$$\frac{(N-n)!(N-M)!}{N!(N-M-n)!}F(-n,-M,N-M-n+1,1)$$

where

$$F(\alpha,\beta,\gamma,t) = 1 + \frac{\alpha\beta}{\gamma}\frac{t}{1!} + \frac{\alpha(\alpha+1)\beta(\beta+1)}{\gamma(\gamma+1)}\frac{t^2}{2!} + \cdots$$

is a hypergeometric function. That is why the distribution is called hypergeometric.

The distribution (8.15) arises in the following circumstances. Suppose an urn contains M white balls and $N-M$ black balls. From this n balls are selected without replacement. The probability that the sample contains x white balls and $(n-x)$ black balls is given by (8.15). The hypergeometric distribution, therefore; arises in sampling from a dichotomous population (a population consisting only of two categories of objects) when sampling is without replacement whereas a binomial distribution arises when sampling is with replacement (vide example 8.3).

We have

$$H(x;N,M,n) = \binom{n}{x}\frac{(M)_x(N-M)_{(n-x)}}{(N)_n} \quad \text{for } x = 0,1,\cdots,n.$$

The rth factorial moment about origin is

$$\begin{aligned}
\mu_{(r)} &= \sum_0^n (x)_r \frac{\binom{n}{x}(M)_x(N-M)_{(n-x)}}{(N)_n} \\
&= \frac{1}{(N)_n}\sum_{y=0}^{n-r}(n-y)_r\binom{n}{y}(M)_{(n-y)}(N-M)_y \quad \text{writing } n-x=y \\
&= \frac{(M)_r(n)_r}{(N)_n}\sum_{y=0}^{n-r}\binom{n-r}{y}(M-r)_{(n-r-y)}(N-M)_y \\
&= \frac{(M)_r(n)_r}{(N)_n}\quad (N-r)_{(n-r)} \quad \left[\because \sum_{x=0}^{n} H(x;N,M,n) = 1\right] \\
&= \frac{(M)_r(n)_r}{(N)_r}.
\end{aligned}$$

Hence

$$E(X) = \frac{Mn}{N} = np \left(\text{writing } p = \frac{M}{N}, q = 1-p\right)$$

$$V(x) = \frac{N-n}{N-1}npq$$

$$\sqrt{\beta_1} = \frac{q-p}{\left[\left(\frac{N-n}{N-1}\right)npq\right]^{1/2}} \frac{N-2n}{N-2}$$

$$\beta_2 = \frac{3(N-1)(N+6)}{(N-2)(N-3)} + \frac{N(N^2-1)}{(N-n)(N^2-5N+6)} \left[1 - \frac{6N}{N+1}\left(pq + \frac{n(N-n)}{N^2}\right)\right] \cdot \frac{1}{npq}$$

The following recurrence relations hold :

$$H(x+1; N, M, n) = \frac{(M-x)(n-x)}{(x+1)(N-M-n+x+1)} H(x; N, M, n) \tag{8.16}$$

$$H(x; N, M+1, n) = \frac{(M+1)(N-M-n+x)}{(N-M)(M+1-x)} H(x; N, M, n) \tag{8.17}$$

$$H(x; N, M, n+1) = \frac{(N-M-n+x)(n+1)}{(n+1-x)(N-n)} H(x; N, M, n) \tag{8.18}$$

$$H(x; N+1, M, n) = \frac{(N+1-n)(N+1-M)}{(N+1-n-M+x)(N+1)} H(x; N, M, n) \tag{8.19}$$

From (8.16) it follows that $H(x+1; N, M, n) \gtrless H(x; N, M, n)$ according as

$$x \lessgtr \frac{(n+1)(M+1)}{N+2} - 1.$$

Hence the distribution has mode at k^* where k^* is the greatest integer which does not exceed $\frac{(n+1)(M+1)}{N+2}$. If $\frac{(n+1)(M+1)}{N+2}$ is an integer, the distribution is binomial having modes at

$$x = \frac{(n+1)(M+1)}{N+2} - 1 \text{ and } x = \frac{(n+1)(M+1)}{N+2}.$$

The following binomial approximation to a hypergeometric distribution holds.

$$H(x; N, M, n) \simeq \binom{n}{x} p^x (1-p)^{n-x} \tag{8.20}$$

where $p = \frac{M}{N}$, when $N \to \infty$, provided $\frac{n}{N}$ is negligible.

If $\frac{M}{N}$ is small but n is large, the Poisson approximation

$$H(x; N, M, n) \simeq \frac{(np)^x}{x!} e^{-np} \tag{8.21}$$

where $p = \frac{M}{N}$, holds.

8.51. The Positive Hypergeometric distribution

This is a truncated hypergeometric distribution where the zero observed value is omitted. However, if $n > (N-M)$, there is no zero observed value. If $n \leq (N-M)$, the distribution is

$$P(X=x) = \frac{\binom{M}{x}\binom{N-M}{n-x}}{\left[\binom{N}{n} - \binom{N-M}{n}\right]} \qquad x = 1, \cdots, \min(n, M).$$

8.5.2. The Negative Hypergeometric distribution

This distribution arises in the following circumstances. Suppose we have b black balls and w white balls in an urn. Balls are drawn one a time, without replacement. We are interested in the probability distribution of the number R of draws needed to obtain a specific number k of black balls.

$$\begin{aligned} P(R=r) &= P\big((k-1) \text{ black balls are drawn in the first } (r-1) \text{ draws}\big) \times \\ &\qquad P\big(\text{a black ball is drawn from an urn containing } (b-k+1) \\ &\qquad\qquad \text{black balls and } (w-r+k) \text{ white balls}\big) \\ &= \binom{r-1}{k-1} \frac{b(b-1)\cdots(b-k+2)w(w-1)\cdots(w-r+k+1)}{(b+w)(b+w-1)\cdots(b+w-r+2)} \cdot \\ &\qquad\qquad \frac{b-k+1}{b+w-r+1} \\ &= \binom{r-1}{k-1} \frac{(b)_k (w)_{r-k}}{(b+w)_{(r)}}, \quad r = k, k+1, \cdots \end{aligned}$$

The distribution of R is a negative hypergeometric distribution with parameters b, w, k.

The jth ascending factorial moment of R is

$$E[R(R+1)\cdots(R+j-1)] = \frac{k^{(j)}(b+w+1)^{(j)}}{(b+1)^{(j)}}$$

where $n^{(t)} = n(n+1)\cdots(n+t-1)$.

In particular

$$E(R) = \frac{k(b+w+1)}{b+1}, \quad V(R) = \frac{k(b+w+1)w(b+1-k)}{(b+1)^2(b+2)}.$$

8.6. The Poisson Distribution

A r.v. X is said to have a Poisson distribution with parameter $\lambda(>0)$ if its *pmf* can be written as

$$P(X=x) = P(x;\lambda) = \begin{cases} e^{-\lambda}\dfrac{\lambda^x}{x!}, & x = 0, 1, 2, \cdots \\ 0 & \text{otherwise} \end{cases}$$

The pgf is $P_X(t) = e^{-\lambda(1-t)}$, $|t| \leq 1$ **(vide example 7.3)**

$$\mu_1' = \sum_0^\infty x e^{-\lambda} \frac{\lambda^x}{x!} = \lambda e^{-\lambda}\left(1 + \frac{\lambda}{1!} + \frac{\lambda^2}{2!} + \cdots\right) = \lambda e^{-\lambda} e^{\lambda} = \lambda$$

$$\mu_2' = \mu_1' + \mu_{(2)} = \lambda + \sum_0^\infty x(x-1)e^{-\lambda}\frac{\lambda^x}{x!} = \lambda + \lambda^2$$

Hence $\mu_2 = \lambda$.

The mgf is about zero is

$$M_X(t) = e^{\lambda(e^t - 1)} \quad \text{(vide example 7.11)}$$

Hence

$$\mu_3' = \lambda + 3\lambda^2 + \lambda^3, \quad \mu_4' = \lambda + 7\lambda^2 + 6\lambda^3 + \lambda^4, \quad \mu_3 = \lambda, \quad \mu_4 = \lambda + 3\lambda^2,$$
$$\sqrt{\beta_1} = \lambda^{-\frac{1}{2}}, \quad \beta_2 = 3 + \frac{1}{\lambda}$$

Thus for any Poisson distribution,

$$\mu_1' = \mu_2 \quad \text{and} \quad \beta_2 - \beta_1 - 3 = 0.$$

Lemma 8.2. The central moments of a Poisson distribution satisfy the following recurrence relation

$$\mu_{r+1} = \lambda\left(r\mu_{r-1} + \frac{d\mu_r}{d\lambda}\right) \tag{8.22}$$

Proof. $\mu_r = \sum_{x=0}^\infty (x-\lambda)^r e^{-\lambda}\frac{\lambda^x}{x!}$

$$\frac{d\mu_r}{d\lambda} = \sum_{x=0}^\infty \left[-r(x-\lambda)^{r-1} + (x-\lambda)^r\left(\frac{x}{\lambda} - 1\right)\right] e^{-\lambda}\frac{\lambda^x}{x!} = -r\mu_{r-1} + \frac{1}{\lambda}\lambda_{r+1}.$$

Hence the proof.

The ratio $\dfrac{P(x;\lambda)}{P(x-1;\lambda)} = \dfrac{\lambda}{x}$ **shows that $P(x;\lambda) > (<)P(x;\lambda)$ according as $x < (>)\lambda$. Hence, if λ is not an integer, the distribution has mode at the integral value x^* satisfying**

$$\lambda - 1 < x^* < \lambda.$$

If λ is an integer, $P[X = \lambda - 1] = P[X = \lambda]$ and the distribution has modes at $x = \lambda - 1$ and $x = \lambda$.

The mean deviation of the Poisson distribution is

$$E[|X - \lambda|] = \frac{2e^{-\lambda}\lambda^{[\lambda]+1}}{[\lambda]!}$$

The figures 8.2(a), (b) and (c) represent Poisson distributions with $\lambda = .75$, 1.5 and 3 respectively.

We have seen in example 7.16, if $X_1, \cdots, X_n$ are n independent Poisson variates, X_i being Poisson with parameter λ_i, the sum $S_n = X_1 + \cdots + X_n$ has mgf, $M_{S_n}(t) = e^{\Sigma_1^n \lambda_i (e^t - 1)}$. It, therefore, follows by the uniqueness property of mgf (theorem 7.5) that S_n follows a Poisson distribution with parameter $\sum_{i=1}^{n} \lambda_i$. This is called the reproductive property of the Poisson distribution.

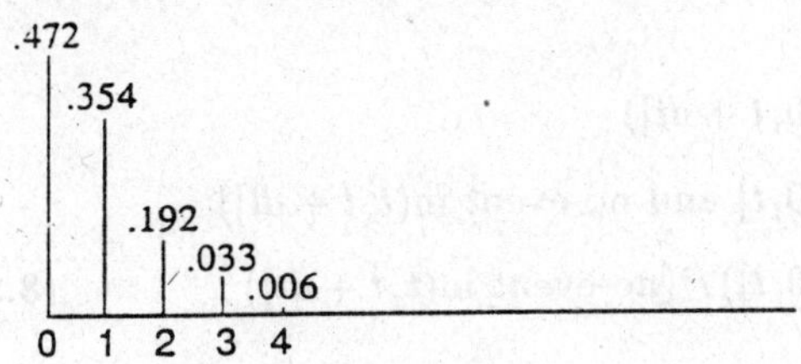

λ = .75

Fig. 8.2(a)

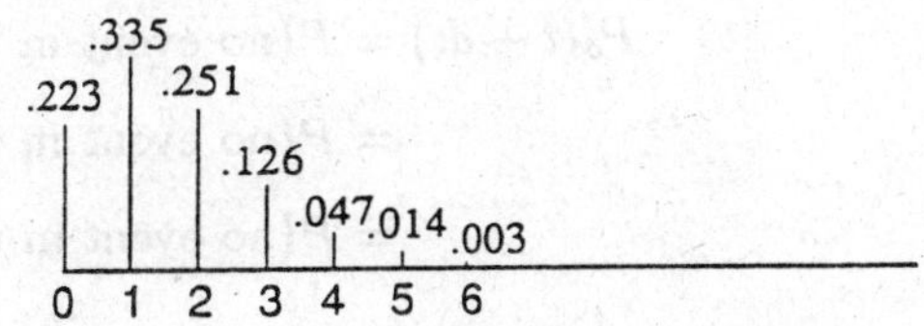

λ = 1.5

Fig. 8.2(b)

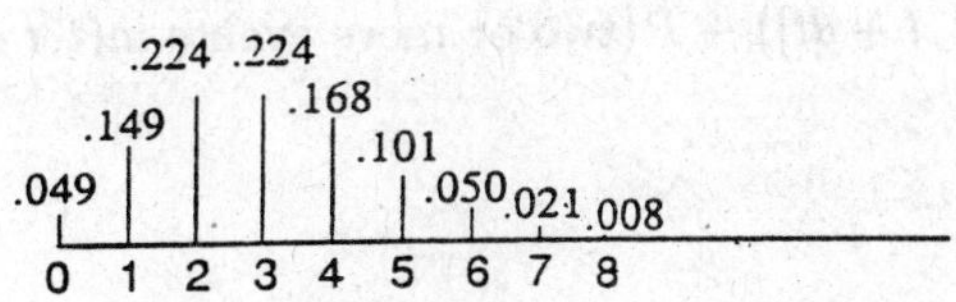

λ = 3

Fig. 8.2(c)

Poisson Probabilities

The Poisson distribution provides a realistic model for many random phenomena. Since the values of a Poisson variable are non-negative integers, the distribution may be used to describe experiments in which observed variable is a count. The Poisson distribution may also be used to describe events occurring randomly and independently in time (space) e.g. number of telephone calls comming into the telephone exchange in time of length t, number of fatal traffic accidents occurring per week in a given region, number of cars passing through a street in time t, number of α particles emitted by a radio-active substance reaching a given portion of sapce during time t, number of flying-bomb hits over a region divided into several small areas. In all such cases we may assume that the conditions of the experiment remain constant in time (space) and that the nonoverlapping intervals of time (space) are stochastically independent in the sense that the information concerning the number of events in one interval does not give any information about the other. These two assumptions provide the basis for a random variable to obey Poisson law.

More specifically let us assume the following :

(i) The probability that exactly one event occurs during a small time interval of length dt is $\lambda dt + o(dt)$ [$o(dt)$ denotes an unspecified function which is of smaller order than dt ; $\lim_{dt \to 0} \frac{o(dt)}{dt} = 0$].

(ii) The probability that more than one event occur during a small interval of length dt is $o(dt)$.

(iii) Number of occurances in non-overlapping time intervals are independent.

The quantity λ may be called the mean rate of occurance.

- **THEOREM 8.2.** *Let X be a r.v. denoting the number of occurances of an event in a time interval of length t. If the assumptions (i) - (iii) above are satisfied, X follows a Poisson distribution with parameter λt.*

Proof. Let $P_x(t)$= Probability that x events occur during the time interval $(0, t]$, $x = 0, 1, 2, \cdots$.

Hence

$$\begin{aligned} P_0(t+dt) &= P\big(\text{no event in } (0, t+dt]\big) \\ &= P\big(\text{no event in } (0,t] \text{ and no event in} (t, t+dt]\big) \\ &= P\big(\text{no event in } (0,t]\big) P\big(\text{no event in} (t, t+dt]\big) \end{aligned} \tag{8.23}$$

Now

$$\begin{aligned} &P\big(\text{no event in}(t, t+dt]\big) \\ &= 1 - P\big(\text{one event in}(t, t+dt]\big) - P\big(\text{two or more events in}(t, t+dt]\big) \\ &= 1 - \lambda dt - o(dt) \end{aligned}$$

Hence from (8.23)

$$P_0(t+dt) = P_0(t)(1 - \lambda dt - o(dt)) \quad \text{or} \quad \frac{P_0(t+dt) - P_0(t)}{dt} = -\lambda P_0(t) + \frac{o(dt)}{dt} P_0(t)$$

Taking limit as $dt \to 0$

$$P_0'(t) = -\lambda P_0(t)$$

The solution of this differential equation is

$$P_0(t) = e^{-\lambda t} \quad \text{since } P_0(0) = 1$$

For $x = 1, 2, \cdots$

$$\begin{aligned} P_x(t+dt) &= P\big(x \text{ events in } (0,t] \text{ and no event in } (t, t+dt]\big) \\ &\quad + P\big(x-1 \text{ events in } (0,t] \text{ and one event in } (t, t+dt]\big) \\ &\quad + \sum_{h=2}^{x} P\big(x-h \text{ events in } (0,t] \text{ and } h \text{ events in } (t, t+dt]\big) \\ &= P_x(t)(1 - \lambda dt - o(dt)) + P_{x-1}(t)(\lambda dt + o(dt)) + \sum_{h=2}^{x} P_{x-h}(t)(o(dt)) \end{aligned}$$

$$\therefore \quad P_x(t+dt) = P_x(t)(1-\lambda dt) + P_{x-1}(t)\lambda dt + o(dt)$$

$$\text{or} \quad \frac{P_x(t+dt)-P_x(t)}{dt} = -\lambda[P_x(t)-P_{x-1}(t)] + \frac{o(dt)}{dt}$$

Taking limit as $dt \to 0$

$$P_x'(t) = \lambda[P_{x-1}(t) - P_x(t)] \tag{8.24}$$

Hence, For $x = 1$

$$P_1'(t) + \lambda P_1(t) = \lambda e^{-\lambda t}$$

Multiplying both sides by $e^{\lambda t}$ and integrating, we obtain

$$e^{\lambda t} P_1(t)\lambda = t + K$$

where K is a constant.

The initial conditions $P_1(0) = 0$ implies

$$P_1(t) = \lambda t e^{-\lambda t} \tag{8.25}$$

Similarly $P_2(t)$ is found from (8.24) by using (8.25). Induction on n shows that

$$P_n(t) = e^{-\lambda t}\frac{(\lambda t)^n}{n!}.$$

❑ **EXAMPLE 8.4.** *A telephone switch board receives 20 calls on an average during an hour. Find the probability that during a period of 5 minutes (a) no call is received , (b) exactly 3 calls are received (c) more that 5 calls are received. Assume that the time is measured in minutes.*

We assume that the number of incoming calls during any time period obeys a Poisson law (which follows on the basis of the assumptions (i) – (iii) above). 20 calles per hour is equivalent to .33 calls per minute, which is the mean rate of occurance. Hence number of calls in a 5 minute period follows a Poisson distribution with parameter $\lambda = 1.65$.

(a) P (no call in a 5-minute period) $=e^{-1.65} = .192$

(b) P (3 calls in a 5-minute period)$= e^{-1.65}\dfrac{(1.65)^3}{3!} = .144$

(c) P (more than 5 calls in a 5 minute period)$=\displaystyle\sum_{x=6}^{\infty} e^{-1.65}\frac{(1.65)^x}{x!} = .007.$

❑ **EXAMPLE 8.5.** *A traffic light has a constant probability λdt of changing from a red light to a green light or from a green light to a red light in an infinitesimal interval of length dt. Prove that a car arriving at a random instant has a probability $\frac{1}{2}$ of passing through without waiting and a proabbility element $\frac{1}{2}\lambda e^{-\lambda w}$ of waiting for a time $w(>0)$ before passing.*

A car would pass at any moment with or without waiting. Since both the cases are mutually exclusive and equally likely the probability of each case is $\frac{1}{2}$.

The car waits for a time period w if (i) the traffic light is red when it arrives and (ii) does not change to green within the next time period w and (iii) thereafter (within time dt) changes. The three events are mutually independent and probability of (i) and (iii) are $\frac{1}{2}$ and λdt respectively.

Let $\frac{w}{dt} = n$. The chance that the traffic light does not change in time w is

$$(1-\lambda dt)^n = (1-\lambda dt)^{\frac{w}{dt}} = (1-x)^{\frac{w\lambda}{x}} \quad \text{where } x = \lambda dt,$$

which on taking limit as $dt \to 0$ (i.e. $x \to 0$) reduce to $e^{-w\lambda}$.

The required probability is, therefore, $\frac{1}{2}e^{-w\lambda}\lambda dt$.

❑ **EXAMPLE 8.6.** *Let X_1, X_2 be two independent Poisson variates with parameters λ_1 and λ_2 respectively. Show that the conditional distribution of X_1 given $X_1 + X_2$ is binomial.*

For positive integers $k, n (k < n)$,

$$P(X_1 = k | X_1 + X_2 = n) = \frac{P(X_1 = k, X_2 = n-k)}{P(X_1 + X_2 = n)}$$

$$= \frac{\left(e^{-\lambda_1}\frac{\lambda_1^k}{k!}\right)\left(e^{-\lambda_2}\frac{\lambda_2^{n-k}}{(n-k)!}\right)}{e^{-(\lambda_1+\lambda_2)}\frac{(\lambda_1+\lambda_2)^n}{n!}}$$

(since X_1, X_2 are independent and $X_1 + X_2$ is a Poisson $(\lambda_1 + \lambda_2)$).

$$= \binom{n}{k} p^k (1-p)^{n-k} \quad \text{where } p = \frac{\lambda_1}{\lambda_1+\lambda_2}; \qquad k = 0, 1, \cdots, n$$

Hence the conditional distribution of X_1 given $X_1 + X_2 = n$, is binomial with parameters n and $p = \frac{\lambda_1}{\lambda_1 + \lambda_2}$.

❑ **EXAMPLE 8.7.** *Suppose that the number of eggs laid by an insect follows a poisson distribution with parameter λ and the probablity of an egg developing is p. Assuming mutual independence of the events, show that the number of surviving insects follows a Poisson distribution with parameter λp.*

Let X, Y be r.v.'s denoting respectively the number of eggs laid and number of survivors.

$$P(X = n, Y = k) = \left(e^{-\lambda}\frac{\lambda^n}{n!}\right)\left(\binom{n}{k} p^k q^{n-k}\right), q = 1-p, k \le n$$

(by independence)

$$P(Y=k)=\sum_{n=k}^{\infty}P(X=n,Y=k)=e^{-\lambda}\left(\frac{p}{q}\right)^k\sum_{n=k}^{\infty}\frac{(q\lambda)^n}{k!(n-k)!}$$

$$=e^{-\lambda}\left(\frac{p}{q}\right)^k(q\lambda)^k\sum_{n-k=0}^{\infty}\frac{(q\lambda)^{n-k}}{k!(n-k)!}=e^{-\lambda}\left(\frac{p}{q}\right)^k\frac{(q\lambda)^k}{k!}e^{q\lambda}$$

$$=e^{-\lambda p}\frac{(\lambda p)^k}{k!},\quad k=0,1,\cdots$$

❑ **EXAMPLE 8.8.** ***The number of female insects in a given region follows a Poisson distribution with parameter λ. The number of eggs laid by each female insect follows a Poisson distribution with mean μ. Find the probability distribution of number of eggs in the region.***

Let X denote the number of female insects and Y the total number of eggs laid down in the region.

$$P(X=t)=e^{-\lambda}\frac{\lambda^t}{t!}$$

Let Y_i denote the number of eggs laid down by the ith insect. Hence

$$P(Y=s|X=t)=P\left(Y_1=s_1,Y_2=s_2,\cdots,Y_t=s_t;\sum_1^t s_i=s\right)$$

$$=\sum_{\substack{s_1,\cdots,s_t:\\ \Sigma_1^t s_i=s}}\left(e^{-\mu}\frac{\mu^{s_1}}{s_1!}\right)\cdots\left(e^{-\mu}\frac{\mu^{s_t}}{s_t!}\right)=\sum{}' e^{-\mu t}\frac{\mu^s}{s_1!\cdots s_t!}\quad\text{(say)}$$

$$=\frac{1}{s!}e^{-\mu t}(\mu t)^s.$$

Hence

$$P(X=t;Y=s)=\left(e^{-\lambda}\frac{\lambda^t}{t!}\right)\left(e^{-\mu t}\frac{(\mu t)^s}{s!}\right).$$

$$P(Y=s)=\sum_{t=0}^{\infty}P(X=t;Y=s)=\frac{e^{-\lambda}\mu^s}{s!}\sum_{t=0}^{\infty}\frac{\lambda^t t^s}{t!}e^{-\mu t}$$

which is the distribution of Y. The mgf of Y is

$$M_Y(h)=E(e^{hY})=e^{-\lambda}\sum_{s=0}^{\infty}\frac{e^{hs}e^{-\lambda}\mu^s}{s!}\sum_{t=0}^{\infty}\frac{\lambda^t t^s}{t!}e^{-\mu t}=e^{-\lambda}\sum_{t=0}^{\infty}\sum_{s=0}^{\infty}e^{-\mu t}\frac{(\mu t e^h)^s}{s!}\frac{\lambda^t}{t!}$$

$$=e^{-\lambda}\sum_{t=0}^{\infty}\frac{\{\lambda e^{\mu(e^h-1)}\}^t}{t!}=e^{-\lambda}e^{\lambda e^{\mu(e^h-1)}}=e^{\lambda(e^{\mu(e^h-1)}-1)}.$$

$$=\exp\left[\lambda\left\{e^{\mu(h+\frac{h^2}{2!}+\cdots)}-1\right\}\right]$$

$$=\exp\left[\lambda\left\{\mu\left(h+\frac{h^2}{2!}+\cdots\right)+\frac{\mu^2}{2!}\left(h+\frac{h^2}{2!}+\cdots\right)^2+\cdots\right\}\right]$$

$$= 1 + \lambda\left[\mu\left(h + \frac{h^2}{2!} + \cdots\right) + \frac{\mu^2}{2!}\left(h + \frac{h^2}{2!} + \cdots\right)^2 + \cdots\right]$$

$$+ \frac{\lambda^2}{2!}\left[\mu\left(h + \frac{h^2}{2!} + \cdots\right) + \frac{\mu^2}{2!}\left(h + \frac{h^2}{2!} + \cdots\right)^2 + \cdots\right]^2 + \cdots$$

Hence

$$\mu_1' = \text{coefficient of } h = \mu\lambda, \quad \mu_2' = \text{coefficient of } \frac{h^2}{2!} = \lambda(\mu^2 + \mu + \lambda\mu^2)$$

$$\mu_2 = \lambda\mu(\mu + \lambda).$$

The following lemma shows that a binomial distribution tends to a Poisson distribution under certain conditions.

Lemma 8.3. Let X be a binomial r.v. with parameters n and p. If $n \to \infty$ and $p \to 0$ in such a way that $np = \lambda$ remains constant then

$$b(x; n, p) \to e^{-\lambda}\frac{\lambda^x}{x!}$$

for fixed integer x.

Proof.

$$b(x; n, p) = \binom{n}{x} p^x q^{n-x} = \frac{n!}{x!(n-x)!}\left(\frac{\lambda}{n}\right)^x \left(1 - \frac{\lambda}{n}\right)^{n-x}$$

$$= \left\{\frac{\lambda^x}{x!}\left(1 - \frac{\lambda}{x}\right)^n \left(1 - \frac{\lambda}{n}\right)^{-x}\right\}\left\{\frac{n!}{n^x(n-x)!}\right\}$$

Hence

$$\lim_{n\to\infty} b(x; n, p) = e^{-\lambda}\frac{\lambda^x}{x!}\left[\lim_{n\to\infty}\frac{n!}{n^x(n-x)!}\right] \tag{8.26}$$

Using Stirling's approximation for $n!$,

$$n! \simeq \sqrt{2\pi} e^{-n} n^{n+\frac{1}{2}}$$

the quantity in [ ] in (8.26) is

$$\lim_{n\to\infty}\left[\frac{e^{-x} n^{n+\frac{1}{2}-x}}{(n-x)^{n-x+\frac{1}{2}}}\right] = \lim_{n\to\infty}\left[e^x\left(1 - \frac{x}{n}\right)^n \left(1 - \frac{x}{n}\right)^{-x+\frac{1}{2}}\right]^{-1} = e^x \cdot e^{-x} = 1$$

Hence the proof.

COROLLERY 8.1.

$$\lim_{\substack{n\to\infty \\ np=\lambda}} \sum_A b(x; n, p) = \sum_A e^{-\lambda}\frac{\lambda^k}{k!}$$

where $\sum_A$ denotes summation over any (finite or infinite) subset A of the non-negative integer $0, 1, 2, \cdots$.

8.6.1. The Truncated Poisson distribution

Consider a truncated Poisson variable, truncated at $x = 0$, whose pmf is

$$P(X = x) = e^{-\lambda}\frac{\lambda^x}{x!}(1 - e^{-\lambda})^{-1} = \frac{\lambda^x}{x!(e^{\lambda} - 1)}, \quad x = 1, 2, \cdots$$

$$E(X) = \frac{\lambda}{1 - e^{-\lambda}}, \quad V(X) = (1 - e^{-\lambda})^{-1}\left[1 - \lambda e^{-\lambda}(1 - e^{-\lambda})^{-1}\right]$$

The rth factorial moment is

$$\mu_{(r)} = \frac{\lambda^r}{(1 - e^{-\lambda})}$$

8.7. The Geometric Distribution

A r.v. X is said to have a geometric distribution with parameter p if its pmf is

$$P(X = x) = g(x; p) = \begin{cases} p(1 - p)^x, & x = 0, 1. \cdots \\ 0 & \text{otherwise} \end{cases} \tag{8.27}$$

The parameter p satisfies $0 < p < 1$.

The geometric distribution arises in the following circumstances. Bernoulli trials with constant probability p of success are performed until the first success occurs. The probability that exactly x failures precede the first success is given by (8.27).

The pgf is

$$P_X(t) = p(1 - tq)^{-1} \quad \text{(vide example 7.2)}$$

The mgf about origin is

$$M_X(t) = p(1 - qe^t)^{-1} \quad \text{(vide example 7.12)}$$

Thus

$$\mu_1' = \frac{q}{p}, \quad \mu_2' = \frac{q}{p} + 2\frac{q^2}{p^2}, \quad \mu_2 = \frac{q}{p^2}$$

$$\sqrt{\beta_1} = \frac{1 + q}{\sqrt{q}}, \quad \beta_2 = 9 + \frac{p^2}{q}, \quad \beta_2 - \beta_1 = 5.$$

The factorial moment of order r is

$$\mu_{(r)} = r!\left(\frac{q}{p}\right)^r.$$

A very intersting property of the geometric distribution is that the conditional probability that a geometric r.v. X is geater than or equal to $j + k$ given that X is greater than j is equal to the unconditional probability that it will be greater than or equal to k. Thus if a sequence of repeated, independent Bernoulli trials, each with probability p

of success, is performed until the first success occurs, the probability that the number of failures X preceding the first success will be $\geq j+k$ given that $X \geq j$, is equal to the unconditional probability that $X \geq k$. The event that the waiting time for the first success has already been at least j, does not affect the probabiliy of the future waiting time. As a crude example if the arrival of a bus of a particular route at a specified bus stop in each minute (unit of time) is considered as a Bernoulli trial with constant probability p and if these trials are assumed independent, the probability that a person has to wait, say, at least for 5 minutes, remains independent of whether he has already waited for 3 minutes or is just coming. This is called the "lack of memory" property of the geometric distribution. This is proved in

• **THEOREM 8.3.** *If X is a geometric distribution then for any two positive integers j and k,*

$$P\{X \geq j+k | X \geq j\} = P\{X \geq k\}$$

Proof.

$$P\{X \geq j+k | X \geq j\} = \frac{p\{X \geq j+k\}}{P\{X \geq j\}} = \frac{\sum_{x=j+k}^{\infty} pq^x}{\sum_{y=j}^{\infty} pq^y} = q^k$$

Again $P\{X \geq k\} = p\sum_{y=k}^{\infty} q^y = q^k$.

Some authors define a geometric random variable as the waiting time till the first success, including the trial on which the success occurs. Here

$$P(X=x) = pq^{x-1}, \quad x = 1, 2, \cdots$$

Here $E(X) = \frac{1}{p}$, $V(X) = \frac{q}{p^2}$, $M_X(t) = \frac{pe^t}{1-qe^t}$

❑ **EXAMPLE 8.9.** *Let X_1, X_2 be independent r.v.'s each having geometric distribution $q^x p$, $x = 0, 1, \cdots$. Show that the conditional distribution of X_1 given $X_1 + X_2$ is rectangular.*

$$P(X_1 = r | X_1 + X_2 = n) = \frac{P(X_1 = r, X_2 = n-r)}{P(X_1 + X_2 = n)}$$

$$= \frac{P(X_1 = r)P(X_2 = n-r)}{P(X_1 + X_2 = n)} \quad \text{(by independence)} \tag{8.28}$$

Now

$$P(X_1 + X_2 = n) = \sum_{x=0}^{n} P(X_1 = x)P(X_2 = n-x) = \sum_{x=0}^{n} q^x pq^{n-x}p = p^2q^n(n+1).$$

Hence (8.28) $= \frac{1}{n+1}$.

❑ **EXAMPLE 8.10.** *Let X_1, X_2 be two independent r.v.'s each having geometric distribution $q^k p$, $k = 0, 1, \cdots$. Find the probability distribution of $Z = \max(X_1, X_2)$.*

$$\begin{aligned} P(Z = j) &= P(Z = j, X_1 = j, X_2 < j) + P(Z = j, X_2 = j, X_1 < j) \\ &\qquad + P(Z = j, X_1 = j, X_2 = j) \\ &= 2q^j p \sum_{k=0}^{j-1} q^k p + (q^j p)^2 = 2q^j p^2 \left(\frac{1 - q^j}{1 - q}\right) + q^{2j} p^2 \\ &= 2q^j p(1 - q^j) + q^{2j} p^2 = 2q^j p - q^{2j} p - q^{2j+1} p. \end{aligned}$$

8.8. The Negative Binomial Distribution

A r.v. X is said to have a negative binomial distribution with parameters k and p if its *pmf* is given by

$$P(X = x) = b'(x; k, p) = \begin{cases} \binom{k + x - 1}{x} p^k q^x = \binom{-k}{x} p^k (-q)^x, & x = 0, 1, 2, \cdots \\ 0 & \text{otherwise} \end{cases} \tag{8.29}$$

where the parameters satisfy $0 < p < 1$, $k = 1, 2, 3, \cdots$

A negative binomial distribution arises in the following circumstances. An urn contains np white and nq black balls. Balls are drawn at random with replacement. The probability that exactly $x + k$ trials will be required to produce k white balls is

$$\begin{aligned} \binom{x + k - 1}{k - 1} q^x p^k &= \frac{(x + k - 1)(x + k - 2) \cdots k}{x!} q^x p^k \\ &= (-1)^x \frac{(-k)(-k - 1) \cdots (-k - x + 1)}{x!} q^x p^k = (-1)^x \binom{-k}{x} p^k q^x. \end{aligned}$$

Clearly

$$\sum_{x=0}^{\infty} b'(x; k, p) = p^k \sum_{0}^{\infty} \binom{-k}{x} (-q)^x = p^k (1 - q)^{-k} = 1.$$

The distribution (8.29) ramains meaningful even when k is not an integer. When k is an integer the distribution is sometimes called a Pascal distribution. In this case (8.29) is also called a discrete waiting time distribution as it gives the probability that the waiting time is x (in terms of failures) to attain the kth success in an unending sequence of Bernoulli trials. For $k = 1$, the distribution reduces to the geometric distribution.

It is conventional to express the distribution in terms of parameters $Q = \dfrac{1}{p}$, $P = \dfrac{1 - p}{p}$ so that $Q - P = 1$. The expression (8.29) reduces to

$$\binom{k+x-1}{x}\left(\frac{P}{1+P}\right)^x(1+P)^{-k} = Q^{-k}\binom{k+x-1}{x}\left(\frac{P}{Q}\right)^x$$
$$= \binom{k+x-1}{x}\left(\frac{P}{Q}\right)^x\left(1-\frac{P}{Q}\right)^k, x = 0, 1, 2, \cdots \tag{8.30}$$

which is the $(x+1)$th term in the expansion of $(Q-P)^{-k}$.

The pgf is

$$E(t^X) = (Q-P_t)^{-k} = p^k(1-tq)^{-k}, |t| \leq 1 \qquad \text{(vide example 7.5.)}$$

The mgf about origin is

$$M_X(t) = E(e^{tX}) = \sum_{x=0}^{\infty}\binom{k+x-1}{x}\left(\frac{Pe^t}{Q}\right)^x\left(1-\frac{P}{Q}\right)^k$$
$$= Q^{-k}\sum_{x=0}^{\infty}\binom{-k}{x}\left(-\frac{Pe^t}{Q}\right)^x = (Q-Pe^t)^{-k}$$
$$= p^k(1-qe^t)^{-k} \quad \text{for} \quad qe^t < 1$$

Thus

$$\mu_1' = M_X'(t)\Big]_{t=0} = \left[kPe^t(Q-Pe^t)^{-(k+1)}\right]_{t=0} = kP$$

$$\mu_2' = M_x''(t)\Big]_{t=0}$$
$$= \left[kPe^t(Q-Pe^t)^{-(k+1)} + k(k+1)P^2e^{2t}(Q-Pe^t)^{-(k+2)}\right]_{t=0} = kP + k(k+1)P^2.$$

Hence

$$\mu_2 = kPQ = k\frac{q}{p^2}(> \mu_1')$$

Note that the variance of the negative binomial distribution is greater than its mean (for binomial distribution, variance is less than its mean).

The jth factorial moment about zero is

$$\mu_{(j)} = \sum_{x=j}^{\infty}(x)_j\binom{k+x-1}{k-1}\left(\frac{P}{Q}\right)^x\left(1-\frac{P}{Q}\right)^k$$
$$= (k+j-1)_{(j)}\left(\frac{P}{Q}\right)^j Q^{-k}\sum_{x=j}^{\infty}\binom{k+j+x-j-1}{k+j-1}\left(\frac{P}{Q}\right)^{x-j}$$
$$= (k+j-1)_{(j)}P^j.$$

From these it follows

$$\sqrt{\beta_1} = \frac{Q+P}{\sqrt{kPQ}}, \quad \beta_2 = 3 + \frac{1+6PQ}{kPQ}$$

The points (β_1, β_2) lie on the straight line

$$\beta_2 - \beta_1 - 3 = \frac{2}{k}.$$

The cumulants satisfies the followint recurrence relation

$$k_{r+1} = PQ\frac{\partial k_r}{\partial Q}, \quad r \geq 1.$$

The ratio

$$\frac{b'(x+1;k,p)}{b'(x;k,p)} = \frac{(k+x)P}{(x+1)Q}$$

shows that $b'(x+1;k,p) \gtrless b'(x;k,p)$ according as $x \lessgtr kP - Q$. Hence if $kP > Q$ there is a mode at x^* where x^* satisfies $kP - Q - 1 < x^* < kP - Q$. If $kP < Q$ the mode is at $x = 0$.

Lemma 8.4. As $k \to \infty$ and $P \to 0$ with kP remaining fixed at λ, $b'(x;k,p) \to e^{-\lambda}\frac{\lambda^x}{x!}$.

Proof.

$$\begin{aligned}
\lim_{\substack{k\to\infty \\ P\to 0}} b'(x;k,p) &= \lim_{\substack{k\to\infty \\ P\to 0}} \frac{(k+x-1)\cdots k}{x!}\left(\frac{P}{1+P}\right)^x (1+P)^{-k} \\
&= \lim_{\substack{k\to\infty \\ P\to 0}} \frac{1}{x!}\left(1+\frac{x-1}{k}\right)\left(1+\frac{x-2}{k}\right)\cdots 1\cdot\lambda^x\left(1+\frac{\lambda}{k}\right)^{-x}\left(1+\frac{\lambda}{k}\right)^{-k} \\
&= e^{-\lambda}\frac{\lambda^x}{x!}
\end{aligned}$$

- **THEOREM 8.4.** *Let $X_1, \cdots X_n$ be independently distributed negative binomial variables, X_i having parameters $(k_i, p), i = 1, \cdots, n$. Then $S_n = X_1 + \cdots + X_n$ has a negative binomial distribution with parameters (k, p) when $k = \sum_{i=1}^n k_i$.*

Proof. mgf of S_n is

$$M_{Sn}(t) = \prod_{i=1}^n M_{X_i}(t) = \prod_{i=1}^n p^{k_i}(1-qe^t)^{-k_i} = p^k(1-qe^t)^{-k}$$

which is the mgf of a negative binomial distribution with parameters k, p. Hence the result follows by the uniqueness property of mgf (theorem 7.5).

COROLLERY 8.2. Let $X_1, \cdots, X_n$ be independent geometric variables each with parameter p. Then S_n has a negative binomial distribution with parameters (n, p)

8.8.1. The Truncated Negative Binomial Distribution

Consider a negative binomial distribution truncated at $x = 0$. The pmf of the truncated distribution is

$$\begin{aligned}
b'_t(x;k,p) &= \binom{k+x-1}{x}\frac{p^k q^x}{1-p^k} \\
&= \binom{k+x-1}{x}(1-Q^{-k})^{-1}\left(\frac{P}{Q}\right)^x\left(1-\frac{P}{Q}\right)^k \quad x = 1, 2, \cdots
\end{aligned}$$

The moments about zero of this distribution are obtained by multiplying the raw moments of the untruncated distribution by $(1-Q^{-k})^{-1}$. Thus for $b_t'(x;k,p)$,

$$E(X)=\frac{kP}{1-Q^{-k}},\quad E(X^2)=\frac{kPQ+(KP)^2}{1-Q^{-k}}$$

$$V(X)=\frac{kPQ}{1-Q^{-k}}\left[1-k\left(\frac{P}{Q}\right)\left\{\left(1-Q^{-k}\right)^{-1}-1\right\}\right].$$

8.9. The Power Series Distribution

Consider the power series

$$f(z)=\sum_{x=0}^{\infty}a_x z^x$$

where a_x's are real non-negative constants, $0<z<r$ (radius of convergence of $f(z)$) so that $f(z)$ is finite. Define a r.v. X with *pmf*

$$P\{X=x\}=P_x=\frac{a_x z^x}{f(z)},\quad x=0,1,2,\cdots \tag{8.31}$$

Clearly $P_x \geq 0, \sum_0^{\infty} P_x = 1$ so that X has a probability distribution P_x. The distribution of X is called the power series distribution (*psd*).

$$\mu_1'=E(X)=\frac{1}{f(z)}\sum_0^{\infty}xa_x z^x$$

Now

$$f'(z)=\frac{1}{z}\sum_1^{\infty}xa_x z^x$$

Hence

$$\mu_1'=\frac{zf'(z)}{f(z)}=z\frac{d\log f(z)}{dz},\quad \mu_r'=\frac{1}{f(z)}\sum_0^{\infty}x^r a_x z^x,\quad r=1,2,\cdots \tag{8.32}$$

Now

$$\frac{d\mu_r'}{dz}=\frac{1}{f(z)}\sum_x a_x x^{r+1}z^{x-1}-\left(\sum_x x^r a_x z^x\right)\frac{f'(z)}{(f(z))^2}.$$

Thus

$$z\frac{d\mu_r'}{dz}=\frac{1}{f(z)}\sum_x a_x x^{r+1}z^x-\left(\sum_x x^r a_x z^x\right)\frac{zf'(z)}{(f(z))^2}.$$

i.e. $z\dfrac{d\mu_r'}{dz}=\mu_{r+1}'-\mu_r'\mu_1'$.

Thus we have the recurrence relation

$$\mu'_{r+1} = z\frac{d\mu'_r}{dz} + \mu'_r\mu'_1. \tag{8.33}$$

The central moments are

$$\mu_r = \frac{1}{f(z)}\sum_{0}^{\infty}(x-\mu'_1)^r a_x z^x.$$

Now

$$z\frac{d\mu_r}{dz} = \frac{d}{dz}\left\{\sum_{0}^{\infty}(x-\mu'_1)^r a_x z^x\right\}\frac{z}{f(z)} - \left\{\sum_{0}^{\infty}(x-\mu'_1)^r a_x z^{x+1}\right\}\frac{f'(z)}{(f(z))^2}.$$

Again,

$$\frac{d}{dz}\left\{\sum_x(x-\mu'_1)^r a_x z^x\right\} = \sum_x a_x\left[z^x r(x-\mu'_1)^{r-1}\left(-\frac{d\mu'_1}{dz}\right) + (x-\mu'_1)^r x z^{x-1}\right].$$

Thus

$$z\frac{d\mu_r}{dz} = \frac{1}{f(z)}\sum_x a_x\cdot x(x-\mu'_1)^r z^x - zr\frac{d\mu'_1}{dz}\frac{1}{f(z)}\sum_x a_x z^x$$

$$(x-\mu'_1)^{r-1} - \frac{zf'(z)}{(f(z))^2}\sum_x(x-\mu'_1)^r a_x z^x \tag{8.34}$$

The first and the last term in the rhs of (8.34) together simplify to, by virtue of (8.31),

$$\frac{1}{f(z)}\sum_x a_x z^x(x-\mu'_1)^{r+1} = \mu_{r+1}.$$

Hence we can write (8.34) as

$$z\frac{d\mu_r}{dz} = \mu_{r+1} - rz\frac{d\mu'_1}{dz}\mu_{r-1}$$

when we have the recurrence relation between the central moments,

$$\mu_{r+1} = z\left[\frac{d\mu_r}{dz} + r\frac{d\mu'_1}{dz}\mu_{r-1}\right] \tag{8.35}$$

Putting $r = 1$, $\mu_o = 1$, $\mu_1 = 0$ we get the variance of X as

$$\mu_2 = z\frac{d\mu'_1}{dz} = z\frac{d}{dz}\left\{z\frac{d}{dz}\log f(z)\right\} = z^2\frac{d^2\log f(z)}{dz^2} + \mu'_1 \tag{8.36}$$

Now

$$\frac{d^2 \log f(z)}{dz^2} = \frac{f''(z)f(z) - (f'(z))^2}{(f(z))^2}$$

Hence

$$\mu_2 = z^2 \frac{f''(z)}{f(z)} - z^2 \left(\frac{f'(z)}{f(z)}\right)^2 + z\frac{f'(z)}{f(z)}. \tag{8.37}$$

From (8.37), (8.35) can be written as

$$\mu_{r+1} = z\frac{d\mu_r}{dz} + r\mu_2\mu_{r-1} \tag{8.38}$$

The characteristic function of X is

$$\psi_X(t) = \sum_x e^{itx} P_x = \frac{1}{f(z)} \sum_x e^{itx} a_x z^x = \frac{f(e^{it}z)}{f(z)} \tag{8.39}$$

Some Special Cases :

(a) Poisson distribution : Let $f(z) = e^z = 1 + z + \frac{z^2}{z!} + \frac{z^3}{3!} + \cdots$ Here $a_x = \frac{1}{x!}$. Thus $P_x = \frac{a_x z^x}{f(z)} = e^{-z}\frac{z^x}{x!}, x = 0, 1, 2, \cdots$, which is the Poisson distribution.

From (8.32) and (8.36) respectively, $\mu_1' = z$, $\mu_2 = z$.
The recurrence relation (8.35) assumes the form

$$\mu_{r+1} = z\left(\frac{d\mu_r}{dz} + r\mu_{r-1}\right).$$

(b) Binomial distribution. Let $f(z) = (1+z)^n = 1 + \binom{n}{1} z + \binom{n}{2} z^2 + \cdots + z^n$.

Here $a_x = \binom{n}{x}, x = 0, 1, \cdots, n$.

$$P_x = \binom{n}{x}\left(\frac{z}{1+z}\right)^x \left(\frac{1}{1+z}\right)^{n-x}$$

when X follows a Binomial distribution with parameter n and $p = \frac{z}{1+z}$.

From (8.32), $\mu_1' = z\frac{f'(z)}{f(z)} = np$

From (8.36) $\mu_2 = z\frac{d\mu_1'}{dz} = \frac{nz}{(1+z)^2} = np(1-p)$.

The recurrence relations (8.33) and (8.35), respectively, assume the form

$$\mu'_{r+1} = z\frac{d\mu'_r}{dz} + n\left(\frac{z}{1+z}\right)\mu'_r = pq\left(\frac{d\mu'_r}{dp} + \frac{n}{q}\mu'_r\right) \qquad \text{(relation 8.5)}$$

and

$$\mu'_{r+1} = \frac{z}{(1+z)^2}\left[(1+z)^2\frac{d\mu_r}{dz} + nr\mu_{r-1}\right] = pq\left[\frac{d\mu_r}{dp} + nr\mu_{r-1}\right] \qquad \text{(relation 8.4)}$$

(c) Negative Binomial distribution. Let $f(z) = (1-z)^{-k}$, $k > 0$, $0 < z < 1$. Hence

$$f(z) = 1 + \binom{k}{1}z + \binom{k+1}{2}z^2 + \binom{k+2}{3}z^3 + \cdots + \binom{k+x-1}{x}z^x + \cdots$$

Here $a_x = \binom{k+x-1}{x}$

Hence

$$P_x = \frac{a_x z^x}{f(z)} = \binom{k+x-1}{x} z^x(1-z)^k$$

which is the negative binomial distribution. By (8.32)

$$E(X) = z\frac{d}{dz}\log\{(1-z)^{-k}\} = \frac{zk}{1-z}.$$

The recurrence relation for the central moments is

$$\mu_{r+1} = z\left[\frac{d\mu_r}{dz} + r\mu_{r-1}\frac{k}{(1-z)^2}\right].$$

Hence

$$\mu_2 = \frac{kz}{(1-z)^2}, \qquad \mu_3 = \frac{k(1+z)z}{(1-z)^3}$$

$$\mu_4 = z\left[\frac{d\mu_3}{dz} + 3\mu_2\frac{k}{(1-z)^2}\right] = \frac{kz(1+4z+3kz+z^2)}{(1-z)^4}.$$

The characteristic function for this distribution is

$$\psi(t) = \frac{f(e^{it}z)}{f(z)} = \left(\frac{1-e^{it}z}{1-z}\right)^{-k}$$

(d) Geometric distribution. Let $f(z) = (1-z)^{-1} = 1 + z + z^2 + \cdots, |z| < 1$. Here $P_x = z^x(1-z)$ which is the geometric distribution with probability of success $(1-z)$.

$$\mu'_1 = z\frac{f'(z)}{f(z)} = \frac{z}{1-z}, \qquad \mu_2 = z\frac{d\mu'_1}{dz} = \frac{z(1-2z)}{(1-z)^2}.$$

(e) Logarithmic series distribution. Let $f(z) = -\log(1-z) = z + \dfrac{z^2}{2} + \dfrac{z^3}{3} + \cdots$, $0 < z < 1$.

Hence $P_x = -\dfrac{z^x}{x\log(1-z)}$, which is the logarithmic series distribution.

$$\mu_1' = z\frac{f'(z)}{f(z)} = -\frac{z}{(1-z)\log(1-z)}$$

$$\frac{d\mu_1'}{dz} = -\frac{\log(1-z)+z}{(1-z)^2\{\log(1-z)\}^2}$$

Hence

$$\mu_2 = -\frac{z^2 + z\log(1-z)}{(1-z)^2\{\log(1-z)\}^2}$$

Thus

$$\mu_{r+1} = z\left[\frac{d\mu_r}{dz} - r\mu_{r-1}\frac{z+\log(1-z)}{(1-z)^2\{\log(1-z)\}^2}\right]$$

Characteristic function is

$$\psi(t) = \frac{f(e^{it}z)}{f(z)} = \frac{\log(1-e^{itz})}{\log(1-z)}.$$

We now consider an interesting theorem characterising the property of a Poisson distribution.

• **THEOREM 8.5.** *The equality of mean and variance is a necessary and sufficient condition for a psd to be Poisson.*

Proof. For any *psd* given by (8.31), $\mu_2 = z\dfrac{d\mu_1'}{dz}$.

When $f(z) = e^z$, the *psd* gives Poisson distribution for which $\mu_1' = z\dfrac{f'(z)}{f(z)} = z$ and hence $\mu_2 = z$. Hence the necessity.

To prove the sufficiency, let $\mu_2 = \mu_1'$ which implies $\dfrac{d\mu_1'}{dz} = \dfrac{\mu_1'}{z}$.

Integrating both sides,

$$\log\mu_1' = \log(cz)$$

where c is a positive constant. Hence

$$\mu_1' = cz.$$

Again, by (8.32)

$$z\frac{d\log f(z)}{dz} = cz \quad \text{or} \quad \frac{d\log f(z)}{dz} = c.$$

Integrating both sides,

$$\log f(z) = cz + \log k, \quad k \text{ a positive constant.}$$

Thus

$$f(z) = ke^{cz}$$

from which it follows

$$P_x = e^{-cz}\frac{(cz)^x}{x!}$$

so that X follows a Poisson distribution with parameter cz.

8.10. Exercises

1. *A die is cast repeatedly until an ace turns up. Find the expected number of throws necessary.*

2. *Find the expected number of aces drawn in a bridge hand of 13 cards. Also find the expected number of face cards drawn.*

3. *Let X have a probability distribution defined by the function $f(x) = \dfrac{1}{\pi(1+x^2)}$.*

Let $Y = 1(0)$ if $X \geq 1(< 1)$. Find probability distribution of Y. $\left[\text{Bernoulli, } p = \dfrac{1}{4}\right]$

4. *Let X, Y be independent geometric r.v.'s. Show that minimum (X, Y) and $X - Y$ are independent.*

5. *Let $X_1, \cdots, X_n$ be independent geometric r.v.'s with parameters $p_1, \cdots p_n$ respectively. Show that $Z = \min(X_1, \cdots, X_n)$ is a geometric r.v. with paramenter $p = 1 - \prod_{i=1}^{n}(1 - p_i)$.*

6. *Let X and Y be two independently and identically distributed r.v.'s with $P\{X = x\} = p_x$, $x = 0, 1, 2, \cdots$.*

Suppose

$$P\{X = s | X + Y = s\} = P\{X = s - 1 | X + Y = s\} = \frac{1}{s+1}, s \geq 0.$$

Show that X and Y are geometric r.v.'s.

7. *Let X be a non-negative integer-valued random variable satisfying*

$$P\{X > j + k | X > j\} = P\{X \geq k\}.$$

for any $j(> 0), k(> 0)$. Then X must have a geometric distribution.

8. *Show that for a binomial distribution* $b(x;n,p) = \binom{n}{x} p^x q^{n-x}$,

$$P\{X \geq j\} \leq b(j,n,p)\frac{q(j+1)}{j+1-(n+1)p} \quad \text{for } j \geq np.$$

$$P\{X \leq j\} \leq b(j,n,p)\frac{(n-j+1)}{(n+1)p-j} \quad \text{for } j \leq np$$

9. *Let* $X_1, \cdots, X_n$ *be* n *independently and identically distributed Bernoulli variates each with parameter* p. *Show that their sum* $S_n = \sum_{i=1}^{n} X_i$ *is distributed as a binomial variate with parameter* (n,p).

10. *Let* $X_1, \cdots X_n$ *be independent Bernoulli variates with common parameter* $p = P\{X_i = 1\}$, $0 < p < 1$, $i = 1, \cdots, n$. *Set* $S_j = X_1 + \cdots + X_j$, $1 \leq j \leq n$.*Show that* $P[S_m = r | S_n = s]$, $m < n$ *does not depend on* p. *Identify the probability distribution and obtain the first two moments of it.*

$\left[\text{Hypergeometric } H(x;n,m,s)\right]$

11. *Let* X_1, X_2 *be independent binomial wih parameters* $\left(n_i, \frac{1}{2}\right)$, $i = 1, 2,$. *Find the pmf of* $X_1 - X_2 + n_2$.

$\Big[(n_2 - X_2)$ is binomial with parameters $(n_2, \frac{1}{2})$.

$$P(X_1 - X_2 + n_2 = k) = \binom{n_1+n_2}{n_2+k} \Big/ 2^{n_1+n_2}, \quad (-n_2 \leq k \leq n_1)\Big]$$

12. *In a population of* n *elements* n_1 *are red and* $n_2 (= n - n_1)$ *are black. A group of* r *elements are chosen at random without replacement. Find out the probability* $q_k (k = 0, 1, \cdots, \min(n_1, r))$ *that the group contains exactly* k *red elements; hence show that if* $np = n_1$, $q = 1 - p$,

$$\binom{r}{k}\left(p - \frac{k}{n}\right)^k \left(q - \frac{r-k}{n}\right)^{r-k} < q_k < \binom{r}{k} p^k q^{r-k} \left(1 - \frac{r}{n}\right)^{-r}$$

Justify how you can approximate this q_k *by a binomial mass function.*

13. *Show that*

$$\frac{\binom{Np}{x}\binom{Nq}{n-x}}{\binom{N}{n}} \to \binom{n}{x} p^x q^{n-x}, \quad q = 1-p, 0 < p < 1$$

as $N \to \infty$ *for any fixed* x, *provided* $\frac{n}{N}$ *is negligible.*

14. *An urn contains a white balls and b black balls. A sample of 2 balls is chosen wihout replacement. Under what conditions will the probability that both the chosen balls are of the same colour be greater than* $\frac{1}{2}$ *?* $[(a-b)^2 > a+b.]$

15. *Show that*

$$\binom{k+x-1}{x} p^k q^x \to e^{-\lambda} \frac{\lambda^x}{x!}$$

as $p \to 1$ *and* $k \to \infty$ *in such a way that* $k(1-p) = \lambda$ *remains constant.*

16. *Obtain by two methods, the probability that in a series of Bernoulli trials (with probability p of success) 'a' successes occur before at most b failures. Hence prove the identity*

$$\sum_{k=0}^{b-1} \binom{a+b-1}{a+k} p^k q^{b-k-1} = \sum_{k=0}^{b-1} \binom{a+k-1}{k} q^k.$$

[Prob. that in $(a+b-1)$ trials, $(a+k)$ successes and $b-k-1$ failures occur is

$$\binom{a+b-1}{a+k} p^{a+k} q^{b-k-1}.$$

Hence probability of getting 'a' successes before 'b' failures is

$$\sum_{k=0}^{b-1} \binom{a+b-1}{a+k} p^{a+k} q^{b-k-1}. \qquad \text{(i)}$$

Alternatively, probability that in $(a+k)$ trials the last trial is a success and the preceding $(a+k-1)$ trials contain $(a-1)$ successes is

$$\binom{a+k-1}{a-1} p^a q^k.$$

The required probability is therefore

$$= p^a \sum_{k=0}^{b-1} \binom{a+k-1}{k} q^k \qquad \text{(ii)}$$

Hence the identity.]

17. *Let* X_1, X_2 *be two independent negative binomial variables,* X_i *having parameters* $k_i, p, i = 1, 2$. *Find the conditional probability that* $X_1 = t$, *given that* $X_1 + X_2 = r$. *Show also that when* $k_1 = k_2 = 1$, *then* $P(X_1 = t | X_1 + X_2 = r) = \frac{1}{r+1}$, $t = 0, 1, \cdots, r; r = 0, 1, 2, \cdots$.

$$\left[\text{Hints : } P(X_1 = t|X_1 + X_2 = r) = \frac{\binom{t + k_1 - 1}{t}\binom{r + k_2 - t - 1}{r - t}}{\binom{r + k_1 + k_2 - 1}{r}}, r = 0, 1, 2, \cdots\right]$$

18. *Consider a r.v. X distributed as $H(x; N, M, n)$. Show that for fixed n and fixed $p = \frac{M}{N}$, the rth factorial moment of this distribution tends to that of a binomial distribution with parameters (n, p) as $N \to \infty$.*

19. *An urn contains N balls numbered $1, \cdots, N$. From these balls n balls are drawn at random without replacement. Let X_i denote the number drawn on the ith ball $(i = 1, \cdots n)$ and $S_n = \sum_{i=1}^{n} X_i$. Find $E(S_n)$, $V(S_n)$.*

[These are $\binom{N}{n}$ possible sets of n sampled balls $s = (i, \cdots, i_n)$, i_t denoting the number marked on a sampled ball. Let $T=$ sum of the numbers on the sampled balls $=\sum_{t=1}^{n} i_t$. Probability of selecting each s is M_o^{-1}, $s \in \mathcal{S}$ (sample sapce), where $M_j = \binom{N-j}{n-j}$, $j = 0, 1, 2, \cdots$.

$$E(T) = \sum_{s \in \mathcal{S}} \left[\sum_{t=1}^{n} i_t\right] \frac{1}{M_o}$$

$$= \frac{1}{M_o} \sum_{i=1}^{N} i \times (\text{Number of samples containing the ball marked } i)$$

$$= \frac{M_1}{M_o} \sum_{i=1}^{N} i = \frac{n(N+1)}{2}.$$

$$E(T^2) = E\left[\sum_{t=1}^{n} i_t^2 + \sum \sum_{t \neq t'=1}^{n} i_t i_{t'}\right]$$

$$E\left(\sum_{t=1}^{n} i_t^2\right) = \sum_{S \in \mathcal{S}} \left[\sum_{t=1}^{n} i_t^2\right] \cdot \frac{1}{M_o} = \frac{M_1}{M_o} \sum_{i=1}^{N} i^2 = \frac{n(N+1)(2N+1)}{6}.$$

$$E\left(\sum \sum_{t \neq t'=1}^{n} i_t i_{t'}\right) = \frac{1}{M_o} \sum_{S \in \mathcal{S}} \left[\sum \sum_{t \neq t'=1}^{n} i_t i_{t'}\right]$$

$$= \frac{1}{M_o} \sum \sum_{t \neq i'=1}^{N} ii' \times (\text{number of samples containing the balls marked } i \text{ and } i')$$

$$= \frac{M_2}{M_o} \frac{N(N+1)(3N+2)(N-1)}{12} = \frac{n(n-1)(N+1)(3N+2)}{12}$$

Hence

$$E(T^2) = \frac{n(N+1)}{6}\left[2N+1+\frac{(n-1)(3N+2)}{2}\right]$$

$$V(T) = E(T^2) - (E(T))^2 = \frac{n(N+1)(N-n)}{12}.$$

20. *Telephone calls arrive at a certain exchange at an average rate of 3 calls per minute. Assuming that the number of incoming calls follow a Poisson distribution, find the probability that 3 or more calls arrive in a 15-second time interval.*

[.0406]

21. *A Geiger Muller counter registrars on an average 50 counts per minute in the neighbourhood of a certain radioactive substance. Find the probability there will be (i) 3 counts in a 6-second period (ii) x counts in a T-second period.*

$$\left[\text{(i) } .140;\ \text{(ii) } e^{-.833T}\frac{(.833T)^x}{x!}\right]$$

22. *Find the probability generating function of the sum S of the numbers obtained when one ball is drawn from each of m urns, each of which contains n balls numbered $1, 2, \cdots, n$. Hence find the probability distribution of S.*

[Hints : Let X_i be the number on the ball drawn from the ith urn ($i = 1, \cdots, m$). Each X_i has a uniform distribution $P(X_i = k) = \frac{1}{n}$, $k = 1, \cdots, n$. Hence proceed as in exercise 9, section 7.6]

9 SOME CONTINUOUS DISTRIBUTIONS ON R^1

9.1. Introduction

We consider some univariate continuous distributions in this chapter. Uniform distribution is discussed in the section 9.2; sections 9.3–9.6 discuss respectively exponential, gamma (chi-square), beta and cauchy distributions. Section 9.7 deals with the celebrated normal distribution. We consider log-normal, Laplace, Pareto, Weibull, extreme-value and logistic distributions in sections 9.8 to 9.13 respectively.

9.2. Continuous Uniform Distribution

A random variable X is said to have a continuous uniform (rectangular) distribution on the interval $[a, b]$ $(-\infty < a < b < \infty)$ if its *pdf* is given by

$$f(x; a, b) = \begin{cases} \dfrac{1}{b-a} & \text{if } a \le x \le b \\ 0 & \text{otherwise,} \end{cases} \tag{9.1}$$

a, b are the parameters of the distribution. The *cdf* is

$$F_X(x) = \begin{cases} 0, & x \le a \\ \dfrac{x-a}{b-a}, & a < x < b \\ 1 & x \ge b \end{cases} \tag{9.2}$$

$$E(X) = \int_a^b \frac{x}{b-a}dx = \frac{a+b}{2}$$

$$E(X^2) = \frac{b^2+ab+a^2}{3}$$

$$V(X) = \frac{(b-a)^2}{12}$$

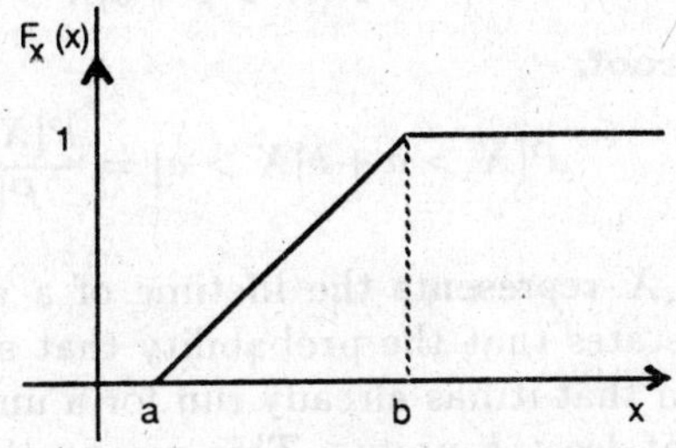

Fig. 9.1 cdf of rectangular distribution

The *mgf* is

$$M_X(t) = \int_a^b \frac{e^{tx}}{b-a}dx = \frac{e^{bt}-e^{at}}{(b-a)t}, \quad t \neq 0.$$

The distribution can also be defined over the open interval (a, b) or semi-closed intervals $[a, b), (a, b]$. The c.d.f. remains (9.2) for all these distributions (fig. 9.1).

9.3. Exponential Distribution

A r.v. X is defined to have an (negative) exponential distribution with parameter $\lambda > 0$ if its *pdf* is given by

$$f(x, \lambda) = \begin{cases} \lambda e^{-\lambda x} & 0 < x < \infty, \lambda > 0 \\ 0 & \text{otherwise} \end{cases} \tag{9.3}$$

The *cdf* is

$$F_X(x) = \begin{cases} 0, & x < 0 \\ 1 - e^{-\lambda x}, & 0 \leq x \end{cases}$$

The *mgf* of the distribution is

$$M_X(t) = \frac{1}{1 - \left(\frac{t}{\lambda}\right)}$$

$$\mu_1' = \frac{1}{\lambda}, \quad \mu_r' = \frac{r!}{\lambda^r}, \quad r = 2, 3, \ldots$$

$$\mu_2 = \frac{1}{\lambda^2}, \quad \mu_3 = \frac{2}{\lambda^3}, \quad \mu_4 = \frac{9}{\lambda^4}$$

$$\sqrt{\beta_1} = 2, \quad \beta_2 = 9.$$

The mean deviation is $\frac{2}{e\lambda}$. The ratio (mean deviation)/s.d. $= \frac{2}{e} = 0.736$. The median of the distribution is $\frac{1}{\lambda}\log_e 2$. The mode is at $x = 0$.

The cumulant generating function is

$$-\log\left(1 - \frac{t}{\lambda}\right),$$

$$k_1 = \frac{1}{\lambda}, \quad k_r = (r-1)!\lambda^{-r}, \quad r > 1.$$

Fig. 9.2 shows an exponential density.

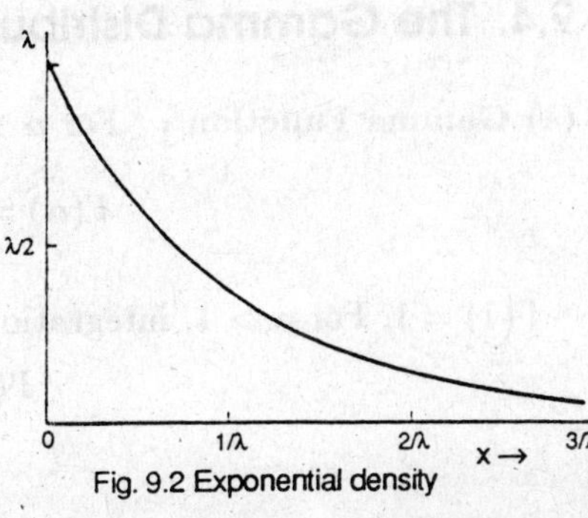

Fig. 9.2 Exponential density

• **THEOREM 9.1.** *For an exponential r.v. X with parameter λ,*

$$P[X > a+b|X > a] = P[X > b] \quad \text{for } a > 0, b > 0.$$

Proof.

$$P[X > a+b|X > a] = \frac{P[X > a+b]}{P[X > a]} = \frac{e^{-\lambda(a+b)}}{e^{-\lambda a}} = e^{-\lambda b} = P[X > b].$$

If X represents the lifetime of a given component, say, an electric bulb, theorem 9.1 states that the probability that a bulb will be operative for at least $(a+b)$ units given that it has already run for a units is the same as its intial probability for lasting for at least b units. This means that the future lifetime of an individual has the same distribution, no matter how old it is at present. This is the "lack of memory" property of the exponential distribution. A similar property was noted for the geometric distribution (vide section 8.7).

Suppose that the events occur randomly in time in such a way that the number of happenings in a fixed time interval $[0, t]$ has a poisson distribution with parameter λt. Suppose one happening has just occured. Probability that the next happening will occur not before time T is P [waiting time $> T$] $= P$ [no happening in time interval of length T] $= e^{-\lambda T}$. Hence $P(X \leq T) = 1 - e^{-\lambda T} (T > 0)$ so that X, the waiting time has an exponential distribution.

Thus exponential distribution and poisson distribution are related (vide also example 9.1).

Because of the simplicity in mathematics associated with it, a negative exponential distribution is often used as an approximate representation of more complex models suitable for a particular application. The lifetime of an component can often be represented by an exponential random variable. The distribution is often used in the field of life-testing.

An exponential distribution truncated from below is still an exponential distribution, with the same scale parameter.

An exponential distribution truncated by exclusion of values exceeding A has the *pdf*.

$$f_t(x; \lambda) = \lambda\{1 - e^{-\lambda A}\}^{-1} e^{-\lambda x}, \quad 0 < x < A \tag{9.4}$$

The expected value of the distribution (9.4) is

$$\frac{1}{\lambda}\left[1 - A\lambda(e^{\lambda A} - 1)^{-1}\right].$$

9.4. The Gamma Distribution and Chi-squared Distribution

(a) Gamma Function : For $\alpha > 0$, the gamma function is defined as

$$\Gamma(\alpha) = \int_0^\infty t^{\alpha-1} e^{-t} dt \quad (\alpha > 0). \tag{9.5}$$

$\Gamma(1) = 1$. For $\alpha > 1$, integration by parts yields

$$\Gamma(\alpha) = (\alpha - 1)\Gamma(\alpha - 1) \tag{9.6}$$

If $\alpha = n$, a positive integer

$$\Gamma(n) = (n-1)! \tag{9.7}$$

$$\Gamma\left(\frac{1}{2}\right) = \sqrt{\pi} \tag{9.8}$$

Gauss' multiplication formula :

$$\Gamma(n\alpha) = (2\pi)^{-\frac{n-1}{2}} n^{n\alpha-\frac{1}{2}} \prod_{j=0}^{n-1} (\alpha + jn^{-1}) \tag{9.9}$$

Duplication formula ($n = 2$) :

$$\Gamma(2\alpha) = (2\pi)^{-\frac{1}{2}} 2^{2\alpha-\frac{1}{2}} \Gamma(\alpha)\, \Gamma(\alpha + \tfrac{1}{2}) \tag{9.10}$$

The incomplete gamma function :

$$\Gamma_y(\alpha) = \int_0^y e^{-t} t^{\alpha-1} dt \tag{9.11}$$

The incomplete gamma function ratio :

$$\frac{\Gamma_y(\alpha)}{\Gamma(\alpha)} \tag{9.12}$$

Stirling's formula :

$$\Gamma(\alpha+1) = \sqrt{2\pi}\alpha^{\alpha+\frac{1}{2}} e^{-(\alpha+\frac{\theta_\alpha}{12\alpha})} \tag{9.13}$$

for some $\theta_\alpha, 0 < \theta_\alpha < 1$.

(b) Gamma Distribution : A r.v. X is defined to have a gamma distribution with parameters $(\lambda, p)(\lambda > 0, p > 0)$ if its *pdf* is given

$$f_G(x; \lambda, p) = \begin{cases} \dfrac{\lambda}{\Gamma(p)} \quad e^{-\lambda x}(\lambda x)^{p-1}, & 0 < x < \infty \\ 0 \quad \text{otherwise} & \end{cases} \tag{9.14}$$

The mgf is

$$M_X(t) = E(e^{tX}) = \int_0^\infty \frac{\lambda^p}{\Gamma(p)} e^{tx} e^{-\lambda x} x^{p-1} dx$$

$$= \left(\frac{\lambda}{\lambda - t}\right)^p \int_0^\infty \frac{(\lambda - t)^p}{\Gamma(p)} x^{p-1} e^{-(\lambda-t)x} dx = \left(\frac{\lambda}{\lambda - t}\right)^p \tag{9.15}$$

Hence

$$\mu_1' = \frac{p}{\lambda}, \quad \mu_2' = \frac{p(p+1)}{\lambda^2}, \quad \mu_r' = \frac{\Gamma(p+r)}{\Gamma(p)\,\lambda^r}$$

$$\mu_2 = \frac{p}{\lambda^2}, \quad \mu_3 = \frac{2p}{\lambda^3}, \quad \mu_4 = \frac{3p^2 + 6p}{\lambda^4}$$

$$\sqrt{\beta_1} = \frac{2}{\sqrt{p}}, \quad \beta_2 = 3 + \frac{6}{p}$$

The cumulative distribution function is given by the incomplete gamma function ratio

$$F(x) = P(X < x) = \frac{\Gamma_x(p)}{\Gamma(p)}.$$

The distribution has a mode at $x = \frac{1}{\lambda}(p-1)$. There are points of inflexion, equidistant from the mode, at $x = p-1 \pm \sqrt{p-1}$ (provided the values are real and positive). When $p = 1$, (9.14) reduces to the exponential distribution (9.3).

The standard form of the gamma distribution is obtained by putting $\lambda = 1$, giving

$$f_G(x;p) = \frac{e^{-x}x^{p-1}}{\Gamma(p)} \quad (x \geq 0) \tag{9.16}$$

The standard distribution (9.16) has a single mode at $x = p - 1$ if $p \geq 1$. If $p < 1, f_G(x;p) \to \infty$ as $x \to 0$.

Some typical standard gamma distributions are shown in fig. 9.3. It is seen as p increases the shape of the curve becomes similar to that of the normal propability curve (discussed in section 9.7).

(c) χ^2-distribution A r.v. X is said to have a chi-square distribution with n degrees of freedom if its *pdf* is given by

$$f(x) = \frac{1}{2^{\frac{n}{2}}\Gamma(\frac{n}{2})}e^{-\frac{x}{2}}x^{\frac{n}{2}-1}, \quad 0 < x < \infty \tag{9.17}$$

Clearly $\frac{x}{2}$ has a standard gamma distribution (9.16) with parameter $p = \frac{n}{2}$.

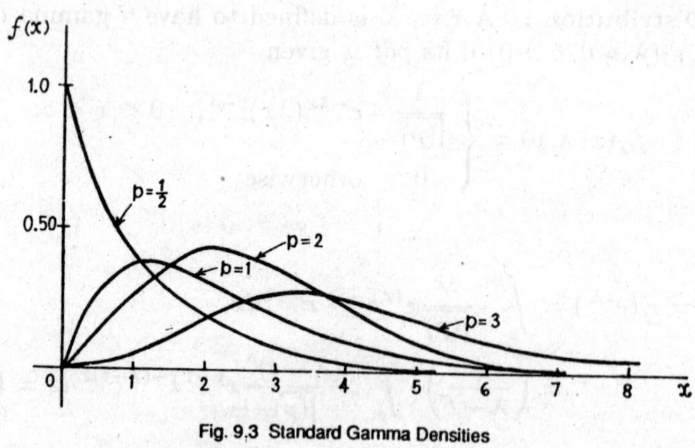

Fig. 9.3 Standard Gamma Densities

• **THEOREM 9.2.** *If $X_1, \cdots, X_n$ are independent r.v.'s such that X_j is distributed as gamma with parameters (λ, p_j) then their sum $S_n = \sum_{j=1}^{n} X_j$ is distributed as gamma with parameters (λ, p) where $p = \sum_{j=1}^{n} p_j$.*

Proof. The *mgf* of $\mathcal{S}_n = M_{\mathcal{S}_n}(t) = \prod_{j=1}^{n} M_{X_j}(t)$

$$= \prod_{j=1}^{n} \left(\frac{\lambda}{\lambda - t}\right)^{p_j} = \left(\frac{\lambda}{\lambda - t}\right)^{p}.$$

The result follows by the uniqueness property of the *mgf*.

COROLLARY 9.1. *If $X_1, \cdots, X_n$ are independent exponential variables each with parameter λ their sum $\mathcal{S}_n = \sum_{j=1}^{n} X_j$ is a gamma variable with parameters (λ, n).*

COROLLARY 9.2. *If $X_1, \cdots, X_n$ are independent chi-square variables, X_j having degree of freedom n_j, their sum $\mathcal{S}_n = \sum_{j=1}^{n} X_j$ has a chi-square distribution with degree of freedom $n = \sum_{j=1}^{n} n_j$.*

• **THEOREM 9.3.** *Let X be a poisson r.v. with parameter λ. Then*

$$P\{X \le t\} = \frac{1}{\overline{|(t+1)}} \int_{\lambda}^{\infty} e^{-x} x^t dx \tag{9.18}$$

(9.18) expresses the d.f. of X in terms of an incomplete gamma function.

Proof. Integrating by parts

$$\frac{1}{t!} \int_{\lambda}^{\infty} e^{-x} x^t dx = \frac{1}{t!} \lambda^t e^{-\lambda} + \int_{\lambda}^{\infty} \frac{1}{(t-1)!} x^{t-1} e^{-x} dx = \cdots$$

$$= e^{-\lambda}\left(1 + \lambda + \frac{\lambda^2}{2!} + \cdots + \frac{\lambda^t}{t!}\right)$$

❑ **EXAMPLE 9.1.** *The life-time T of a component has an exponential distribution*

$$f_T(t) = \lambda e^{-\lambda t} \quad (t \ge 0, \lambda > 0).$$

Assume that as soon as a component fails it is replaced by another component with exactly the same distribution of life-time. Show that the probability distribution of number of failures X in a period of fixed length τ is poisson with parameter $\lambda\tau$.

Let T_i denote the life-time of the i-th component.

$$P(X = k) = P(T_1 + \cdots + T_{k+1} > \tau) - P(T_1 + \cdots + T_k > \tau) \tag{i}$$

For a fixed m, $\sum_{j=1}^{m} T_j$ follows a gamma distribution with parameter (λ, m). Hence

$$P(T_1+\cdots+T_{k+1}>\tau)=\int_\tau^\infty \frac{\lambda}{\Gamma(k+1)}e^{-\lambda y}(\lambda y)^k\,dy$$
$$=\frac{1}{\Gamma(k+1)}\int_{\lambda\tau}^\infty e^{-z}z^k\,dz \qquad \text{where } z=\lambda y$$
$$=\sum_{j=0}^{k}e^{-\lambda\tau}\frac{(\lambda\tau)^j}{j!} \qquad \text{(by theorem 9.3)}$$

Hence from (i),

$$P(X=k)=\sum_{j=0}^{k}e^{-\lambda\tau}\frac{(\lambda\tau)^j}{j!}-\sum_{j=0}^{k-1}e^{-\lambda\tau}\frac{(\lambda\tau)^j}{j!}=e^{-\lambda\tau}\frac{(\lambda\tau)^k}{k!}.$$

9.5. Beta Distribution

(a) The Beta function

$$B(p,q)=\int_0^1 t^{p-1}(1-t)^{q-1}dt \qquad (p,q>0) \tag{9.19}$$
$$=2\int_0^{\frac{\pi}{2}}\sin^{2p-1}\theta\cos^{2q-1}\theta\,d\theta \tag{9.20}$$

In terms of gamma function

$$B(p,q)=\frac{\Gamma(p)\,\Gamma(q)}{\Gamma(p+q)}=B(q,p) \tag{9.21}$$

The incomplete beta function

$$B_y(p,q)=\int_0^y t^{p-1}(1-t)^{q-1}dt \tag{9.22}$$

The incomplete beta function ratio

$$I_y(p,q)=\frac{B_y(p,q)}{B(p,q)} \tag{9.23}$$

(b) Beta distribution A r.v. X is said to have a beta distribution with parameters $a,b(a>0,b>0)$ if its *pdf* is given by

$$f(x;a,b)=\begin{cases}\dfrac{1}{B(a,b)}x^{a-1}(1-x)^{b-1}, & 0<x<1, \quad a>0,b>0\\ 0 & \text{otherwise}\end{cases} \tag{9.24}$$

The rth raw moment

$$\mu_r'=\frac{1}{B(a,b)}\int_0^1 x^{r+a-1}(1-x)^{b-1}dx=\frac{B(r+a,b)}{B(a,b)}=\frac{\Gamma(r+a)\,\Gamma(a+b)}{\Gamma(a)\,\Gamma(r+a+b)}$$

In particular,

$$\mu_1' = \frac{a}{a+b}, \qquad \mu_2 = \frac{ab}{(a+b)^2(a+b+1)},$$

$$\mu_3 = \frac{2ab(b-a)}{(a+b)^3(a+b+1)(a+b+2)} \qquad \mu_4 = \frac{3ab\{ab(a+b-6)+2(a+b)^2\}}{(a+b)^4(a+b+1)(a+b+2)(a+b+3)}$$

$$\sqrt{\beta_1} = \frac{2(b-a)\sqrt{a+b+1}}{\sqrt{ab}(a+b+2)} \qquad \beta_2 = \frac{3(a+b+1)\{ab(a+b-6)+2(a+b)^2\}}{ab(a+b+2)(a+b+3)}$$

The mgf of the distribution is

$$M_X(t) = E(e^{tX}) = \frac{1}{B(a,b)}\int_0^1 e^{tx}x^{a-1}(1-x)^{b-1}dx.$$

The mgf does not have a simple form. However the moments of all orders exist and hence one can write

$$M_X(t) = \sum_{j=0}^{\infty}\mu_j'\frac{t^j}{j!} = \sum_{j=0}^{\infty}\frac{t^j}{j!}\cdot\frac{\Gamma(a+j)\,\Gamma(a+b)}{\Gamma(a)\,\Gamma(a+b+j)}$$

The cumulative distribution function is.

$$F_X(u) = P(X \le u) = \begin{cases} 0 & u \le 0 \\ \dfrac{1}{B(a,b)}\displaystyle\int_0^u x^{a-1}(1-x)^{b-1}dx & 0 < u < 1 \\ \dfrac{B_u(a,b)}{B(a,b)} = I_u(a,b) & \\ 1 & u > 1 \end{cases}$$

The incomplete Beta function $I_u(a,b)$ has been tabulated extensively.

If $a = b = 1$, (9.24) reduces to the continuous uniform distribution over (0, 1).

Fig. 9.4 shows some beta density curves for different values of (a, b).

The beta density can assume a variety of shapes depending on different values of parameters and hence this distribution has an wide application to model experiments of various types.

The distribution (9.24) is also called beta distribution of the first type.

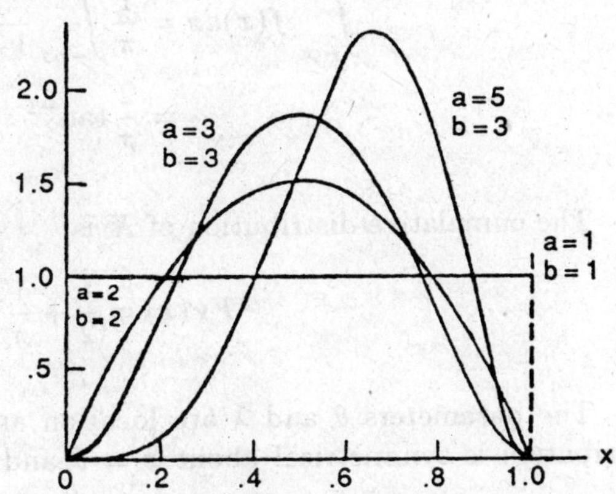

Fig. 9.4 Beta densities

(c) Beta distribution of the second type : A r.v. X is said to have a beta distribution of the second type with parameters $a, b (a > 0, b > 0)$ if its *pdf* is given by

$$f(x; a, b) = \begin{cases} \dfrac{1}{B(a,b)} \dfrac{x^{a-1}}{(1+x)^{a+b}} & x \geq 0, \quad a > 0, b > 0 \\ 0 & \text{otherwise} \end{cases} \tag{9.25}$$

The rth raw moment is,

$$\mu_r' = \frac{1}{B(a,b)} \int_0^\infty \frac{x^{a+r-1}}{(1+x)^{a+b}} dx.$$

Putting $(1+x) = \dfrac{1}{y}$

$$= \frac{1}{B(a,b)} \int_0^1 y^{b-r-1}(1-y)^{a+r-1} dy = \frac{B(b-r, a+r)}{B(a,b)} = \frac{\Gamma(b-r)\,\Gamma(a+r)}{\Gamma(a)\,\Gamma(b)}.$$

In particular,

$$\mu_1' = \frac{a}{b-1}, \quad \mu_2 = \frac{a(a+b-1)}{(b-1)^2(b-2)}$$

9.6. Cauchy Distribution

A r.v. X is said to have a Cauchy distribution with parameter θ and $\lambda (\lambda > 0)$ if its *pdf* is given by

$$f(x; \theta, \lambda) = \frac{\lambda}{\pi} \cdot \frac{1}{\lambda^2 + (x-\theta)^2}, \quad -\infty < x < \infty, \quad \lambda > 0 \tag{9.26}$$

We check that

$$\int_{-\infty}^{\infty} f(x) dx = \frac{1}{\pi} \int_{-\infty}^{\infty} \frac{dz}{1+z^2} \quad \text{putting } z = \frac{x-\theta}{\lambda}$$

$$= \frac{2}{\pi} \tan^{-1} z \Big|_0^\infty = 1.$$

The cumulative distribution of X is

$$F_X(x) = \frac{1}{2} + \frac{1}{\pi} \tan^{-1}\left(\frac{x-\theta}{\lambda}\right).$$

The parameters θ and λ are location and scale parameters respectively. The distribution is symmetrical about $x = \theta$ and hence the median is at $x = \theta$; the upper and lower quantiles are $\theta \pm \lambda$. the *pdf* has points of inflexion at $\theta \pm \dfrac{\lambda}{\sqrt{3}}$. The density function is maximum at $x = \theta$, hence $x = \mu$ is mode of the distribution.

The distribution does not possess finite moments of order greater than or equal to 1. Hence the distribution does not possess finite expected value or standard deviation (see example 6.4).

Considering,

$$\mu_1' = \frac{\lambda}{\pi}\int_{-\infty}^{\infty} \frac{x dx}{\lambda^2 + (x-\theta)^2} dx = \frac{1}{\pi}\int_{-\infty}^{\infty} \frac{(\theta + \lambda z)}{1+z^2} dz \quad \left(\text{putting } \frac{x-\theta}{\lambda} = z\right)$$

$$= \theta + \frac{\lambda}{\pi}\int_{-\infty}^{\infty} \frac{z dz}{1+z^2} = \theta + \frac{\lambda}{\pi} \lim_{\substack{A\to\infty \\ B\to\infty}} \int_{-A}^{B} \frac{z dz}{1+z^2} \tag{9.27}$$

The integral in (9.27) does not exist if A, B approach infinity independently. Consequently the mean of the distribution does not exist. However, when $A = B$, the limit of this integral exists and the limit is zero. Hence if we take the principal value of this integral, the value of the expression in (9.27) is θ.

The characteristic function of the distribution is

$$\Psi_X(t) = E(E^{itX}) = e^{it\theta - |t|\lambda} \qquad \text{(vide example 7.23)}$$

The standard form of the distribution, represented in fig. 9.5 is obtained by putting $\theta = 0, \lambda = 1$.

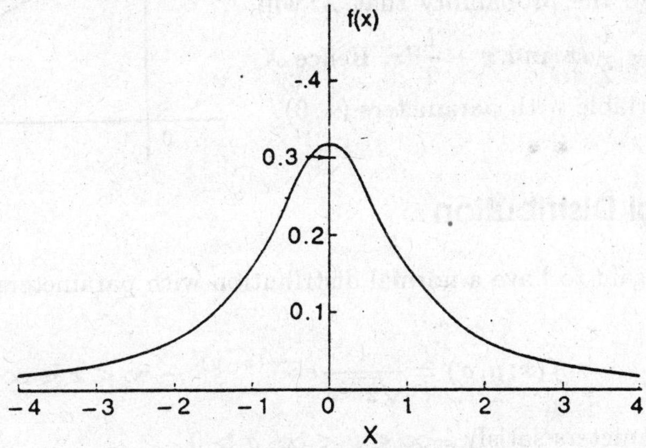

Fig. 9.5 Cauchy density

• **THEOREM 9.4.** *If $X_1, \cdots, X_n$ are independent cauchy variables, X_j having parameters $(\lambda_j, \theta_j), j = 1, \cdots, n$ then the sum $\mathcal{S}_n = \sum_{j=1}^{n} X_j$ is a cauchy variable with parameters $\lambda' = \sum_{j=1}^{n} \lambda_j, \theta' = \sum_{j=1}^{n} \theta_j$.*

Proof. The characteristic function of $\mathcal{S}_n$ is

$$\Psi_{\mathcal{S}_n}(t) = \prod_{i=1}^{n} \Psi_{X_i}(t) = \exp\left[it\sum_{j=1}^{n}\theta_j - |t|\sum_{j=1}^{n}\lambda_j\right]$$

which is the characteristic function of a cauchy variable with parameters λ', θ'. Hence the theorem follows by the uniqueness property of the characteristic function (theorem 7.7).

COROLLARY 9.3. *The arithmatic mean* $\bar{X} = \frac{1}{n}\sum_{i=1}^{n} X_i$ *of n independently and identically distributed cauchy variables with parameters* (λ, θ) *is a cauchy variable with the same set of parameters* (λ, θ).

❑ **EXAMPLE 9.2.** *A straight line AB is allowed to move through a fixed point A with coordinates* $(0, \lambda)$ *and the length of the intersection X it makes with the x-axis is noted. AB makes an angle* $\theta (= \angle OAP)$ *with the y-axis.. Assuming that* θ *has an uniform distribution between* $-\frac{\pi}{2}$ *(corresponding to* $OP = -\infty$*) and* $\frac{\pi}{2}$ *(corresponding to* $OP = \infty$*), find the probability density function of X.*

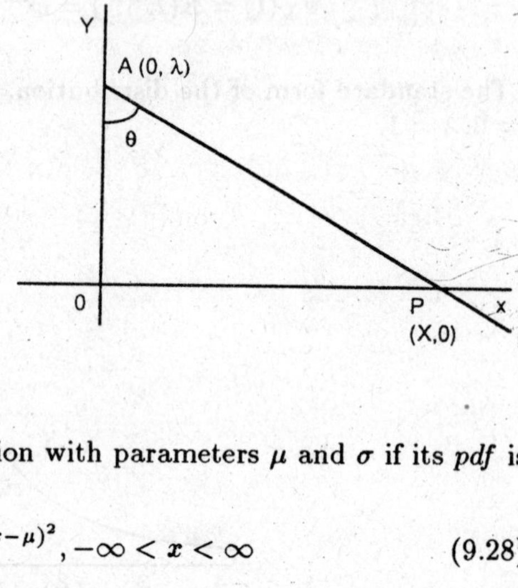

Relation between θ and X is $\theta = \tan^{-1}\left(\frac{X}{\lambda}\right)$. Probability that θ takes value between $\theta - \frac{1}{2}d\theta$ and $\theta + \frac{1}{2}d\theta$ is $\frac{d\theta}{\pi} = \frac{\lambda dx}{\pi(\lambda^2 + x^2)}$ and this is also the probability that X will lie between $x - \frac{1}{2}dx$ and $x + \frac{1}{2}dx$. Hence X is a cauchy variable with parameters $(\lambda, 0)$.

9.7. Normal Distribution

A r.v. X is said to have a normal distribution with parameters μ and σ if its *pdf* is given by

$$f(x; \mu, \sigma) = \frac{1}{\sqrt{2\pi}\,\sigma} e^{\frac{1}{2\sigma^2}(x-\mu)^2}, -\infty < x < \infty \tag{9.28}$$

where the parameters satisfy $-\infty < \mu < \infty$, $\sigma > 0$.

To show that

$$\int_{-\infty}^{\infty} f(x; \mu, \sigma) dx = 1 \tag{9.29}$$

we make the transformation $\frac{x - \mu}{\sigma} = z$. The lhs of (9.29) is

$$\frac{1}{\sqrt{2\pi}} \int_{-\infty}^{\infty} e^{-\frac{z^2}{2}} dz = I \quad \text{(say)}$$

$$I^2 = \frac{1}{2\pi} \int_{-\infty}^{\infty} e^{-\frac{z^2}{2}} dz \int_{-\infty}^{\infty} e^{-\frac{y^2}{2}} dy = \frac{1}{2\pi} \int_{-\infty}^{\infty} \int_{-\infty}^{\infty} e^{-\frac{y^2+z^2}{2}} dydz.$$

We make now polar transformation $y = r\cos\theta$, $z = r\sin\theta$,

$$0 < r < \infty, 0 < \theta < 2\pi$$

$$I^2 = \frac{1}{2\pi}\int_0^\infty \int_0^{2\pi} re^{-\frac{1}{2}r^2} dr d\theta = \int_0^\infty re^{-\frac{1}{2}r^2} dr = 1 \quad \text{when } I = 1.$$

The probability curve (9.28) is symmetrical about $x = \mu$, because $f(\mu+\lambda) = f(\mu-\lambda)$ for all values of λ. The ordinate decreases rapidly as $(x-\mu)$ increases numerically. The curve extends to infinity on either side of the ordinate at $x = \mu$. The distribution is unimodal with mode at $x = \mu$. The modal value of the probability density function is $(\sigma\sqrt{2\pi})^{-1} = \frac{.3979}{\sigma}$. Since the ordinate at $x = \mu$ divides the area under the normal curve into two equal parts, the median is also μ. The mean (= μ, shown below), median and the mode coincide for this distribution. The density function has points of inflexion at $x = \mu \pm \sigma$. $\left(\text{The points of inflexion are obtained by solving the equation } \frac{d^2 f(x)}{dx^2} = 0\right.$ and verifying that at these points $\left.\frac{d^3 f(x)}{dx^3} \neq 0\right)$. These are the points at which the curve of a normal distribution changes from concave upwords to convex downwards.

The figures 9.6 show some normal density curves for various values of μ and σ.

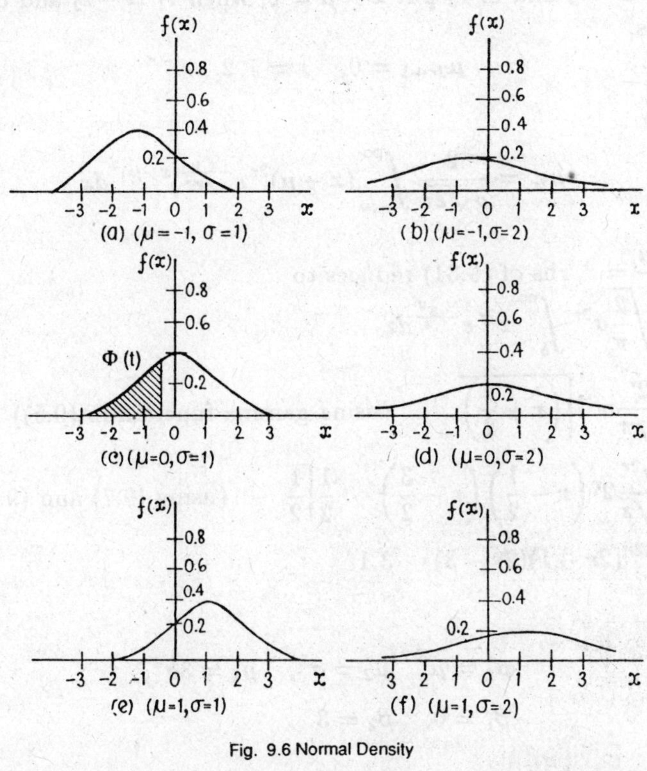

Fig. 9.6 Normal Density

The mean deviation about mean μ is

$$E|X-\mu| = \frac{1}{\sigma\sqrt{2\pi}}\int_{-\infty}^{\infty}|x-\mu|e^{-\frac{1}{2\sigma^2}(x-\mu)^2}dx$$

$$= \frac{\sigma}{\sqrt{2\pi}}\int_{-\infty}^{\infty}|z|e^{-\frac{z^2}{2}}dz \qquad \left(\text{putting } \frac{x-\mu}{\sigma}=z\right)$$

$$= \frac{\sigma}{\sqrt{2\pi}}\left[-\int_{-\infty}^{0} ze^{-\frac{z^2}{2}}dz + \int_{0}^{\infty} ze^{-\frac{z^2}{2}}dz\right]$$

$$= \sigma\sqrt{\frac{2}{\pi}}\int_{0}^{\infty} ze^{-\frac{z^2}{2}}dz = \sigma\sqrt{\frac{2}{\pi}} = .7979\sigma$$

Considering the odd order central moments

$$\mu_{2r+1} = \frac{1}{\sigma\sqrt{2\pi}}\int_{-\infty}^{\infty}(x-\mu)^{2r+1}e^{-\frac{1}{2\sigma^2}(x-\mu)^2}dx$$

$$= \frac{1}{\sigma\sqrt{2\pi}}\left[\int_{-\infty}^{\mu}(x-\mu)^{2r+1}e^{-\frac{1}{2\sigma^2}(x-\mu)^2}dx + \int_{\mu}^{\infty}(x-\mu)^{2r+1}e^{-\frac{1}{2\sigma^2}(x-\mu)^2}dx\right]$$

$$= \frac{1}{\sigma\sqrt{2\pi}}[I_1+I_2] \qquad \text{(say)}$$

In I_1 put $\mu - x = y$ and in I_2 put $x-\mu = y$, when $I_1 = -I_2$ and central moments of all odd orders,

$$\mu_{2r+1} = 0, \quad r = 1, 2, \cdots \tag{9.30}$$

Similarly,

$$\mu_{2r} = \frac{2}{\sigma\sqrt{2\pi}}\int_{-\infty}^{\infty}(x-\mu)^{2r}e^{-\frac{1}{2\sigma^2}(x-\mu)^2}dx \tag{9.31}$$

Putting $\frac{x-\mu}{\sigma} = z$, rhs of (9.31) reduces to

$$\sqrt{\frac{2}{\pi}}\sigma^{2r}\int_{0}^{\infty} z^{2r}e^{-\frac{z^2}{2}}dz$$

$$= \frac{2^r}{\sqrt{\pi}}\sigma^{2r}\Gamma\left(r+\frac{1}{2}\right) \qquad \text{(using gamma function in (9.5))}$$

$$= \frac{\sigma^{2r}}{\sqrt{\pi}}2^r\left(r-\frac{1}{2}\right)\left(r-\frac{3}{2}\right)\cdots\frac{1}{2}\Gamma\frac{1}{2} \qquad \text{(using (9.7) and (9.8))}$$

$$= \sigma^{2r}(2r-1)(2r-3)\cdots 3.1.$$

In particular,

$$\mu_1' = \mu, \quad \mu_2 = \sigma^2, \quad \mu_4 = 3\sigma^4,$$

$$\beta_1 = 0, \quad \beta_2 = 3.$$

The moment generating function about $x = 0$ is

$$M_X(t) = e^{\mu t+\frac{1}{2}t^2\sigma^2} \quad \text{(vide example 7.14)} \tag{9.32}$$

The mgf about mean, μ is

$$M_{X-\mu}(t) = e^{\frac{1}{2}t^2\sigma^2} \tag{9.33}$$

The cumulant generating function about $x = 0$ is

$$K_X(t) = \log_e M_X(t) = \mu t + \frac{1}{2}t^2\sigma^2,$$

Consequently

$$k_1 = \mu, \quad k_2 = \sigma^2, \quad k_r = 0, \qquad r \geq 3.$$

The characteristic function about $x = 0$ is

$$\psi_X(t) = e^{i\mu t-\frac{1}{2}t^2\sigma^2} \tag{9.34}$$

The c.f. about mean is

$$\psi_{X-\mu}(t) = e^{-\frac{1}{2}t^2\sigma^2}. \tag{9.35}$$

When X is normally distributed with mean μ and standard deviation σ it is customary to write X is distributed as $N(\mu, \sigma)$ or $X \sim N(\mu, \sigma)$.

- **THEOREM 9.5.** *If $X_1, \cdots, X_n$ are independent normal variables, X_i being $N(\mu_i, \sigma_i)$, then the variate $Y = \sum_{i=1}^{n} a_i X_i$ is an $N(\mu', \sigma')$ r.v. with $\mu' = \sum_{i=1}^{n} a_i\mu_i$, $\sigma' = \sqrt{\sum_{i=1}^{n} a_i^2\sigma_i^2}$, where a_i's are constants.*

Proof. The characteristic function of Y is

$$\begin{aligned}
\psi_Y(t) &= E(e^{itY}) = E(e^{it\Sigma_{i=1}^{n} a_i x_i}) \\
&= E\left\{\prod_{i=1}^{n} e^{ita_iX_i}\right\} = \prod_{i=1}^{n} E(e^{ita_iX_i}) \qquad \text{(by virtue of independence)} \\
&= \prod_{i=1}^{n} \psi_{a_iX_i}(t) = \prod_{i=1}^{n}(e^{a_i\mu_i t-\frac{1}{2}t^2\sigma_i^2a_i^2}) = e^{t\mu'-\frac{1}{2}t^2\sigma'^2}
\end{aligned}$$

which is the characteristic function of an $N(\mu', \sigma')$ r.v. Hence the result follows by the uniqueness property of characteristic function (theorem 7.7).

COROLLARY 9.4. *Let $X_1, \cdots, X_n$ be independent normal variables, X_i being $N(\mu_i, \sigma_i)$. Then the sum $S_n = \sum_{i=1}^{n} X_i$ is an $N(\mu', \sigma')$ variable where $\mu' = \sum_{i=1}^{n} \mu_i$ and $\sigma' = \sqrt{\sum_{i=1}^{n} \sigma_i^2}$.*

COROLLARY 9.5. *If $X_1, \cdots, X_n$ are independently and identically distributed $N(\mu, \sigma)$ variables, the arithmatic mean $\bar{X} = \frac{S_n}{n}$ is an $N(\mu, \frac{\sigma}{\sqrt{n}})$ r.v.*

The variable $\frac{X-\mu}{\sigma} = \tau$ (say) is called a standard or unit normal variable. Its *pdf* is

$$\frac{1}{\sqrt{2\pi}} e^{-\frac{\tau^2}{2}} \tag{9.36}$$

Clearly $E(\tau) = 0, V(\tau) = 1$,

$$M_\tau(t) = e^{\frac{t^2}{2}}, \quad \psi_\tau(t) = e^{-\frac{t^2}{2}} \tag{9.37}$$

The ordinate (9.36) of the unit normal variate is denoted as $\phi(\tau)$. Clearly $\phi(-t) = \phi(t)$, for any $t > 0$. Cumulative distribution function of τ,

$$F_\tau(u) = P[\tau \le u] = \frac{1}{\sqrt{2\pi}} \int_{-\infty}^{u} e^{-\frac{\tau^2}{2}} d\tau \tag{9.38}$$

is denoted as $\Phi(u)$, $\Phi(u)$ represents the area of a standard normal curve to the left of the ordinate at $\tau = u$ (fig. 9.6(c)). Clearly $\Phi(0) = .5, \Phi(\infty) = 1, \Phi(-t) = 1 - \Phi(t), P[|\tau| < t] = 2\Phi(t) - 1$ for any $t > 0$. The ordinates $\phi(t)$ and the area $\Phi(t)$ under the curve (9.38) upto the left of the ordinate at t have been extensively tabulated and are reproduced in tables 1 and 2 respectively at the appendix. It is seen

$$P\{|\tau| > 1\} = .318, \quad P\{|\tau| > 2\} = .046, \quad P\{|\tau| > 3\} = .0027$$

Hence for a $N(\mu, \sigma)$ variable X, $P\{|X - \mu| > 3\sigma\} = .002$. Thus practically a normal distribution is concentrated within three standard deviation units from the mean. It is also seen

$$\Phi(1.96) = .9750, \quad \Phi(2.58) = .9951.$$

If X is $N(\mu, \sigma)$, for any two integers $a < b$,

$$P(a \le X \le b) = \frac{1}{\sigma\sqrt{2\pi}} \int_a^b e^{-\frac{(x-\mu)^2}{2\sigma^2}} dx = \frac{1}{\sqrt{2\pi}} \int_{\frac{a-\mu}{\sigma}}^{\frac{b-\mu}{\sigma}} e^{-\frac{\tau^2}{2}} d\tau, \quad \tau = \frac{x-\mu}{\sigma}$$

$$= \Phi\left(\frac{b-\mu}{\sigma}\right) - \Phi\left(\frac{a-\mu}{\sigma}\right) \tag{9.39}$$

Normal distribution arises as a reasonable approximation to many random phenomena happening in nature. Many distributions such as those relating to measurements on plant and animal, gradation of students in test are approximately of the normal form. Many theoretical reasons for the use of the normal distribution are based on the forms of *central limit theorems* (discussed in chapter 12). These theorems state conditions under which the distribution of the standardised sum of random variables tends to a unit normal distribution as the number of variables in the sum increases. Many theoretical distributions, often of a complex nature, are approximated by a nomal distribution with the same expected value and standard deviation.

❑ **EXAMPLE 9.3.** *The results of a particular examination are as follows :*

RESULT	PERCENTAGE OF STUDENTS
Passed with distinction	15
Passed without distinction	42
Failed	43

It is known a candiate fails if he obtains less than 40% marks while he must obtain at least 75% marks in order to pass with distinction. Hence determine the mean and standard deviation of the distribution of marks assuming it to be normal.

Let μ and σ be the mean and standard deviation of the distribution of X, the marks of a student. Hence

$$F_X(40) = \Phi\left(\frac{40-\mu}{\sigma}\right) = .43 = \Phi(-.18), \quad F_X(75) = \Phi\left(\frac{75-\mu}{\sigma}\right) = .85 = \Phi(1.04)$$

because 43% of students secured less than 40 meakrs, 85% of students secured less than 75 marks. Hence

$$\frac{40-\mu}{\sigma} = -.18, \quad \frac{75-\mu}{\sigma} = 1.04 \qquad \text{when } \mu = 45.16, \sigma = 28.68.$$

• **THEOREM 9.6.** *(DeMoivre-Laplace theorem). Let X be a binomial variate with mean np and variance npq. If $n \to \infty$, the probability distribution of the standardised binomial variable $z = \dfrac{(X-np)}{\sqrt{npq}}$ tends to that of a unit normal variate.*

Proof.

$$P(X=r) = P(n,r) = \frac{n!}{r!(n-r)!}p^r q^{n-r}$$

Applying Stirling's approximation (9.13) to factorials,

$$\lim_{n\to\infty} P(n,r) = \frac{e^{-n}n^{n+\frac{1}{2}}p^r q^{n-r}}{\sqrt{2\pi}e^{-r}r^{r+\frac{1}{2}}e^{-(n-r)}(n-r)^{n-r+\frac{1}{2}}}$$

$$= \frac{A}{\sqrt{2\pi npq}} \quad \text{(say)}$$

where $A = \left(\frac{np}{r}\right)^{r+\frac{1}{2}} \left(\frac{nq}{n-r}\right)^{n-r+\frac{1}{2}} = \left(1 + z_r\sqrt{\frac{q}{np}}\right)^{-(np+z_r\sqrt{npq}+\frac{1}{2})}$
$\left(1 - z_r\sqrt{\frac{p}{nq}}\right)^{-(nq-z_r\sqrt{npq}+\frac{1}{2})}$

where $z_r = \frac{(r-np)}{\sqrt{npq}}$.

$$\log_e A = -\left(np + z_r\sqrt{npq} + \frac{1}{2}\right)\log_e\left(1 + z_r\sqrt{\frac{q}{np}}\right)$$
$$-\left(nq - z_r\sqrt{npq} + \frac{1}{2}\right)\log_e\left(1 - z_r\sqrt{\frac{p}{nq}}\right)$$

Assuming $|z_r| < \min\left(\sqrt{\frac{np}{q}}, \sqrt{\frac{nq}{p}}\right)$,

$$\log_e A = -\left(np + z_r\sqrt{npq} + \frac{1}{2}\right)\left[z_r\sqrt{\frac{q}{np}} - \frac{z_r^2 q}{2np} + \frac{z_r^3 q^{\frac{3}{2}}}{3(np)^{\frac{3}{2}}} + 0(n^{-2})\right]$$
$$+\left(nq - z_r\sqrt{npq} + \frac{1}{2}\right)\left[z_r\sqrt{\frac{p}{nq}} + \frac{z_r^2 p}{2nq} + \frac{z_r^3 p^{\frac{3}{2}}}{3(nq)^{\frac{3}{2}}} + 0(n^{-2})\right]$$
$$= \frac{z^2}{2} + \frac{z(q-p)}{2\sqrt{npq}} - \frac{z^2}{4npq}(p^2+q^2) - \frac{z^3}{6\sqrt{npq}}(q-p) + \frac{z^3}{6(npq)^{\frac{3}{2}}}(q^3-p^3) \tag{9.40}$$

As $n \to \infty$, (9.40) $\to \frac{z^2}{2}$.

As r takes values $0, 1, \cdots, n$ and $n \to \infty$, z takes values between $-\infty$ and ∞. Also as r increases by value 1, z increased by the amount $\frac{1}{\sqrt{npq}}$, which for large n, would be taken as dz. Hence z can be taken as a continuous variable taking vaues between $-\infty$ and ∞ and

$$\lim_{n\to\infty} P(n,r) = \frac{1}{\sqrt{2\pi}} e^{-\frac{z_r^2}{2}} dz, \quad -\infty \le z_r \le \infty \tag{9.41}$$

For an alternative proof see section 12.2.

COROLLARY 9.6. *Let X be a binomial variable with parameters (n,p). For a fixed a and b,*

$$P[a \le X \le b] \to \Phi\left(z_b + \frac{1}{2}\right) - \Phi\left(z_a - \frac{1}{2}\right) \tag{9.42}$$

as $n \to \infty$.

Proof. By (9.41)

$$P[a \le X \le b] \simeq \sum_{r=a}^{b} \phi(z_r) \tag{9.43}$$

as $n \to \infty$.

By mean value theorem of integral calculus $\left[\int_a^b f(x)dx = (b-a)f(x_0)\right.$ for an $\left. x_0 \epsilon [a, b]\right]$,

$$\Phi\left(z_r + \frac{1}{2}\right) - \Phi\left(z_r - \frac{1}{2}\right) = dz\phi(z_r') \qquad \text{for some}$$

$$z_r - \frac{1}{2} < z_r' < z_r + \frac{1}{2}.$$

Hence

$$\phi(z_r)dz = dz\phi(z_r')e^{\frac{1}{2}(z_r'^2 - z_r^2)} = \left[\Phi\left(z_r + \frac{1}{2}\right) - \Phi\left(z_r - \frac{1}{2}\right)\right] e^{\frac{1}{2}(z_r'^2 - z_o^2)}$$

Now

$$\frac{1}{2}|z_r'^2 - z_r^2| = \frac{1}{2}|z_r' - z_r||z_r' + z_r| < dz\left[|z_r| + \frac{dz}{4}\right] < \epsilon$$

for sufficiently large n. Hence

$$\phi(z_r)dz \simeq \Phi\left(z_r + \frac{1}{2}\right) - \Phi\left(z_r - \frac{1}{2}\right)$$

when the result follows from (9.43).

(9.42) shows that for large n an approximate value for the probability that a binomial random variable falls in an interval can be obtained from the standard normal distribution.

The following approximations hold for $\Phi(x)$.

(a) $\frac{1}{2}\left[1 + \sqrt{1 - e^{\frac{-x^2}{2}}}\right] \le \Phi(x) \le \frac{1}{2}\left[1 + \sqrt{1 - e^{-x^2}}\right]$

(b) $\Phi(x) \simeq \frac{1}{2}\left[1 + \left\{1 - \exp\left(-\frac{2x^2}{\pi}\right)\right\}^{\frac{1}{2}}\right]$

9.7.1. Truncated Normal Distribution

A r.v. X has a doubly truncated normal distribution if its *pdf* is given by

$$f_t(x; \mu, \sigma) = \frac{\frac{1}{\sigma\sqrt{2\pi}} e^{-\frac{1}{2\sigma^2}(x-\mu)^2}}{\int_A^B \frac{1}{\sigma\sqrt{2\pi}} e^{-\frac{1}{2\sigma^2}(x-\mu)^2} dx}, \quad (A \le x \le B) \tag{9.44}$$

$$= \left[\frac{1}{\sigma}\phi\left(\frac{x-\mu}{\sigma}\right)\right] \Big/ T$$

where $T = \Phi\left(\frac{B-\mu}{\sigma}\right) - \Phi\left(\frac{A-\mu}{\sigma}\right)$.

The distribution is singly truncated from below or above if B is replaced by ∞ or A is replaced by $-\infty$ respectively. The expected value and variance of the distribution in (9.44) is

$$E(X) = \mu + \frac{\sigma}{T}[\phi_1 - \phi_2]$$

$$V(X) = \sigma^2 \left[1 + \frac{\left(\frac{A-\mu}{\sigma}\right)\phi_1 - \left(\frac{B-\mu}{\sigma}\right)\phi_2}{T} - \left\{\frac{\phi_1 - \phi_2}{T}\right\}^2 \right]$$

where $\phi_1 = \phi\left(\frac{A-\mu}{\sigma}\right), \phi_2 = \phi\left(\frac{B-\mu}{\sigma}\right)$.

❑ **EXAMPLE 9.4.** *A normal distribution* $f(x) = \frac{1}{\sigma\sqrt{2\pi}} e^{-\frac{1}{2\sigma^2}(x-\mu)^2}, -\infty < x < \infty$ *is truncated at* $x = A$, *values* $x \le A$ *are discarded. Find the mean and the variance of the truncated distribution.*

$$\mu_1' = \int_A^\infty x f(x) dx \bigg/ \int_A^\infty f(x) dx$$

$$= \int_{A_\circ}^\infty (\mu + z\sigma) e^{-\frac{z^2}{2}} dz \bigg/ \int_{A_\circ}^\infty e^{-\frac{z^2}{2}} dz \quad \left(\text{where } z = \frac{x-\mu}{\sigma}, A_\circ = \frac{A-\mu}{\sigma}\right)$$

$$= \mu + \frac{\sigma}{M\sqrt{2\pi}} e^{-\frac{A_\circ^2}{2}} \quad \text{where } M = 1 - \Phi(A_\circ)$$

$$\mu_2' = \int_A^\infty x^2 f(x) dx \bigg/ \int_A^\infty f(x) dx = \frac{1}{M\sqrt{2\pi}} \int_{A_\circ}^\infty (\mu + \sigma_z)^2 e^{-\frac{z^2}{2}} dz$$

$$= \mu^2 + \frac{2\mu\sigma}{M\sqrt{2\pi}} e^{-\frac{A_\circ^2}{2}} + \frac{\sigma^2}{M\sqrt{2\pi}} \int_{A_\circ}^\infty z^2 e^{-\frac{z^2}{2}} dz$$

Now $$\int_{A_\circ}^\infty z^2 e^{\frac{-z^2}{2}} dz = z \int e^{\frac{-z^2}{2}} d\left(\frac{z^2}{2}\right)\bigg|_{A_\circ}^{\alpha} - \int_{A_\circ}^\infty \left[\int e^{\frac{-z^2}{2}} d\left(\frac{z^2}{2}\right)\right] dz$$

$$= -ze^{\frac{-z^2}{2}}\bigg|_{A_\circ}^\infty + \int_{A_\circ}^\infty e^{\frac{-z^2}{2}} dz = A_\circ e^{\frac{-A_\circ^2}{2}} + M\sqrt{2\pi}$$

Hence $$\mu_2 = \mu_2' - \mu_1'^2 = \sigma^2 \left[1 + \frac{A_\circ e^{\frac{-A_\circ^2}{2}}}{M\sqrt{2\pi}} - \frac{1}{2\pi M^2} e^{-A_\circ^2}\right].$$

9.8. Log-Normal Distribution

A r.v. X is said to have a log-normal distribution if $\log X$ has a normal distribution $(-\infty < \log X < \infty)$. If $\log X$ has mean μ and s.d. σ, the probability differential of $\log X$ is

$$f(\log x) d(\log x) = \frac{1}{\sigma\sqrt{2\pi}} e^{-\frac{1}{2\sigma^2}(\log x - \mu)^2} d(\log x), \quad -\infty < \log x < \infty \tag{9.45}$$

and hence

$$f(x)dx = \frac{1}{\sigma^x\sqrt{2\pi}} e^{-\frac{1}{2\sigma^2}(\log x-\mu)^2} dx, \quad 0 < x < \infty \tag{9.46}$$

$$\mu_r' = \int_0^\infty \frac{x^r}{\sigma^x\sqrt{2\pi}} e^{-\frac{1}{2\sigma^2}(\log x-\mu)^2} dx \tag{9.47}$$

Substituting $\dfrac{(\log x - \mu)}{\sigma} = z$

$$\mu_r' = \frac{1}{\sqrt{2\pi}} \int_{-\infty}^{\infty} e^{r\sigma z+\mu r-\frac{z^2}{2}} dz = e^{\mu r+\frac{r^2\sigma^2}{2}} \cdot \frac{1}{\sqrt{2\pi}} \int_0^\infty e^{-\frac{1}{2}(z-r\sigma)^2} dz$$

$$= e^{\mu r+\frac{r^2\sigma^2}{2}}; \tag{9.48}$$

μ_r', therefore, increases very rapidly with r.

In particular,

$$\mu_1' = e^{\mu+\frac{\sigma^2}{2}}, \quad \mu_2' = e^{2\mu+2\sigma^2}, \quad \mu_3' = e^{3\mu+\frac{9}{2}\sigma^2}, \quad \mu_4' = e^{4\mu+8\sigma^2}$$

Denoting $e^{\sigma^2} = \alpha$,

$$\mu_2 = \alpha(\alpha-1)e^{2\mu}, \quad \mu_3 = \alpha^{\frac{3}{2}}(\alpha-1)^2(\alpha+2)e^{3\mu}, \mu_4 = \alpha^2(\alpha-1)^2(\alpha^4+2\alpha^3+3\alpha^2-3)e^{4\mu}$$

In general

$$\mu_r = \frac{\mu_2^{\frac{r}{2}}}{(\alpha-1)^{\frac{r}{2}}} \sum_{j=0}^{r} (-1)^j \binom{r}{j} \alpha^{\frac{(r-j)(r-j-1)}{2}} \tag{9.49}$$

$$\sqrt{\beta_1} = \sqrt{(\alpha-1)}\,(\alpha+2), \quad \beta_2 = \alpha^4 + 2\alpha^3 + 3\alpha^2 - 3.$$

The curve is therefore positively skewed. Note that neither $\sqrt{\beta_1}$, nor β_2 depend on μ.

The coefficient of variation is $\sqrt{(\alpha-1)}$ and does not depend on μ.

The distribution is unimodal because $\dfrac{df(x)}{dx} = 0$ gives

$$-\frac{1}{\sigma x^2\sqrt{2\pi}} e^{-\frac{1}{2\sigma^2}(\log x-\mu)^2} \left[\frac{1}{\sigma^2}(\log x - \mu) + 1\right] = 0 \tag{9.50}$$

when the mode is $x_m = e^{(\mu-\sigma^2)}$.

The value X_α such that $P[X \le X_\alpha] = \alpha$ is related to the corresponding percentile z_α of the unit normal distribution

$$z = \frac{\log X - \mu}{\sigma} \tag{9.51}$$

as follows :

$$P[X \le X_\alpha] = \alpha \quad \Rightarrow \quad P[\log X \le \log X_\alpha] = \alpha$$

$$\Rightarrow \quad P\left[\frac{(\log X - \mu)}{\sigma} \le \frac{(\log X_\alpha - \mu)}{\sigma}\right] = \alpha$$

when

$$z_\alpha = \frac{\log X_\alpha - \mu}{\sigma} \quad \text{or} \quad X_\alpha = e^{(\mu+\sigma z_\alpha)}.$$

In particular for $\alpha = .50, z_\alpha = 0$, when the median of the distribution is e^μ. For this distribution, therefore,

$$\mu_1' > \text{median} > \text{mode} \quad \text{and} \quad \frac{\text{mode}}{\mu_1'} = \left(\frac{\text{median}}{\mu_1'}\right)^3.$$

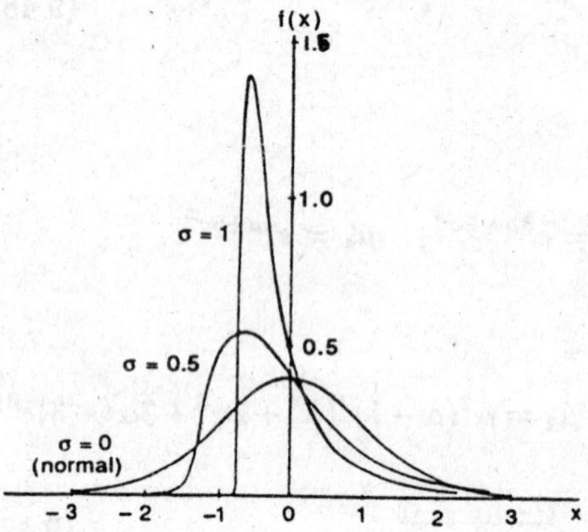

Fig. 9.7 Standardized Lognormal Distribution

The standardised log-normal distribution is

$$\xi = \frac{X - \mu_1'}{\sqrt{\mu_2}} = \frac{e^{z\sigma} - e^{\frac{\sigma^2}{2}}}{\left[e^{\sigma^2}(e^{\sigma^2} - 1)\right]^{\frac{1}{2}}} \tag{9.52}$$

where z is as defined in (9.51). As $\sigma \to 0$, distribution of ξ tends to a unit normal distribution.

Fig. 9.7 shows some standardised log-normal distributions.

A log-normal distribution often gives a good approximation to a normal distribution which has a small absolute value of coefficient of variation.

9.9. Double Exponential (Laplace) Distribution

A r.v. X is defined to have a double exponential distribution if its *pdf* is given by

$$f(x; \mu, \sigma) = \frac{1}{2\sigma} e^{-\frac{|x-\mu|}{\sigma}} \qquad (\sigma > 0) \tag{9.53}$$

when $-\infty < \mu < \infty$ and $\sigma > 0$. Here $E(X) = \mu, V(X) = 2\sigma^2$.

The distribution is symmetrical about $x = 0. \sqrt{\beta_1} = 0, \beta_2 = 6$. The mean deviation is $E(|X|) = 1$.

The mode is at $x = \mu$. The cumulative distribution function is

$$F_X(x) = \frac{1}{2}\exp\left[-\frac{(\mu - x)}{\sigma}\right] \quad (x \le \mu), \qquad 1 - \frac{1}{2}\exp\left[-\frac{(x-\mu)}{\sigma}\right] \quad (x \ge \mu).$$

The lower and upper quartiles are $\mu \pm \sigma \log_e 2$. Fig. 9.8 shows the form of the distribution.

A standard from of the distribution is obtained by putting $\mu = 0, \sigma = 1$ when the *pdf* is

$$f(x) = \frac{1}{2} e^{-|x|} \tag{9.54}$$

The characteristic function corresponding to (9.54) is

$$\psi_X(t) = E(e^{itX}) = \frac{1}{1+t^2}$$

(vide example 7.2)

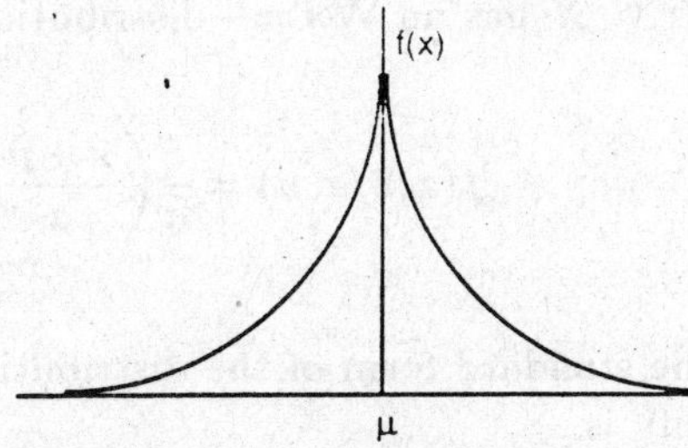

Laplace Density
Fig. 9.8

The mgf is

$$M_X(t) = \frac{1}{1-t^2}.$$

The central moments are

$$\mu_r = 0, \quad \text{if } r \text{ is odd; } \mu_r = r! \text{ if } r \text{ is even.}$$

The *cgf* is $K_X(t) = -\log(1-t^2)$. Thus

$$k_r = 0 \quad \text{if } r \text{ is odd; } k_r = 2[(r-1)!] \text{ if } r \text{ is even.}$$

9.10. Pareto Distribution

A r.v. X is said to have a Pareto distribution with parameters α, β if its *pdf* is given by

$$f(x) = \frac{\alpha\beta^\alpha}{x^{\alpha+1}} \quad \alpha > 0, x \geq \beta > 0 \tag{9.55}$$

The *cdf* is

$$F_X(x) = 1 - \left(\frac{\beta}{x}\right)^\alpha.$$

Provided $r < \alpha$,

$$\mu_r' = \frac{\alpha\beta^r}{\alpha - r}.$$

In particular,

$$\mu_1' = \frac{\alpha\beta}{\alpha-1}(\alpha > 1), \quad \mu_2 = \frac{\alpha\beta^2}{(\alpha-1)^2(\alpha-2)} \quad (\alpha > 2).$$

The mode of the distribution is at $x = \beta$. The median is $x_m = \beta 2^{\frac{1}{\alpha}}$. The population geometric mean is $\beta \cdot e^{\frac{1}{\alpha}}$. The population harmonic mean is $\beta\left(1+\dfrac{1}{\alpha}\right)$.

In many socio-economic investigations, like distribution of population sizes, occurrence of natural resources, size of firms, distribution of income over population, pareto curve gives a good fit to the data towards the extremes of the range, while log-normal distribution fits well over the other part of the data.

9.11. Weibull Distribution

A r.v. X has an Weibull distribution with parameters $c(>0), a(>0), \mu$ if its *pdf* is

$$f(x; c, a, \mu) = \frac{c}{a}\left(\frac{x-\mu}{a}\right)^{c-1} \exp\left[-\left(\frac{x-\mu}{a}\right)^{c}\right], \quad x > \mu \tag{9.56}$$

The standard form of the distribution is obtained by putting $\mu = 0, a = 1$ when the density is

$$f(x) = cx^{c-1}e^{-x^c} \quad (x > 0) \tag{9.57}$$

$$E(X) = \Gamma\left(\tfrac{1}{c}+1\right), \quad V(X) = \Gamma\left(\tfrac{2}{c}+1\right) - \left(\Gamma\left(\tfrac{1}{c}+1\right)\right)^2.$$

For $c = 1$ (9.57) reduces to the exponential density.

The distribution finds frequent use in the theory of reliability.

9.12. Extreme Value Distribution

The *pdf* for this distribution with parameters $\theta(>0), \mu$ is

$$f(x) = \theta^{-1}e^{-\frac{(x-\mu)}{\theta}} \exp[-e^{-\frac{(x-\mu)}{\theta}}] \tag{9.58}$$

$$E(X) = \mu - .57721\sigma, \quad V(X) = \pi^2\frac{\theta^2}{6}.$$

mode $= \mu$, median $= \mu + \sigma \log\log 2$

9.13. Logistic Distribution

The *cdf* for this distribution is

$$F_X(x) = \frac{1}{1+e^{-\frac{(x-\mu)}{\sigma}}}, \quad \sigma > 0 \tag{9.59}$$

The corresponding *pdf* is

$$f_X(x) = \frac{e^{-(x-\mu)/\sigma}}{\sigma[1+e^{-\frac{(x-\mu)}{\sigma}}]^2}, \quad -\infty < x < \infty, \sigma > 0 \tag{9.60}$$

$$E(X) = \mu, \quad V(X) = \frac{\sigma^2\pi^2}{3}.$$

The function (9.59) is has been widely used to represent growth functions with x representing time.

9.14. Exercise

1. Show that for the rectangular distribution

$$f(x) = \frac{1}{b}, \quad a \le x \le a+b$$

the mgf is $e^{at}\frac{[e^{bt}-1]}{bt}, \mu_1' = a + \frac{b}{2}$ (= median), $\mu_2 = \frac{b^2}{12}, \mu_r = 0$ for r odd, $= \left(\frac{b}{2}\right)^r \Big/ (r+1)$ for r even.

2. *X has a negative exponential distribution with parameter λ. If $P[X \le 1] = P[X > 1]$, find var (X).*

3. *$\bar{X}$ is mean of n independent $N(\mu, \sigma^2)$-variables. Show that for any*

$\epsilon > 0, \lim_{n\to\infty} P(|\bar{X} - \mu| \ge \epsilon) = 0.$

$$\left[\bar{X} \sim N\left(\mu, \frac{\sigma}{\sqrt{n}}\right); \quad P(|\bar{X} - \mu| \ge \epsilon) = 1 - \frac{1}{\sqrt{2\pi}} \int_{-\epsilon\sqrt{n}/\sigma}^{\epsilon\sqrt{n}/\sigma} e^{-\frac{t^2}{2}} dt \left(t = \frac{(\bar{x}-\mu)\sqrt{n}}{\sigma}\right).\right.$$

$$\left.\therefore \quad \lim_{n\to\infty} P|\bar{X} - \mu| \ge \epsilon) = 1 - \frac{1}{\sqrt{2\pi}} \int_{-\infty}^{\infty} e^{-\frac{t^2}{2}} dt = 0\right]$$

4. *Show that for the triangular distribution (fig. 9.9)*

$$f(x) = \begin{cases} \dfrac{2x}{A}, & 0 \le x \le A \\ \dfrac{2(1-x)}{(1-A)}, & A \le x \le 1 \end{cases}$$

$$\mu_1' = \frac{A+1}{3}, \quad \mu_2 = \frac{1}{18}(1 - A + A^2)$$

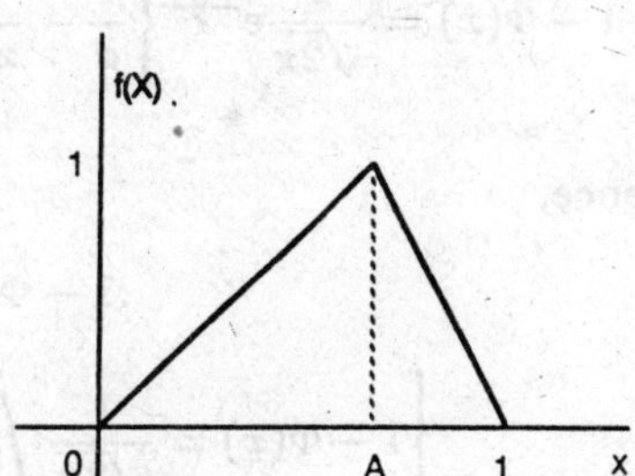

Fig. 9.9 Triangular Distribution

5. *Show that for Erlang distribution*

$$f(x) = \frac{\left(\frac{x}{b}\right)^{c-1} e^{-\frac{x}{b}}}{b[(c-1)!]}, \quad 0 \le x \le \infty, b > 0, c \text{ (integer) } > 0$$

mgf is $(1-bt)^{-c}, t > \frac{1}{b}; \mu_1' = bc, \mu_2 = b^2c, \sqrt{\beta_1} = \frac{2}{\sqrt{c}}, \beta_2 = 3 + \frac{6}{c}$, mode is $b(c-1)$.

6. *For a beta distribution with density*

$$f(x) = \frac{1}{B(p,q)}(1-x)^{p-1}x^{q-1}, \quad 0 < x < 1, p > 0, q > 0$$

verify that the harmonic mean is less than the arthmatic mean.

7. Show that for the power function distribution

$$f(x) = cx^{c-1}, \quad 0 \le x \le 1,$$

$$\mu'_r = \frac{c}{c+r}, \quad \mu_2 = \frac{c}{(c+2)(c+1)^2},$$

mode = 1 for $c > 1$, 0 for $c < 1$, median = $\left(\frac{1}{2}\right)^{\frac{1}{c}}$.

8. *Let N be a non-negative random integer with geometric distribution, $P(N = n) = (1 - e^{-\mu})e^{-n\mu}, n = 0, 1, 2, \cdots$ Let M be a positive integer with pmf*

$$P(M = m) = \frac{e^{-\mu}\mu^m}{m!(1 - e^{-\mu})}, \quad m = 1, 2, \cdots$$

Let $(X_1, \cdots, X_M)$ be M independent variables, each uniform over (0, 1). Show that

$$Y = \mu\{N + \min(X_1, \cdots, X_N)\}$$

has the standard exponential distribution.

9. *Show that for $x > 0$.*

$$1 - \Phi(x) = \frac{1}{\sqrt{2\pi}} e^{\frac{-x^2}{2}} \left\{ \frac{1}{x} - \frac{1}{x^3} + \frac{1.3}{x^5} - \frac{1.3.5}{x^7} + \cdots + (-1)^k \frac{1.3\cdots(2k-1)}{x^{2k+1}} \right\} \qquad \text{(i)}$$

Hence,

$$1 - \Phi(x) < \frac{\phi(x)}{x} \quad \text{for } x > 0. \qquad \text{(ii)}$$

$$\left[1 - \Phi(x) = \frac{1}{\sqrt{2\pi}} \int_x^\infty e^{\frac{-u^2}{2}} du \right.$$

$$\int_x^\infty e^{\frac{-u^2}{2}} du = \int_x^\infty \frac{1}{u}(ue^{\frac{-u^2}{2}}) du = \frac{1}{x} e^{\frac{-x^2}{2}} - \int_x^\infty \frac{1}{u^3}(ue^{\frac{-u^2}{2}}) du \qquad \text{(a)}$$

[Integrating by parts, taking $\frac{1}{u}$ as the first factor]

Similarly, evaluating $\int_x^\infty \frac{1}{u^3}(ue^{\frac{-u^2}{2}}) du$ by integration by parts (with $\frac{1}{u^3}$ as the first factor) and proceeding similarly the first part follows.

From (a) $\int_x^\infty e^{\frac{-u^2}{2}} du < \frac{1}{x} e^{-\frac{x^2}{2}}$ for $x > 0$. Hence the second part.]

10 PROBABILITY DISTRIBUTION ON R^n

10.1. Introduction

In chapter 5 we considered a r.v. X and its probability distribution on R^1. In many experiments, however, observations are expressed as a number of several numerical quantities. In tossing a pair of coins, for example, observations are (x, y), $x(y)$ denoting the face value turned up by the first (second) die. In such cases we have, therefore, a vector of random variables $\boldsymbol{X} = (X_1, \cdots, X_n)$ where each X_i is a random variable. In this chapter we shall study the probability distribution of n-dimensional random vector $\boldsymbol{X}$ on R^n. Section 10.2 deals with the concept of joint densities, marginal densities and conditional densities and the respective distribution functions. The concept of stochastic independence is also considered. Section 10.3 considers expectation of functions of random variables and moments, section 10.4 the generating functions. Conditional expectation, conditional variance, regression are considered in 10.5. The remaining two sections deal with multinomial distribution and bivariate normal distribution respectively.

10.2. Probability Distribution of a Random Vector

Definition 10.1. *Random Vector. Consider the probability space $(\mathcal{S}, \mathcal{A}, P)$ arising out of a random experiment. A vector of functions $(X_1, \cdots, X_n)$ which maps $\mathcal{S}$ into R^n is called a random vector if and only if for each $a_i \in R^1 (i = 1, \cdots, n)$,*

$$\{w : X_1(w) \leq a_1, \cdots, X_n(w) \leq a_n\} \subset \mathcal{A} \tag{10.1}$$

i.e.

$$\boldsymbol{X}^{-1}[I = \{(x_1, \cdots, x_n) : -\infty < x_i \leq a_i, \quad i = 1, \cdots, n\}] \subset \mathcal{A} \tag{10.2}$$

i.e. if $\boldsymbol{X}$ is a $\mathcal{A}$-measureable function. Here I is an n-dimensional interval.

❑ **EXAMPLE 10.1.** *A fair coin is tossed twice. Let us assign the score 1(2) for $H(T)$. Let X_1, X_2 denote respectively the number at the first toss, sum of scores at the two tosses. Here $\boldsymbol{X} = (X_1, X_2)$ and*

$$\boldsymbol{X}^{-1}\{(-\infty, a_1], (-\infty, a_2]\} = \begin{cases} \phi, & a_1 < 0, a_2 < 1 \\ (HH), & 0 < a_1 \leq 1, 1 < a_2 \leq 2 \\ (HT), & 0 < a_1 \leq 1; 2 < a_2 \leq 3 \\ (TH), & 1 < a_1 \leq 2, 2 < a_2 \leq 3 \\ (TT), & 1 < a_1 \leq 2, 3 < a_2 \leq 4 \\ \mathcal{S}, & a_1 > 2, a_2 > 4. \end{cases}$$

Thus X is a random vector.

We shall first consider in details the two dimensional random vector (X, Y). The same abbreviation r.v. will be used to denote a random vector as well a a random variable.

Definition 10.2. *The joint cumulative distribution function of (X, Y) is defined as*

$$F_{X,Y}(x, y) = P[X \leq x, Y \leq y] \quad \text{for all } (x, y) \epsilon R^2. \tag{10.3}$$

As in the case of a single r.v., the following properties hold for $F_{X,Y}(.,.)$.

(i) (a) $F_{X,Y}(-\infty, y) = \lim_{x \to -\infty} F(x, y) = 0$ for all y

(b) $F_{X,Y}(x, -\infty) = \lim_{y \to -\infty} F(x, y) = 0$ for all x

(ii) $F_{X,Y}(\infty, \infty) = \lim_{\substack{x \to \infty \\ y \to \infty}} F(x, y) = 1$

(iii) if $x_1 < x_2, y_1 < y_2$, $P[x_1 < X \leq x_2 \quad y_1 < Y \leq y_2]$

$$= F(x_2, y_2) - F(x_2, y_1) - F(x_1, y_2) + F(x_1, y_1) \geq 0$$

(iv) $F(x, y)$ is continuous from the right in each argument; that is, $\lim_{\epsilon \downarrow 0} F(x + \epsilon, y) = \lim_{\epsilon \downarrow 0} F(x, y + \epsilon) = F(x, y)$

Any function satisfying the properties (i)–(iv) above is a bivariate d.f.

Definition 10.3. *If (X, Y) takes only finite or countably infinite number of pair of values (x, y) in R^2 then the r.v. is of discrete type.*

Let $(x_i, y_j), i = 1, 2, \cdots; j = 1, 2, \cdots$ be the values assumed by (X, Y) with probability $P[X = x_i, Y = y_j] = p_{ij}$. p_{ij} is called the joint pmf of (X, Y) at the mass point (x_i, y_j) if

(i) $p_{ij} \geq 0$ for all $i = 1, 2, \cdots; j = 1, 2, \cdots$

(ii) $\sum_i \sum_j p_{ij} = 1$.

The d.f. is given by

$$F_{X,Y}(x, y) = P[X \leq x, Y = y] = \sum_{x_i \leq x} \sum_{y_j \leq y} p_{ij} \tag{10.4}$$

Every point (x_i, y_j) where $p_{ij} > 0$ is a jump-point of $F_{X,Y}(x, y)$ and p_{ij} is its jump at (x_i, y_j).

Definition 10.4. A r.v. (X, Y) is of continuous type if $F_{X,Y}(x, y)$ is continuous i.e., if there exists a function $f(x, y)(\geq 0)$ such that for every $(x, y)\epsilon R^2$,

$$F_{X,Y}(x, y) = \int_{-\infty}^{x} \left[\int_{-\infty}^{y} f(u, v)dv \right] du. \tag{10.5}$$

If $F(x, y)$ has partial derivatives of order up to two at (x, y)

$$\frac{\partial^2 F(x, y)}{\partial x \partial y} = f(x, y). \tag{10.6}$$

The function $f(x, y)$ is called the joint pdf of (X, Y).

Clearly

$$F_{X,Y}(\infty, \infty) = \lim_{\substack{x \to \infty \\ y \to \infty}} \int_{-\infty}^{x} \int_{-\infty}^{y} f(u, v)dudv = \int_{-\infty}^{\infty} \int_{-\infty}^{\infty} f(u, v)dudv = 1.$$

Again

$$\lim_{\substack{\delta x \to 0 \\ \delta y \to 0}} \frac{F(x + \delta x, y + \delta y) - F(x, y + \delta y) - F(x + \delta x, y) + F(x, y)}{\delta x \delta y} = \frac{\partial^2 F(x, y)}{\partial x \partial y}$$

i.e.,

$$\lim_{\substack{\delta x \to 0 \\ \delta y \to 0}} \frac{P[x < X \leq x + \delta x, y < Y \leq y + \delta y]}{\delta x \delta y} = f(x, y)$$

i.e.,

$$P[x < X \leq x + dx, y < Y \leq y + dy] = f(x, y)dxdy \tag{10.7}$$

Also

$$P[x < X \leq x + dx, y \leq Y] = F(x + dx, y) - F(x, y) = \frac{\partial F(x, y)}{\partial x \partial y} dx \tag{10.8}$$

Similarly

$$P[X \leq x, y < Y \leq y + dy] = \frac{\partial F(x, y)}{\partial x \partial y} dy \tag{10.9}$$

For any two pair of real mumbers $(a_1, b_1), (a_2, b_2), a_i < b_i (i = 1, 2)$,

$$P[a_1 < X \leq b_1, a_2 < Y \leq b_2] = \int_{a_2}^{b_2} \left[\int_{a_1}^{b_1} f(u, v)du \right] dv \tag{10.10}$$

❑ **EXAMPLE 10.2.** *A pair of fair dice are tossed. Let X denote the face value of the first die, Y the larger of two face values. Here (X, Y) is a discrete r.v. The joint pmf is given below :*

TABLE 10.1

Y \ X	1	2	3	4	5	6	Total
1	1/36	0	0	0	0	0	1/36
2	1/36	2/36	0	0	0	0	3/36
3	1/36	1/36	3/36	0	0	0	5/36
4	1/36	1/36	1/36	4/36	0	0	7/36
5	1/36	1/36	1/36	1/36	5/36	0	9/36
6	1/36	1/36	1/36	1/36	1/36	1/6	11/36
Total	1/6	1/6	1/6	1/6	1/6	1/6	1

Thus $P(X = 2, Y = 4) = \frac{1}{36}$. The d.f. can be calculated from the above table. For example, $F_{X,Y}(3,4) = \frac{12}{36}$.

Definition 10.5. *Marginal pmf.* If (X, Y) is a discrete r.v. with joint pmf $P(X = x_i, Y = y_j) = p_{ij}, i = 1, 2, \cdots; j = 1, 2, \cdots$ then $\{p_{io}\}$ where $p_{io} = \sum_j p_{ij}$ is called the marginal pmf of X. Clearly $p_{io} = P(X = x_i) \geq 0$ and $\sum_i p_{io} = 1$.

Similarly $\{p_{oj}\}$ where $p_{oj} = \sum_i p_{ij}$ is the marginal pmf of Y. $p_{.j} = P(Y = y_j)$.

For example 10.2, the column (row) totals give the marginal pmf of $X(Y)$. Thus $p_{io} = \frac{1}{6}, i = 1, \cdots, 6; p_{01} = \frac{1}{36}, p_{02} = \frac{3}{36}, p_{03} = \frac{5}{36}, p_{04} = \frac{7}{36}, p_{05} = \frac{9}{36}, p_{06} = \frac{11}{36}$.

Definition 10.6. *Marginal pdf.* If (X, Y) is a continuous r.v. with joint pdf $f(x, y)$, then

$$f_X(x) = \int_{-\infty}^{\infty} f(x, y)dy \tag{10.11}$$

$$f_Y(y) = \int_{-\infty}^{\infty} f(x, y)dx \tag{10.12}$$

are called the marginal pdf of X and the marginal pdf of Y respectively.

Clearly $f_X(x) \geq 0, f_Y(y) \geq 0, \int_{-\infty}^{\infty} f_X(x)dx = \int_{-\infty}^{\infty} f_Y(y)dy = 1$.

❑ **EXAMPLE 10.3.** *Let (X, Y) have joint pdf*

$$f_{X,Y}(x, y) = 2, \quad 0 < x < y < 1.$$

$$f_X(x) = \int_x^1 f(x, y)dy = 2(1 - x), \quad 0 < x < 1$$

$$f_Y(y) = \int_0^y 2dx = 2y, \quad 0 < y < 1.$$

❑ **EXAMPLE 10.4.** ***(X,Y) is distributed with a constant density inside a square R of side b. Find $f(x,y), F(x,y)$ and the marginal pdfs.***

$$f(x,y) = \frac{1}{b^2} \quad \text{for } (x,y) \in R$$
$$= 0 \quad \text{otherwise}$$
$$F(x,y) = 0 \quad \text{for } x \le 0, y \le 0.$$
$$F(x,y) = \int_0^x \int_0^y \frac{dx dy}{b^2} = \frac{xy}{b^2} \quad \text{for } 0 < x \le b, 0 < y \le b,$$
$$F(x,y) = \frac{by}{b^2} = \frac{y}{b} \quad \text{for } x > b, 0 < y \le b$$
$$F(x,y) = \frac{x}{b} \quad 1 \text{ for } 0 < x \le b, y > b$$
$$F(x,y) = 1 \text{ for } x > b, y > b.$$
$$f_X(x) = \int_0^b \frac{dy}{b^2} = \frac{1}{b} \quad \text{for } x \in (0,b)$$

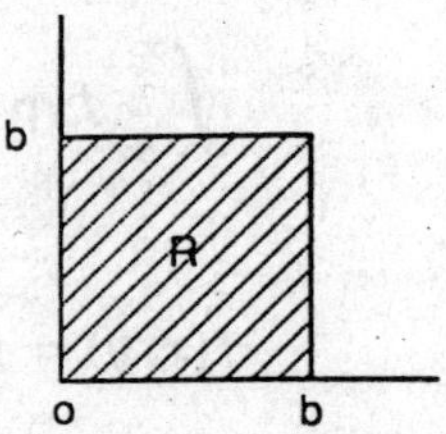

Fig. 10.1

Similarly, $f_Y(y) = \dfrac{1}{b}$ for $y \in (0,b)$.

Definition 10.7. *Conditional pmf.* If (X,Y) is a discrete r.v., with $P(Y = y_j) > 0$ then

$$P(X = x_i | Y = y_j) = \frac{P(X = x_i, Y = y_j)}{P(Y = y_j)} = \frac{p_{ij}}{p_{.j}} = p_{X|Y}(i|j) \quad \text{(say)}. \tag{10.13}$$

The set of values $\left\{\dfrac{p_{ij}}{p_{.j}}\right\}$ is called the conditional pmf of X given that $Y = y_j$.

Clearly $\dfrac{p_{ij}}{p_{.j}} \ge 0, \sum_j \left(\dfrac{p_{ij}}{p_{.j}}\right) = 1.$

Similarly the set of values $\left\{\dfrac{p_{ij}}{p_{i.}}\right\}$ is the conditional pmf of Y given that $X = x_i$. We shall denote $\dfrac{p_{ij}}{p_{i.}}$ as $P_{Y|X}(j|i)$.

Note that joint pmf

$$p_{ij} = p_{X|Y}(i|j)p_{.j} = p_{i.}p_{Y|X}(j|i) \tag{10.14}$$

Definition 10.8. *Conditional d.f.* (discrete case) If X, Y are discrete r.v.'s. with $P(Y = y_j) > 0$, the conditional d.f. of X given that $Y = y_j$ is

$$F_{X|Y}(x|y_j) = P(X \le x | Y = y_j) = \frac{1}{p_{.j}} \sum_{x_i \le x} p_{ij}. \tag{10.15}$$

Similarly the conditional d.f. of Y given that $X = x_i$ is

$$F_{Y|X}(y|x_i) = P(Y \le y | X = x_i) = \frac{1}{p_{i.}} \sum_{y_j \le y} p_{ij} \tag{10.16}$$

provided $p_{i.} > 0$.

Definition 10.9. *Conditional pdf.* Let X, Y be continuous r.v.'s. At every point (x, y) where f is continuous and $f_Y(y) > 0$, and is continuous, the conditional pdf of X given $Y = y$ is

$$f_{X|Y}(x|y) = \frac{f_{X,Y}(x,y)}{f_Y(y)} \quad (\geq 0). \tag{10.17}$$

(10.17) remains undefined at points where $f_Y(y) = 0$. Similarly the conditional pdf of Y given $X = x$ is

$$f_{Y|X}(y|x) = \frac{f_{X,Y}(x,y)}{f_X(x)} \quad (\geq 0)$$

provided $f_X(x) > 0$. Clearly

$$\int_{-\infty}^{\infty} f_{X|Y}(x|y)dx = \int_{-\infty}^{\infty} f_{Y|X}(y|x)dy = 1.$$

Note that

$$f(x,y) = f_{X|Y}(x|y)f_Y(y) = f_{Y|X}(y|x)f_X(x) \tag{10.18}$$

Definition 10.10 *Conditional d.f.* (continuous case). Let (X, Y) be a continous r.v. The conditional d.f. of X given Y is defined as

$$F_{X|Y}(x|y) = \int_{\infty}^{x} f_{X|Y}(u|y)du \tag{10.19}$$

for all y for which $f_Y(y) > 0$. Similarly

$$F_{Y|X}(y|x) = \int_{-\infty}^{y} f_{Y|X}(v|x)dv$$

for all x for which $f_X(x) > 0$.

Note : If one variable, say, X is discrete and Y is continuous, then also the joint density, marginal densities, conditional densities are similarly defined.

❑ **EXAMPLE 10.5.** *X, Y have joint pdf $f(x,y) = xe^{-x(y+1)} (x \geq 0, y \geq 0)$. Find marginal and conditinal pdf's.*

$$f_X(x) = xe^{-x}\int_0^{\infty} e^{-xy}dy = e^{-x},$$

$$f_Y(y) = \int_0^{\infty} xe^{-x(y+1)}dx = \frac{1}{(y+1)^2}\int_0^{\infty} ze^{-z}dz \quad \text{(putting } x(y+1) = z\text{)}$$

$$= \frac{1}{(y+1)^2}$$

$$f_X(x|y) = (y+1)^2 xe^{-x(y+1)}$$

$$f_Y(y|x) = xe^{-xy}.$$

❑ **EXAMPLE 10.6.** *For a given λ, X follows poisson with parameter λ. λ follows a gamma distribution*

$$f_\lambda(\lambda) = \frac{r^p}{\Gamma(p)} e^{-r\lambda} \lambda^{p-1}, \quad \lambda > 0, r > 0, p \text{ (integer) } > 0.$$

Find the overall distribution of X, its mean and variance.

Here (X, λ) is a r.v. with

$$f_X(x|\lambda) = e^{-\lambda}\frac{\lambda^x}{x!}$$

The joint pdf

$$f_{X,\lambda}(x,\lambda) = \frac{r^p}{\Gamma(p)\, x!} \lambda^{x+p-1} e^{-\lambda(r+1)}$$

$$f_X(x) = \int_0^\infty f_{X,\lambda}(x,\lambda) d\lambda = \frac{r^p \Gamma(x+p)}{x! \Gamma(p)\ (r+1)^{x+p}}$$

$$= \left(\frac{r}{r+1}\right)^p \frac{1}{(r+1)^x} \frac{(x+p-1)!}{(p-1)! x!} = \binom{x+p-1}{x}\left(\frac{1}{r+1}\right)^x \left(\frac{r}{r+1}\right)^p,$$

a negative binomial distribution with parameters p and $\left(\frac{r}{r+1}\right)$.

$$E(X) = \frac{p}{r}, \quad V(X) = \frac{p(r+1)}{r^2}.$$

❑ EXAMPLE 10.7. ***Suppose the conditional density (pmf)***

$$f_{X|Y}(x|y) = e^{-y}\frac{y^x}{x!}, \quad x = 0, 1, 2, \cdots$$

a poisson with parameter y. The r.v. Y is continuous with marginal pdf

$$f_Y(y) = e^{-y}, \quad y \geq 0.$$

Show that $f_X(x) = 2^{-(x+1)}, x = 0, 1, \cdots$

Joint pdf is

$$f_{X,Y}(x,y) = f_Y(y) f_{X|Y}(x|y) = e^{-2y}\frac{y^x}{x!}.$$

Hence

$$f_X(x) = \int_0^\infty f(x,y)dy = \int_0^\infty e^{-2y}\frac{y^x dy}{x!}$$

$$= \frac{\Gamma(x+1)}{x! 2^{x+1}} = \frac{1}{2^{x+1}} \qquad \text{(using gamma function)}$$

Consider now the conditional probability of an event A given the value of a r.v. X, both A and X being defined on the same probability space $(\mathcal{S}, \mathcal{A}, P)$.

If X is discrete

$$P(A|X = x) = \frac{P[A \cap (X = x)]}{P(X = x)} \tag{10.20}$$

Hence

$$P(A) = \sum_x P(A|X = x)P(X = x) \qquad \text{(by theorem 2.8)} \qquad (10.21)$$

Also the joint probability

$$P[A; X \in B] = \sum_{x \in B} P(A|X = x)P(X = x).$$

If X is continuous $P(X = x) = 0$. In this case

$$P(A|X = x) = \lim_{\epsilon \downarrow 0} P[A|x - \epsilon < X < x + \epsilon] \qquad (10.22)$$

provided $P[x - \epsilon < X < x + \epsilon] > 0$ for every ϵ and the limit in (10.22) exists. Thus

$$P(A) = \int_{-\infty}^{\infty} P[A|X = x]f_X(x)dx \qquad (10.23)$$

Again

$$P(A; X \in B) = \int_{x \in B} P[A|X = x]f_X(x)dx.$$

❑ **EXAMPLE 10.8.** *What is the probability that three points selected randomly on the circumference of a circule will lie on a semi-circle.*

Without loss of generality assume that the first point P_1 falls on the positive x-axis. The position of the second point P_2 is X (radian) measured from OP_1 in the anti-clockwise direction (Fig. 10.2). The pdf of X is

$$f(x) = \frac{1}{2\pi}, \quad 0 < x \leq 2\pi$$

since points are located at random. Let A be the event, P_1, P_2, P_3 are on a semi-circle.

Now

$$P(A|x) = \begin{cases} \dfrac{2\pi - x}{2\pi}, & (0 < x < \pi), \text{ since } A \text{ occurs if and only if } P_3 \text{ falls between } P_1' \text{ and } P_2' \text{ i.e., } P_3 \text{ falls on an arc of length } \pi - (x - \pi) = 2\pi - x. \\ \dfrac{x}{2\pi}, & (\pi < x < 2\pi) \end{cases}$$

Hence

$$P(A) = \int_0^{2\pi} P(A|x)f(x)dx$$

$$= \frac{1}{2\pi}\left[\int_0^{\pi} \frac{2\pi - x}{2\pi}dx + \int_{\pi}^{2\pi} \frac{x}{2\pi}dx\right]$$

$$= \frac{3}{4}$$

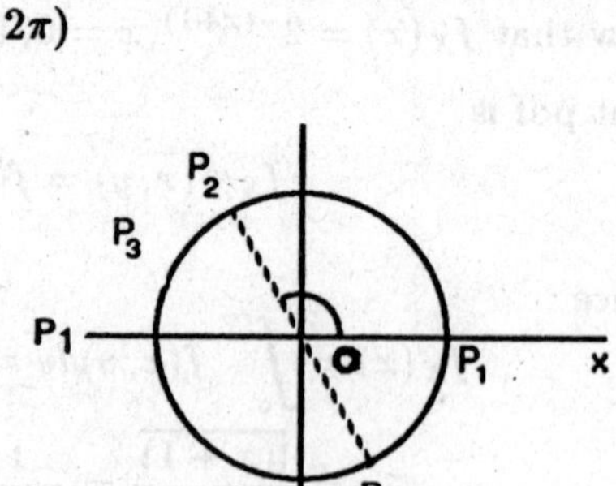

Fig. 10.2

We now consider distributions of a n-dimensional r.v. $\boldsymbol{X} = (X_1, \cdots, X_n)$.

Let $X_1, \cdots, X_n$ be n r.v.'s, all defined on the same probability space $(\mathcal{S}, \mathcal{A}, P)$. The joint cdf of $\boldsymbol{X}$ is

$$F_{\boldsymbol{X}}(x_1, \cdots, x_n) = P[X_1 \le x_1, \cdots, X_n \le x_n] \tag{10.24}$$

for all $(x_1, \cdots, x_n)$.

$F_{\boldsymbol{X}}(x_1, \cdots, x_n) = 0$ if at least one of x_i's is equal to $-\infty$ and $F(\infty, \cdots, \infty) = 1$. $\boldsymbol{X}$ is a n-dim. discrete r.v. if $\boldsymbol{X}$ assumes only a countable number of values $\boldsymbol{X} = (x_1, \cdots, x_n)$ in R^n. In this case the joint pmf of $X_1, \cdots, X_n$ is

$$p_{1,2,\cdots,n} = P[X_1 = x_1, \cdots, X_n = x_n] \ge 0 \tag{10.25}$$

satisfying $\sum_{i_1} \cdots \sum_{i_n} p_{i_1 \cdots i_n} = 1$.

$\boldsymbol{X}$ is n-dimensional continuous r.v. if the corresponding d.f. is continuous and if there exists a $f_{\boldsymbol{X}}(x_1, \cdots, x_n) (\ge 0)$ such that for all $(x_1, \cdots, x_n) \in R^n$,

$$F_{\boldsymbol{X}}(x_1, \cdots, x_n) = \int_{-\infty}^{x_n} \cdots \int_{-\infty}^{x_1} f(u_1, \cdots, u_n) du_1, \cdots du_n \tag{10.26}$$

If F has partial derivatives upto order n,

$$\frac{\partial^n F_{\boldsymbol{X}}(x_1, \cdots, x_n)}{\partial x_1 \cdots \partial x_n} = f_{\boldsymbol{X}}(x_1, \cdots, x_n)$$

$f_{\boldsymbol{X}}(x_1, \cdots, x_n)$ is defined to be the joint pdf of $X_1, \cdots, X_n$ and satisfies

$$\int_{-\infty}^{\infty} \cdots \int_{-\infty}^{\infty} f_{\boldsymbol{X}}(x_1, \cdots, x_n) dx_1 \cdots dx_n = 1. \tag{10.27}$$

For any set A

$$P\{\boldsymbol{X} \in A\} = \int_A \int f_{\boldsymbol{X}}(x_1, \cdots, x_n) dx_1 \cdots dx_n \tag{10.28}$$

The function $F_{\boldsymbol{X}}(\infty, \cdots, \infty, x_i, \infty, \cdots, \infty)$

$$= F_{X_i}(x_i) = P(X_i \le x_i) \tag{10.29}$$

is called the marginal d.f. of X_i. Similarly the marginal d.f. of any subset $(X_{i_1}, \cdots, X_{i_m})$ is obtained by setting all the remaining variables to be ∞.

For discrete $\boldsymbol{X}$ marginal pmf of a subset $(X_{i_1}, \cdots, X_{i_m})$ of $\boldsymbol{X}$ is

$$\begin{aligned} p_{i_1 \cdots, i_m} &= P[X_{i_1} = x_{i_1}, \cdots, X_{i_m} = x_{i_m}, \quad X_j < \infty, \quad j \ne (i_1, \cdots, i_m)] \\ &= \sum_{\text{all values of } X_j} \cdots \sum P[X_{i_1} = x_{i_1}, \cdots, X_{i_m} = x_{i_m}, X_j, j \ne (i_1, \cdots, i_m)] \end{aligned} \tag{10.30}$$

For continuous $\boldsymbol{X}$, marginal pdf of a subset $(X_{i_1}, \cdots, X_{i_m})$ of $\boldsymbol{X}$ is

$$f_{X_{i_1}, \cdots, X_{i_m}}(x_{i_1}, \cdots, x_{i_m}) = \int_{-\infty}^{\infty} \cdots \int_{-\infty}^{\infty} f_{\boldsymbol{X}}(x_1, \cdots, x_n) \prod_{\substack{j \\ (\ne i_1, \cdots, i_m)}} dx_j \tag{10.31}$$

For discrete $\boldsymbol{X}$, the conditional pmf of $X_{i_1}, \cdots, X_{i_k}$ given $X_{j_1} = x_{j_1}, \cdots, X_{j_{n-k}} = x_{j_{n-k}}$ is

$$P\{X_{i_1} = x_{i_1}, \cdots, X_{i_k} = x_{i_k} | X_{j_1} = x_{j_1}, \cdots, X_{j_{n-k}} = x_{j_{n-k}}\}$$

$$= \frac{p_{12\cdots n}}{p_{j_1,\cdots,j_{n-k}}} = p_{X_{i_1}\cdots X_{i_k}|X_{j_1}\cdots X_{j_{n-k}}} \quad (i_1, \cdots, i_k | j_1, \cdots, j_{n-k}) \tag{10.32}$$

The quantities in (10.32) are non-negative and their summation over all values of $X_{i_1}, \cdots, X_{i_k}$ is unity.

For continuous $\boldsymbol{X}$, the conditional *pdf* of $X_{i_1}, \cdots, X_{i_k}$ given $X_{j_1}, \cdots, X_{j_{n-k}}$ is

$$f_{X_{i_1},\cdots,X_{i_k}|X_{j_1}\cdots,X_{j_{n-k}}}(x_{i_1}, \cdots, x_{i_k} | x_{j_1} \cdots, x_{j_{n-k}}) = \frac{f_{X_1,\cdots,X_n}(x_1, \cdots, x_n)}{f_{X_{j_1},\cdots,X_{j_{n-k}}}(x_{j_1}, \cdots, x_{j_{n-k}})} \tag{10.33}$$

Definition 10.11. *Stochastic Independence.* The r.v.'s. $X_1, \cdots, X_n$ are said to be stochastically independent if and only if

$$F_{\boldsymbol{X}}(x_1, \cdots, x_n) = F_{X_1}(x_1)F_{X_2}(x_2) \cdots, F_{X_n}(x_n) \tag{10.34}$$

for all $(x_1, \cdots, x_n) \in R^n$.

If $\boldsymbol{X}$ is a n-dimensional discrete r.v., $X_1, \cdots, X_n$ are stochastically independent if

$$P(X_1 = x_1, \cdots, X_n = x_n) = P(X_1 = x_1) \cdots P(X_n = x_n) \tag{10.35}$$

for all $(x_1, \cdots, x_n) \in R^n$.

If $\boldsymbol{X}$ is a n-dimensional continuous r.v., $X_1, \cdots, X_n$ are stochastically independent if

$$f_{\boldsymbol{X}}(x_1, \cdots, x_n) = f_{X_1}(x_1) \cdots f_{X_n}(x_n) \tag{10.36}$$

for all $(x_1, \cdots, x_n) \in R^n$.

In case of two random variables, (10.35) implies that X, Y are independent if

$$\left.\begin{aligned} P(X = x|Y = y) = P(X = x) \\ P(Y = y|X = x) = P(Y = y) \end{aligned}\right\} \quad \text{for all } (x, y).$$

when X, Y are discrete. For continuous r.v.'s., (10.36) implies X, Y are inpendent if

$$\left.\begin{aligned} f_{X|Y}(x|y) = f_X(x) \\ f_{Y|X}(y|x) = f_Y(y) \end{aligned}\right\} \quad \text{for all } (x, y)$$

(10.35) is equivalent to the following definition : $X_1, \cdots, X_n$ are independent if

$$P[X_1 \in A_1, X_n \in A_n] = P(X_1 \in A_1) \cdots P(X_n \in A_n) \tag{10.37}$$

for all sets $A_1, \cdots, A_n \in \mathcal{B}_1$

The following important theorem holds :

• **THEOREM 10.1.** *Let $X_1, \cdots, X_n$ be independent r.v.'s. Let $g_i(X_i)$ be a function of X_i such that $g_i(X_i)$ is a r.v. $(i = 1, \cdots, n)$. Then $g_1(X_1), \cdots, g_n(X_n)$ are independent r.v.s.*

Proof. Let $A_1, \cdots, A_n$ be some borel sets in $\mathcal{B}_1$. For the sample space $(\mathcal{S}, \mathcal{A})$

$$\{w : g_i(X_i(w)) \in A_i\} \equiv \{w : X_i(w) \in g_i^{-1}(A_i)\}.$$

Hence

$$\begin{aligned} & P[g_1(X_1) \in A_1, \cdots, g_n(X_n) \in A_n] \\ &= P[X_1 \in g_1^{-1}(A_1), \cdots, X_n \in g_n^{-1}(A_n)] \\ &= P[X_1 \in g_1^{-1}(A_1)] \cdots, P[X_n \in g_n^{-1}(A_n)] \\ & \qquad \text{(by independence of } X_1, \cdots, X_n) \\ &= P[g_1(X_1) \in A_1] \cdots P[g_n(X_n) \in A_n]. \end{aligned}$$

❑ **EXAMPLE 10.9.** (X, Y) *have joint d.f.*

$$F_{X,Y}(x,y) = \begin{cases} (1-e^{-ax})y^2 & (0 \le x < \infty, 0 \le y < 1) \\ 1-e^{-ax} & (0 \le x < \infty, 1 < y < \infty) \\ 0 & \text{(otherwise)} \end{cases}$$

where a (> 0) is a constant. Here

$$F_X(x) = \begin{cases} 1-e^{-ax} & (0 \le x < \infty) \\ 0 & \text{otherwise} \end{cases}$$

$$F_Y(y) = \begin{cases} y^2 & (0 \le y \le 1) \\ 1 & (1 < y < \infty) \\ 0 & \text{otherwise} \end{cases}$$

$$f(x,y) = \begin{cases} 2aye^{-ax} & (0 \le x < \infty, 0 \le y \le 1) \\ 0 & \text{otherwise} \end{cases}$$

$$f_X(x) = \begin{cases} ae^{-ax} & (0 \le x < \infty) \\ 0 & \text{otherwise} \end{cases}$$

$$f_Y(y) = \begin{cases} 2y & (0 \le y \le 1) \\ 0 & \text{otherwise} \end{cases}$$

Note

$$\left\{ \begin{aligned} F_{X,Y}(x,y) &= F_X(x)F_Y(y) \\ f_{X,Y}(x,y) &= f_X(x)f_Y(y) \end{aligned} \right\} \quad \text{for all } (x,y)$$

Hence X, Y are independent.

❑ **EXAMPLE 10.10.** *Let X_1, X_2 be independently and identically distributed r.v.'s. with $P(X_i = \pm 1) = \frac{1}{2}, i = 1, 2$. Let $X_3 = X_1X_2$. Show that X_1, X_2, X_3 are pairwise independent but not independent.*

$$P(X_3 = 1) = P(X_1 = 1, X_2 = 1) + P(X_1 = -1, X_2 = -1)$$
$$= P(X_1 = 1)P(X_2 = 1) + P(X_1 = -1)P(X_2 = -1) = \frac{1}{4} + \frac{1}{4} = \frac{1}{2}$$

Similarly $P(X_3 = -1) = \frac{1}{2}$.

$$\begin{aligned}
P(X_1 = 1, X_3 = 1) &= P(X_1 = 1)P(X_3 = 1 | X_1 = 1)\\
&= P(X_1 = 1)\frac{P(X_1 = 1, X_3 = 1)}{P(X_1 = 1)}\\
&= P(X_1 = 1)\frac{P(X_1 = 1, X_2 = 1)}{P(X_1 = 1)}\\
&= \frac{1}{4} = P(X_1 = 1)P(X_3 = 1).
\end{aligned}$$

Thus it can be proved X_1, X_3 and X_2, X_3 are independent. However,

$$P(X_1 = 1, X_2 = 1, X_3 = -1) = P(X_1 = 1, X_2 = 1)P(X_3 = -1 | X_1 = 1, X_2 = 1) = 0.$$

Hence X_1, X_2, X_3 are not independent.

❑ **EXAMPLE 10.11.** $f(x, y) = \frac{1}{2}xy, 0 < y < x, 0 < x < 2.$

$$f_X(x) = \frac{1}{2}x\int_0^x y\,dy = \frac{x^3}{4}, \quad 0 < x < 2$$
$$f_Y(y) = \frac{1}{2}y\int_y^2 x\,dx = y - \frac{y^3}{4}, \quad 0 < y < 2$$

Thus $f_{X,Y}(x, y) \neq f_X(x)f_Y(y)$ and hence, X, Y are not stochastically independent.

10.3. Expectation and Moments of Random Variables

Let $g(\boldsymbol{X})$ be a (borel-measurable) function of $\boldsymbol{X}$ such that $g(\boldsymbol{X})$ is a random variable.

$$E\{g(\boldsymbol{X})\} = \int_{-\infty}^{\infty} \cdots \int_{-\infty}^{\infty} g(x_1, \cdots, x_n)dF_{\boldsymbol{X}}(x_1, \cdots, x_n) \tag{10.38}$$

provided $|g(x_1, \cdots, x_n)|$ is integrable (in the Riemann-Stieltjes sense) with respect to $F_{\boldsymbol{X}}(x_1, \cdots, x_n)$.

In particular,

$$E\{X_k\} = \int_{-\infty}^{\infty} \cdots \int_{-\infty}^{\infty} x_k dF_{\boldsymbol{X}}(x_1, \cdots, x_n) \tag{10.39}$$

provided the integral is absolutely convergent.

In case $\boldsymbol{X}$ is discrete

$$E\{g(\boldsymbol{X})\} = \sum_{x_1 \cdots x_n}\sum g(x_1 \cdots x_n) P[X_1 = x_1, \cdots, X_n = x_n] \tag{10.40}$$

provided the series in (10.40) is absolutely convergent.

In case $\boldsymbol{X}$ is continuous

$$E\{g(\boldsymbol{X})\} = \int_{-\infty}^{\infty} \int_{-\infty}^{\infty} g(x_1, \cdots, x_n) f(x_1, \cdots, x_n) dx_1, \cdots, dx_n \tag{10.41}$$

provided

$$\int_{-\infty}^{\infty} \int_{-\infty}^{\infty} |g(x_1, \cdots, x_n)| f(x_1, \cdots, x_n) dx_1, \cdots, dx_n < \infty.$$

When $g(\boldsymbol{X}) = X_1^{k_1} \cdots X_n^{k_n}, E\{g(\boldsymbol{X})\}$ in (10.38) defines the non-central moment

$$\mu'_{k_1, \cdots, k_n} \tag{10.42}$$

of the distribution $F_{\boldsymbol{X}}(x_1, \cdots, x_n)$.

In the case of two variables X, Y we have bivariate non-central moments

$$\mu'_{r,s} = E(X^r Y^s) \tag{10.43}$$

when

$$E(X) = \mu'_{1,0} = \mu_X \quad \text{(say)}, \quad E(Y) = \mu'_{0,1} = \mu_Y \quad \text{(say)} \tag{10.44}$$

which are mean of X and mean of Y respectively.

The bivariate central moments are

$$\mu_{rs} = E(X - \mu_X)^r (Y - \mu_Y)^s \tag{10.45}$$

We have

$$\mu_{20} = E(X - \mu_X)^2 = \sigma_X^2, \quad \mu_{02} = E(Y - \mu_Y)^2 = \sigma_Y^2 \tag{10.46}$$

which are variances of X and Y respectively.

The product moment

$$\mu_{11} = \mu_{X,Y} = E(X - \mu_X)(Y - \mu_Y) = E(XY) - \mu_X \mu_Y \tag{10.47}$$

is called the covariance between X and Y, often denoted as Cov (X, Y) or σ_{XY}.

It has been proved in section 6.3 that

$$\mu_{X,Y}^2 \le \sigma_X^2 \sigma_Y^2.$$

The quantity

$$\rho(X, Y) = \frac{\mu_{X,Y}}{\sigma_X \sigma_Y} \tag{10.48}$$

is defined as the correlation coefficient between X and Y, provided $\mu_{X,Y}, \sigma_X, \sigma_Y$ exist and $\sigma_X > 0, \sigma_Y > 0$.

Hence

$$\rho^2 \le 1 \quad \text{or} \quad -1 \le \rho \le 1, \tag{10.49}$$

$\rho^2 = 1$ if and only if X or Y is an exact linear function of the other (vide theorem 10.13)

❑ **EXAMPLE 10.12.**

$$f_{X,Y}(x,y) = \frac{3x+y}{4}e^{-x-y}, \quad x>0, y>0.$$

Find the marginal densities, conditional densities and $\rho(X,Y)$.

$$f_X(x) = e^{-x}\int_0^\infty \frac{3x+y}{4}e^{-y}dy = \frac{3x+1}{4}e^{-x}.$$

$$f_Y(y) = e^{-y}\int_0^\infty \frac{3x+y}{4}e^{-x}dx = \frac{y+3}{4}e^{-y}.$$

$$f_{X|Y}(x|y) = \frac{3x+y}{y+3}e^{-x}.$$

$$f_{Y|X}(y|x) = \frac{3x+y}{3x+1}e^{-y}, \quad y>0.$$

$$\mu'_{1,0} = \mu_X = \frac{1}{4}\int_0^\infty x(3x+1)e^{-x}dx = \frac{7}{4}$$

$$\mu'_{2,0} = 5, \quad \sigma_X^2 = \frac{31}{16}, \quad \mu'_{0,1} = \frac{5}{4},$$

$$\mu'_{0,2} = 3, \quad \sigma_Y^2 = \frac{23}{16}$$

$$\mu'_{1,1} = \frac{1}{4}\int_0^\infty\int_0^\infty xy(3x+y)e^{-x-y}dxdy = 2$$

$$\mu_{1,1} = -\frac{3}{16},$$

$$\rho(X,Y) = -\frac{3}{\sqrt{31\times 23}} = -.112$$

• **THEOREM 10.2.** *If $X_1, \cdots, X_n$ are any given set of r.v.'s. with finite means and if*

$$Y = c_0 + c_1X_1 + \cdots + c_nX_n$$

where $c_0, c_1, \cdots, c_n$ are any given set of constants, then

$$E(Y) = c_0 + c_1E(X_1) + \cdots + c_nE(X_n) \tag{10.50}$$

Proof.

$$E(Y) = \int_{-\infty}^\infty \cdots \int_{-\infty}^\infty \left(c_0 + \sum_{i=1}^n c_iX_i\right)dF_{\mathbf{X}}(x_1,\cdots,x_n)$$

$$= c_0 + \sum_{i=1}^n c_i \int_{-\infty}^\infty \cdots \int_{-\infty}^\infty x_i dF_{\mathbf{X}}(x_1,\cdots,x_n) = c_0 + \sum_{i=1}^n c_iE(X_i)$$

by (10.39) and by the fact

$$\int_{-\infty}^{\infty}\cdots\int_{-\infty}^{\infty} dF_{\mathbf{X}}(x_1,\cdots,x_n)=1.$$

• **THEOREM 10.3.** *If $X_1,\cdots,X_n$ are independent r.v.'s. with finite means*

$$E(X_1\cdots X_n)=E(X_1)\cdots E(X_n) \tag{10.51}$$

Proof.

$$\begin{aligned}E(X_1\cdots X_n)&=\int_{-\infty}^{\infty}\cdots\int_{-\infty}^{\infty} x_1\cdots x_n dF_{\mathbf{X}}(x_1,\cdots,x_n)\\&=\int_{-\infty}^{\infty}\cdots\int_{-\infty}^{\infty} x_1\cdots x_n dF_{X_1}(x_1)\cdots dF_{X_n}(x_n)\quad\text{(by (10.34))}\\&=\left[\int_{-\infty}^{\infty} x_1 dF_{X_1}(x_1)\right]\cdots\left[\int_{-\infty}^{\infty} x_n dF_{X_n}(x_n)\right]=E(X_1)\cdots E(X_n).\end{aligned}$$

COROLLARY 10.1. *If $X_1,\cdots,X_n$ are independent r.v.'s. and if $g_i(X_i)$ is a (borel-measureable) function of X_i such that $g_i(X_i)$ $(i=1,\cdots,n)$ is a r.v. then*

$$E\{g_1(X_1)\cdots g_n(X_n)\}=E(g_1(X_1))\cdots E(g_n(X_n)) \tag{10.52}$$

Proof follows from the fact that $g_1(X_1),\cdots g_n(X_n)$ are independent r.v.'s.

❑ **EXAMPLE 10.13.**

$$E(e^{tX_1+\cdots+tX_n})=E\left(\prod_{i=1}^{n}e^{tX_i}\right)=\prod_{i=1}^{n}E(e^{tX_i})$$

i.e., the mgf of sum of n independent r.v.'s. is the product of mgf of each r.v.

Two r.v.'s. X,Y are said to be uncorrelated if

$$E(XY)=E(X)E(Y) \tag{10.53}$$

i.e., if Cov $(X,Y)=0$.

It follows from theorem 10.3 that if X,Y are independent with finite means and if $\mu'_{X,Y}$ exists then X,Y are uncorrelated. The converse is not necessarily true (vide example 4.13 and exercises 11, 12 of section 10.8).

If X and Y are uncorrelated then in general

$$E\{g(X)h(Y)\}\neq E\{g(X)\}E\{h(Y)\},$$

whereas, if they are independent

$$E\{g(X)h(Y)\}=E\{g(X)\}E\{h(Y)\}$$

Defition 10.12. Two random variables are orthogonal if

$$E(XY)=0 \tag{10.54}$$

If X, Y are uncorrelated,

$$E(X - E(X))(Y - E(Y)) = 0$$

i.e., $(X - E(X))$ and $(Y - E(Y))$ are orthogonal.

If X, Y are orthogonal,

$$E(X + Y)^2 = E(X^2) + E(Y^2).$$

If X, Y are uncorrelated and mean of at least one of them is zero then they are also orthogonal.

• **THEOREM 10.4.** *(approximate evaluation of expectation and variance of $g(X)$). Let X be a r.v. with $E(X) = \mu, V(X) = \sigma^2$. Then*

$$E\{g(X)\} \simeq g(\mu) + \frac{g''(\mu)}{2}\sigma^2 \tag{10.55}$$

$$V\{g(X)\} \simeq [g'(\mu)]^2\sigma^2 \tag{10.56}$$

provided $g(X)$ is at least twice differentiable at $X = \mu$.

Proof. Expanding $g(X)$ in a Taylor series about $X = \mu$,

$$g(X) = g(\mu) + (X - \mu)g'(\mu) + \frac{(X - \mu)^2}{2!}g''(\mu) + \Delta \tag{10.57}$$

where Δ is a remainder. Discarding Δ and taking expectation of both sides (10.55) follows.

Ignoring the last two terms in the right hand side of (10.57) and taking variance of both sides, (10.56) follows.

• **THEOREM 10.5.** *Let X, Y be r.v.'s. with finite variances σ_X^2, σ_Y^2. Then*

$$V(X + Y) = V(X) + V(Y) + 2 \text{ Cov } (X, Y).$$

Proof.

$$\begin{aligned} V(X + Y) &= \int_{-\infty}^{\infty}\int_{-\infty}^{\infty} \{x + y - E(X) - E(Y)\}^2 dF_{X,Y}(x, y) \\ &= \int_{-\infty}^{\infty}\int_{-\infty}^{\infty} \{(x - E(X)) + (y - E(Y)\}^2 dF_{X,Y}(x, y) \\ &= \int_{-\infty}^{\infty}\int_{-\infty}^{\infty} (x - E(X))^2 dF(x, y) + \int_{-\infty}^{\infty}\int_{-\infty}^{\infty} (y - E(Y))^2 dF(x, y) \\ &\qquad + 2\int_{-\infty}^{\infty}\int_{-\infty}^{\infty} (x - E(X))(y - E(Y)) dF(x, y) \\ &= V(X) + V(Y) + 2 \text{ Cov } (X, Y), \end{aligned}$$

since means and covariances exist as variances exist.

• **THEOREM 10.6.** *Let $X_1, \cdots, X_n$ be r.v.'s. with $\mu_i = E(X_i), V(X_i) = \sigma_i^2$, Cov $(X_i, X_j) = \sigma_{ij}$ $(i \neq j)$, all finite. Let*

$$Y = c_o + \sum_{i=1}^{n} c_i X_i$$

where $c_j (j = 0, \cdots, n)$ are arbitrary constants. Then

$$V(Y) = \sum_{i=1}^{n} c_i^2 \sigma_i^2 + \sum_{i \neq j=1}^{n}\sum c_i c_j \sigma_{ij}.$$

Proof. Follows as in the proof of theorem 10.5.

• **THEOREM 10.7.** *Let*

$$U = \sum_{i=1}^{n} c_i X_i, \quad V = \sum_{i=1}^{n} d_i X_i$$

be two linear functions of $X_1, \cdots, X_n, c_i, d_i$ $(i = 1, \cdots, n)$ being arbitrary constants. Then

$$\text{Cov } (U, V) = \sum_{i=1}^{n}\sum_{j=1}^{n} c_i d_j \sigma_{ij}$$

where $\sigma_{ii} = V(X_i), \sigma_{ij} =$ Cov $(X_i, X_j), i \neq j$, provided the variances and covariances exist. Proof is straightforward.

COROLLARY 10.2. If $U = a + cX, V = b + dY$,

(i) Cov $(U, V) = cd$ Cov (X, Y).

(ii) $\rho(U, V) = \{\text{sign } (cd)\}\rho(X, Y)$.

❑ **EXAMPLE 10.14.** *Let $X_1, \cdots, X_n$ be independent r.v.'s. with $E|X_i^2| < \infty$ and $V(X_i) = \sigma_i^2, i = 1, \cdots, n$. Let $S = \sum_1^n a_i X_i, a_1, \cdots, a_n$ are constants, $\sum_{i=1}^{n} a_i = 1$. Find a_i's so that $V(S)$ is minimum.*

$$V(S) = \sigma \quad \text{(say)} \quad = \sum_{i=1}^{n} a_i^2 \sigma_i^2 = \sum_{j=1}^{n-1} a_j^2 \sigma_j^2 + (1 - a_1 - \cdots - a_{n-1})^2 \sigma_n^2$$

For minimising σ w.r.t. a_i's we have

$$\frac{\partial \sigma}{\partial a_1} = 2a_1\sigma_1^2 - 2(1 - a_1 - \cdots - a_{n-1})\sigma_n^2 = 0$$

$$\cdots\cdots\cdots\cdots\cdots\cdots\cdots\cdots$$

$$\frac{\partial \sigma}{\partial a_{n-1}} = 2a_{n-1}\sigma_{n-1}^2 - 2(1 - a_1 - \cdots - a_{n-1})\sigma_n^2 = 0$$

It follows that

$$a_j \sigma_j^2 = a_n \sigma_n^2, \quad j = 1, \cdots, n-1.$$

i.e. $a_j = \frac{\lambda}{\sigma_j^2}$, λ the constant of proportionality.

$$\sum_1^n a_j = 1 \text{ gives } \lambda = \left(\sum_1^n \frac{1}{\sigma_j^2}\right)^{-1}.$$

The minimum value of σ is $\sigma_{\min} = \lambda^2 \sum_1^n \frac{1}{\sigma_i^2} = \left(\sum_1^n \frac{1}{\sigma_i^2}\right)^{-1}$.

As in the proof of theorem 10.4 the following theorem can be proved.

- **THEOREM 10.8.**

$$E\{g(X,Y)\} \simeq g(\mu_X, \mu_Y) + \frac{1}{2}\left[\mu_{20}\frac{\partial^2 g}{\partial x^2} + 2\mu_{11}\frac{\partial^2 g}{\partial x \partial y} + \mu_{02}\frac{\partial^2 g}{\partial y^2}\right]_{\mu_X, \mu_Y} \tag{10.58}$$

$$V\{g(X,Y)\} \simeq \left[\left(\frac{\partial g}{\partial x}\right)^2 \mu_{20} + \left(\frac{\partial g}{\partial y}\right)^2 \mu_{02} + 2\left(\frac{\partial g}{\partial x}\right)\left(\frac{\partial g}{\partial y}\right)\mu_{11}\right]_{\mu_X, \mu_Y} \tag{10.59}$$

when $\mu_X = E(X), \mu_Y = E(Y)$.

10.4. Generating Functions

Definition 10.12. *Joint moment generating function.* The joint mgf of the r.v. $\boldsymbol{X} = (X_1, \cdots, X_n)$ is defined as

$$M_{\boldsymbol{X}}(t_1, \cdots, t_n) = M_{\boldsymbol{X}}(\boldsymbol{t}) = E\left[e^{\Sigma_1^n t_i X_i}\right] \quad (\text{where } \boldsymbol{t} = (t_1, \cdots, t_n)) \tag{10.60}$$

if the expectation exists for all values of $t_1, \cdots, t_n$ such that $|t_j| < T$ for some $T(> 0), j = 1, \cdots, n$.

The raw moments are obtained as

$$\mu'_{r_1, \cdots, r_n} = \frac{\partial^{r_1 + \cdots + r_n}}{\partial t_1^{r_1} \cdots \partial t_n^{r_n}} M_{\boldsymbol{X}}(\boldsymbol{t})\Big|_{\boldsymbol{t}=\boldsymbol{0}} \tag{10.61}$$

where $\boldsymbol{0} = (0, \cdots, 0)$, provided the required partial derivatives exist and provided the order of differentiation and integration are interchangeable. In particular

$$\mu'_r = \frac{\partial^r}{\partial t_1^r} M_{\boldsymbol{X}}(\boldsymbol{t})\Big|_{\boldsymbol{t}=\boldsymbol{0}}$$

$$\mu'_{r,s} = \frac{\partial^{r+s}}{\partial t_1^r \partial t_2^s} M_{\boldsymbol{X}}(\boldsymbol{t})\Big|_{\boldsymbol{t}=\boldsymbol{0}}$$

Thus expanding $M_{\boldsymbol{X}}(\boldsymbol{t})$ in a Maclaurin series, one obtains

$$M_{\boldsymbol{X}}(\boldsymbol{t}) = \sum_{r_1, \cdots, r_n} \frac{t_1^{r_1} \cdots t_n^{r_n}}{r_1! \cdots r_n!} \mu'_{r_1, \cdots, r_n} \tag{10.62}$$

The marginal mgf's are obtained as

$$M_{X_1}(t_1) = M_{\mathbf{X}}(t_1, 0, \cdots, 0) = \lim_{\substack{t_i \to 0 \\ i=2,\cdots,n}} M_{\mathbf{X}}(t_1, t_2, \cdots, t_n)$$

$$M_{X_2}(t_2) = M_{\mathbf{X}}(0, t_2, 0, \cdots, 0) = \lim_{\substack{t_i \to 0 \\ i(\neq 2)=1,\cdots,n}} M_{\mathbf{X}}(t_1, \cdots, t_n)$$

If the variables $X_1, \cdots, X_n$ are mutually independent,

$$M_{\mathbf{X}}(t) = \prod_{i=1}^{n} E(e^{t_i X_i}) = \prod_{i=1}^{n} M_{X_i}(t_i) \tag{10.63}$$

i.e., the joint mgf is the product of separate mgf's.

- **THEOREM 10.9.** *The mgf $M(t_1, t_2)$, when it exists, determines the joint distribution of (X, Y) and conversely if the mgf exists, it is unique.*

Proof. Omitted.

- **THEOREM 10.10.** *Two jointly distributed r.v.'s. X, Y are independent if and only if*

$$M_{X,Y}(t_1, t_2) = M_X(t_1) M_Y(t_2) \tag{10.63$'$}$$

for all t_1, t_2 such that $|t_i| < T(> 0), i = 1, 2$.

Proof. If X, Y are independent (10.64) follows from (10.63). If (10.63) holds,

$$\iint e^{t_1 X + t_2 Y} dF(x, y) = \int e^{t_1 x} dF(x) \int e^{t_2 y} dF(y) = \iint e^{t_1 x + t_2 y} dF(x) dF(y)$$

Hence

$$dF(x, y) = dF(x) dF(y)$$

by the uniqueness property of mgf (theorem 10.9).

The theorem can be generalised to n r.v.'s.

❑ **EXAMPLE 10.15.** *Let (X, Y) have a bivariate uniform distribution over the rectangle (R) defined by*

$$a \leq x \leq a + k, \quad b \leq y \leq b + l, \quad k > l > 0.$$

The pdf is

$$f_{X,Y}(x, y) = \begin{cases} \dfrac{1}{kl} & \text{for } (x, y) \in R \\ 0 & \text{otherwise} \end{cases}$$

The mgf is

$$M_{X,Y}(t_1, t_2) = \int_a^{a+k} \int_b^{b+l} e^{t_1 x + t_2 y} \cdot \frac{1}{kl} dx dy = -\frac{\exp[at_1 + bt_2]}{klt_1 t_2} (e^{kt_1} - 1)(e^{lt_2} - 1)$$

❑ **EXAMPLE 10.16.** *X_1, X_2, X_3, X_4 are independently and identically distributed with pdf $e^{-x}, x > 0$. Find the mgf of the joint distribution of U, V, W where $U = X_1 + X_2, V = X_3 + X_2, W = X_4 + X_2$.*

$$\begin{aligned}
\text{Mgf} &= \int_0^\infty\int_0^\infty\int_0^\infty e^{ut_1+vt_2+wt_3}e^{-x_1-x_2-x_3-x_4}dx_1\cdots dx_4.\\
&= \int_0^\infty\int_0^\infty\int_0^\infty e^{x_1(t_1-1)+x_2(t_1+t_2+t_3-1)+x_3(t_2-1)+x_4(t_3-1)}dx_1\cdots dx_4\\
&= \int_0^\infty e^{-x_1(1-t_1)}dx_1\int_0^\infty e^{-x_2(1-t_1-t_2-t_3)}dx_2\int_0^\infty e^{-x_3(1-t_2)}dx_3\int_0^\infty e^{-x_4(1-t_3)}dx_4\\
&= \{(1-t_1)(1-t_2)(1-t_3)(1-t_1-t_2-t_3)\}^{-1}
\end{aligned}$$

Definition 10.13. ***Bivariate probability generating function.*** For a pair of integer-valued r.v.'s. X, Y with a joint density function $P\{X=i, Y=j\} = p_{ij}, i, j = 0, 1\cdots$.we define the pgf

$$P_{X,Y}(t_1,t_2) = \sum_i\sum_j p_{ij}t_1^it_2^j, \quad |t_i|<1, i=1,2 \tag{10.64}$$

The pgf of marginal distributions of X, Y are respectively,

$$P_X(t_1) = P(t_1,1), \quad P_Y(t_2) = P(1,t_2).$$

It follows that pgf of $X+Y$ is $P(t,t)$. X, Y are independent if and only if

$$P(t_1,t_2) = P(t_1,1)P(1,t_2) \quad \text{for all } t_1,t_2 \text{ with } |t_i|<1, \quad i=1,2.$$

❑ **EXAMPLE 10.17.** *Consider*

$$P(t_1,t_2) = e^{\mu_1(t_1-1)+\mu_2(t_2-1)+\lambda(t_1t_2-1)}, \mu_1,\mu_2>0$$

which has a power series expansion with positive coefficients adding upto unity. $P(t_1,t_2))$ is the pgf of a bivariate probability distribution. $P(t_1,1)$ (or $P(1,t_2)$) is the pgf of a Poisson distribution with parameter $\mu_1+\lambda$ (or $\mu_2+\lambda$).

❑ **EXAMPLE 10.18.** *$\{X_n\}$ is a sequence of independently and identically distributed (iid) r.v.'s. with pmf $P\{X_k=j\} = p_j, j=1,2,\cdots$ and the pgf $P_X(t) = \sum p_jt^j, |t|<1$. Let*

$$S_N = X_1+\cdots+X_N$$

where N is a r.v. with pmf $P\{N=n\} = g_n$ and pgf $G_N(t) = \sum_0^\infty g_nt^n, |t|<1$. Find the pgf of S_N.

In particular find the pgf of S_N when X_i's are Bernoulli variables with $P(X_i=1) = p, P(X_i=0) = q$ and N is a Poisson variable with mean λ.

$$P(S_N=j) = \sum_{n=0}^\infty P(N=n)P(S_n=j) = h_j \quad \text{(say)}$$

Hence pgf of S_N

$$= H_{S_N}(t) = \sum_{0}^{\infty} h_j t^j = \sum_{j=0}^{\infty}\left\{\sum_{n=0}^{\infty} g_n P(S_n = j)\right\}t^j$$

$$= \sum_{n=0}^{\infty} g_n \left\{\sum_{j=0}^{\infty} P(S_n = j)t^j\right\} = \sum_{n=0}^{\infty} g_n P_{S_n}(t) \qquad \text{(where } P_{S_n}(t) = \text{pgf of } S_n)$$

$$= \sum_{n=0}^{\infty} g_n [P(t)]^n \qquad \text{(since } X_1, \cdots, X_n \text{ are } iid)$$

$$= G_N(P(t))$$

When X is a Bernoulli variable, its pgf $P_X(t) = q + pt, |t| < 1$. N being a poisson variable, its pgf $G_N(t) = e^{-\lambda(1-t)}, |t| < 1$. Hence pgf of S_N is

$$G_N(P(t)) = e^{-\lambda(1-q-pt)} = e^{-\lambda p(1-t)}.$$

Therefore S_N is a poisson variable with mean λp.

The joint cumulant generating function $\log_e M_{\mathbf{X}}(t)$ is defined similarly as in the case a single random variable, details of which are omitted here. The joint characteristic function $\psi_{\mathbf{X}}(t)$ is analogously defined.

10.5. Conditional Expectation, Variance, Regression

Definition 10.14. Let (X, Y) be two r.v.'s. The conditional expected value of $g(X, Y)$ given $X = x$ is defined as

$$E[g(X,Y)|X = x] = \int_{-\infty}^{\infty} g(x,y) f_{Y|X}(y|x)dy \qquad (10.65)$$

if (X, Y) is a continuous r.v. For discrete (X, Y),

$$E[g(X,Y)|X = x] = \sum_{y} g(x,y)P[Y = y|X = x] \qquad (10.66)$$

where the summation is over all possible values of Y.

In particular if $g(X, Y) = Y$, (10.65), (10.66) define respectively the conditional expectation of Y given x,

$$E(Y|x) = \mu_{Y|x} = \int_{-\infty}^{\infty} y f_{Y|x}(y|x)dy \quad \text{or} \quad \sum_{y} yP[Y = y|X = x] \qquad (10.67)$$

according as (X, Y) is continuous or discrete.

Similarly the conditional expectation of X given $Y = y$, $E(X|y) = \mu_{X|y}$ can be defined.

In general $\mu_{Y|x}$ is a function of x and similarly $\mu_{X|y}$ a function of y.

Definition 10.15. The graph of $E(Y|x)$ as a function of x,

$$y = E(Y|x) \tag{10.68}$$

is called the regression curve of Y on x. The graph of $E(X|y)$ as a function of y,

$$x = E(X|y) \tag{10.69}$$

is called the regression curve of X on y.

• **THEOREM 10.11.** *For two r.v.'s.* (X, Y)

$$E[E\{g(X)|Y)\}] = E[g(X)] \tag{10.70}$$

Proof. We prove the result when (X, Y) is jointly continuous. The left hand side

$$= \int_{-\infty}^{\infty}\left[\int_{-\infty}^{\infty} g(x) f_{X|Y}(x|y)dx\right] f_Y(y)dy = \int_{-\infty}^{\infty}\int_{-\infty}^{\infty} g(x) f(x,y)dxdy = E[(g(X)].$$

As a particular case,

$$E[E(X|Y)] = E(X).$$

Similarly

$$E[E(Y|X)] = E(Y).$$

• **THEOREM 10.12.** *If one of the regression curves is a constant,*

$$\sigma_{X,Y} = \rho(X, Y) = 0.$$

Proof. Suppose $E(Y|x) = k$ (a constant). Then by theorem 10.11, $E(Y) = k$. Hence, assuming X, Y are continuous,

$$\begin{aligned}
\sigma_{X,Y} &= E(X - \mu_X)(Y - \mu_Y) = E(XY) - k\mu_X \\
&= \int_{-\infty}^{\infty}\int_{-\infty}^{\infty} xyf(x,y)dxdy - k\mu_X \\
&= \int_{-\infty}^{\infty} x\left[\int_{-\infty}^{\infty} yf_{Y|X}(y|x)\right] f_X(x)dx - k\mu_X = k(\mu_X - \mu_X) = 0
\end{aligned}$$

(The case when X, Y are discrete can be proved similarly).

• **THEOREM 10.13.** $\rho(X, Y) = \pm 1$ *if and only if there exist constants a, b such that*

$$P(Y = aX + b) = 1$$

i.e., $Y = aX + b$ a.e. Further, if $a < 0$, then $\rho = -1$ and if $a > 0, \rho = 1$.

Proof. (i) Let $Y = aX + b$ a.e. Then $V(Y) = a^2V(X)$, Cov $(X, Y) = aV(x)$ and hence $\rho = \dfrac{a}{|a|} = \pm 1$.

(ii) If $\rho = \pm 1$, $[\text{Cov}\,(X, Y)]^2 = V(X)V(Y)$ which holds by the conditions of equality in Cauchy-Schwartz inequality, if and only if $t(X - \mu_X) = Y - \mu_Y$ a.e. for some t (vide section 6.3) i.e., if and only if $Y = aX + b$ a.e. for some constants a and b i.e., X, Y are perfectly linearly related.

Definition 10.16. *Conditional variance.* The conditional variance of Y given $X = x$ is

$$V(Y|X = x) = E[Y^2|X = x] - [E(Y|X = x)]^2$$

• **THEOREM 10.14.**

$$V(Y) = E[V(Y|X)] + V[E(Y|X)] \tag{10.71}$$

Proof.

$$\begin{aligned} E[V(Y|X)] &= E[E(Y^2|X)] - E[E(Y|X)]^2 \\ &= \left\{E(Y^2) - (E(Y))^2\right\} - \left\{E[E(Y|X)]^2 - (E(Y))^2\right\} \\ &= V(Y) - \left[E[E(Y|X)]^2 - \{EE(Y|X)\}^2\right] = V(Y) - V\{E(Y|X)\}. \end{aligned}$$

• **THEOREM 10.15.** *The following theorem is easy to prove.*

(i) $E[g_1(X) + g_2(X)|Y] = E[g_1(X)|Y] + E[g_2(X)|Y]$ (10.72)

(ii) $E[g_1(X)g_2(Y)|X = x] = g_1(x)E[g_2(Y)|X = x]$

The definitions of conditional expectation, conditional variance can be extended to more than two dimensions. For example

$$E[g(X_1, \cdots, X_n)|X_1 = x_1, \cdots, X_m = x_m], \quad m < n$$
$$= \int_{-\infty}^{\infty}\int_{-\infty}^{\infty} g(x_1, \cdots, x_n) f_{X_{m+1},\ldots,X_n|X_1,\ldots,X_m}(x_{m+1},\ldots,x_n|x_1,\ldots,x_m)dx_{m+1}\cdots dx_n$$

in case $(X_1, \cdots, X_n)$ is a n-dimensional random vector.

❑ **EXAMPLE 10.19.** *For the pdf in example 10.3,*

$$f_{X|Y}(x|y) = \frac{1}{y}, \quad 0 < x < 1$$

$$f_{Y|X}(y|x) = \frac{1}{1-x}, \quad 0 < y < 1$$

$$E(X|y) = \int_0^1 \frac{x}{y}dx = \left.\frac{x^2}{2y}\right|_0^1 = \frac{1}{2y}$$

$$E(Y|X) = \int_0^1 y\frac{dy}{1-x} = \left.\frac{y^2}{2(1-x)}\right|_0^1 = \frac{1}{2(1-x)}$$

Definition 10.17. *Linear Regression.* Consider (X, Y). Regression of Y on X is linear if the corresponding regression curve is a straight line i.e. if

$$y = E(Y|x) = \eta_x \text{ (say) } = \alpha + \beta x \tag{10.73}$$

where α, β are suitable constants. Similarly regression of X on Y is linear if

$$x = E(X|y) = \xi_y \text{ (say) } = \alpha' + \beta' y \tag{10.74}$$

α', β' being suitable constants.

• **THEOREM 10.16.** $E(Y-\eta_x) = E(X-\xi_y) = 0$.
Proof follows readily by theorem 10.11.

• **THEOREM 10.17.** *If the regression of Y on X is linear and if $\sigma_X^2, \sigma_Y^2, \rho$ exist then the constants α, β in (10.73) are given by*

$$\alpha = \mu_Y - \rho\frac{\sigma_Y}{\sigma_X}\mu_X, \quad \beta = \rho\frac{\sigma_Y}{\sigma_X} \tag{10.75}$$

Proof. Assume X, Y are continuous.

$$\eta_x = \alpha + \beta x, \quad E(Y|X) = \alpha + \beta X \tag{10.76}$$

Taking expectation of both sides with respect to X, by theorem 10.11,

$$E(Y) = \mu_y = \alpha + \beta E(X) = \alpha + \beta\mu_X \tag{10.77}$$

Again, multiplying both sides of (10.76) by X and taking expectation w.r.t. X, assuming without loss of generality (X, Y) are continuous,

$$\int_{-\infty}^{\infty} x\left[\int_{-\infty}^{\infty} y f_{Y|X}(y|x)dy\right] f_X(x)dx = \alpha E(X) + \beta E(X^2)$$

i.e.,

$$E(XY) = \alpha E(X) + \beta E(X^2) \tag{10.78}$$

Solving equations (10.77) and (10.78)

$$\beta = \frac{\mu_{X,Y}}{\sigma_X^2} = \rho\frac{\sigma_Y}{\sigma_X}, \quad \alpha = \mu_Y - \beta\mu_X = \mu_Y - \rho\frac{\sigma_Y}{\sigma_X}\mu_X$$

COROLLARY 10.3. Similarly, α', β' in (10.74) are given by

$$\alpha' = \mu_X - \rho\frac{\sigma_X}{\sigma_Y}\mu_Y, \quad \beta' = \frac{\mu_{X,Y}}{\sigma_Y^2}$$

• **THEOREM 10.18.** *If regression of Y on X is linear and if the conditional variance $V(Y|X = x)$ is independent of x (i.e., conditional distribution of Y given x is homoscedestic), then*

$$V(Y|x) = \sigma_Y^2(1-\rho^2).$$

Proof.

$$E[Y - E(Y|X)]^2 = \int_{-\infty}^{\infty}\int_{-\infty}^{\infty}\left\{Y - \mu_Y - \rho\frac{\sigma_Y}{\sigma_X}(X-\mu_X)\right\}^2 f_{X,Y}(x,y)dxdy$$
$$= \sigma_Y^2(1-\rho^2)$$

Now $E[Y - E(Y|X)]^2 = E[E\{Y - E(Y|X)\}^2|X] = E[V(Y|X)]$.

Since $V(Y|x)$ is constant for all x, $E[V(Y|X)] = \sigma_Y^2(1-\rho^2)$ implies $V(Y|x) = \sigma_Y^2(1-\rho^2)$ for all x.

We conclude this chapter by considering two important probability distributions, one is a discrete multinomial distribution (on R^n), the other is continuous—a bivariate normal distribution.

10.6. Multinomial Distribution

Suppose an experiment may result in k possible outcomes $0_1, \cdots, 0_k$ with probabilities $p_1, p_2, \cdots, p_k$ respectively, $\sum_1^k p_i = 1$. N such experiments are performed. Let X_i be the number of experiments that result in outcome $O_i (i = 1, \cdots, k), \sum_{i=1}^k X_i = N$.

$$P(X_1 = x_1, \cdots, X_k = x_k) = P(x_1, \cdots, x_k) = \frac{N!}{\prod_{i=1}^k x_i!} \prod_{i=1}^k p_i^{x_i} \tag{10.79}$$

where $x_i = 0, 1, \cdots, N, \sum_{i=1}^k x_i = N$.

By virtue of independence, any specific order of N outcomes with 0_i occuring x_i times $(i = 1, \cdots, k)$ has probability $\prod_{i=1}^k p_i^{x_i}$ and since there are $\frac{N!}{\prod_{i=1}^k x_i!}$ such orderings, (10.79) follows.

The random vector $(X_1, \cdots, X_k)$ is said to follow a multinomial distribution with parameters $N, p_1, \cdots, p_{k-1}$ when the pmf is given by (10.79).

The expression $P(x_1, \cdots, x_k)$ shows that it can be regarded as the coefficient of $\prod_{j=1}^k t_j^{x_j}$ in the expansion of

$$(t_1 p_1 + \cdots + t_N p_N)^N.$$

The moment generating function is

$$M_{X_1, \cdots, X_k}(t_1, \cdots, t_k) = E(e^{t_1 X_1 + \cdots + t_k X_k})$$

$$= \sum_{x_1, \cdots, x_k} \frac{N!}{\prod_{i=1}^k x_i!} \prod_{i=1}^k (p_i e^{t_i})^{x_i} = (p_1 e^{t_1} + \cdots + p_k e^{t_k})^N \tag{10.80}$$

The marginal mgf of X_i, $(i = 1, \cdots, k)$ is

$$M_{X_i}(t_i) = M(0, \cdots, 0, t_i, 0, \cdots 0) = (p_i e^{t_i} + q_i)^N, \quad q_i = 1 - p_i$$

which is the mgf of a binomial distribution with parameters (N, p_i). Hence marginal distribution of X_i is binomial (N, p_i).

The marginal mgf of a set $(X_1, \cdots, X_m)$, $m < k$ is

$$M_{X_1, \cdots, X_m}(t_1, \cdots, t_m) = M(t_1, \cdots, t_m, 0, \cdots, 0) = (p_1 e^{t_1} + \cdots + p_m e^{t_m} + 1 - p_1 - \cdots - p_m)^N \tag{10.81}$$

so that $(X_1, \cdots, X_m)$ follows a $(m+1)$-nomial distribution with parameters $N, p_1, \cdots, p_m$.

Calculating directly, the marginal pmf of a set $(X_1, \cdots, X_m), m < k$ is

$$\begin{aligned} f_{X_1,\cdots,X_m}(x_1,\cdots,x_m) &= \sum_{x_{m+1},\cdots,x_k} \frac{N!}{x_1!\cdots x_k!} p_1^{x_1}\cdots p_k^{x_k} \\ &= \frac{N!}{x_1!\cdots x_m!\left(N-\sum_{j=1}^{m} x_j\right)!} p_1^{x_1}\cdots p_m^{x_m} Q^{N-\sum_{j=1}^{m} x_j} \\ &\times \left[\sum_{x_{m+1}\cdots,x_k} \frac{\left(N-\sum_{j=1}^{m} x_j\right)!}{x_{m+1}!\cdots x_k!}\left(\frac{p_{m+1}}{Q}\right)^{x_{m+1}}\cdots\left(\frac{p_m}{Q}\right)^{x_k}\right] \end{aligned} \tag{10.82}$$

where $Q = 1 - \sum_{j=1}^{m} p_j$. The quantity in [ ] being the sum of the probability of a multinomial distribution with parameter $\left(N - \sum_{j=1}^{m} x_j\right), \frac{p_{m+j}}{Q} (\geq 0), j = 1, \cdots, k-m$, is unity. Hence the marginal distribution of $X_1, \cdots, X_m$ is multinomial with parameters $N, p_1, \cdots, p_m$. We now consider the conditional distribution of $X_{m+1}, \cdots, X_k$ given $(X_1 = x_1, \cdots, X_m = x_m)$.

$$\begin{aligned} & f_{X_{m+1},\cdots,X_k|X_1,\cdots,X_m}(x_{m+1},\cdots,x_k|x_1,\cdots,x_n) \\ &= \frac{f_{X_1,\cdots,X_k}(x_1,\cdots,x_k)}{f_{X_1,\cdots,X_m}(x_1,\cdots,x_m)} \\ &= \frac{\dfrac{N!}{\prod_{i=1}^{k} x_i!}\prod_{i=1}^{k} p_i^{x_i}}{\dfrac{N!}{x_1!\cdots x_m!\left(N-\sum_{j=1}^{m} x_j\right)!}\prod_{j=1}^{m} p_j^{x_j}\left(1-\sum_{j=1}^{m} p_j\right)^{N-\sum_{j=1}^{m} x_j}} \\ &= \frac{\left(N-\sum_{j=1}^{m} x_j\right)!}{x_{m+1}!\cdots x_k!}\left(\frac{p_{m+1}}{Q}\right)^{x_{m+1}}\cdots\left(\frac{p_k}{Q}\right)^{x_k} \end{aligned} \tag{10.83}$$

This is a multinomial distribution with parameters $\left(N - \sum_{j=1}^{m} x_j\right), \frac{p_{m+j}}{Q}, j = 1, \cdots, k-m$.

We now consider the moments of the distribution (10.79) from its mgf (10.80).

$$\frac{\partial M(t_1,\cdots,t_k)}{\partial t_i} = Np_i e^{t_i}(p_1e^{t_1}+\cdots+p_ke^{t_k})^{N-1}$$

$$\frac{\partial^2}{\partial t_i^2}M(t_1,\cdots,t_k) = N(N-1)p_i^2e^{2t_i}(p_1e^{t_1}+\cdots+p_ke^{t_k})^{N-2} + Np_ie^{t_i}(p_1e^{t_1}+\cdots+p_ke^{t_k})^{N-1}$$

$$\frac{\partial^2}{\partial t_i \partial t_j}M(t_1,\cdots,t_k) = N(N-1)p_ip_je^{t_i+t_j}(p_1e^{t_1}+\cdots+p_ke^{t_k})^{N-2}, \quad \text{(for } i \neq j\text{)}.$$

Hence, evaluating the derivatives at $(0,\cdots,0)$

$$E(X_i) = Np_i,\; E(X_i^2) = -Np_i + N(N-1)p_i^2$$
$$E(X_iX_j) = N(N-1)p_ip_j$$
$$V(X_i) = -Np_i(1-p_i)$$
$$\text{Cov}\,(X_i,X_j) = -Np_ip_j \qquad i \neq j$$
$$\rho(X_i,X_j) = -\sqrt{\frac{p_ip_j}{(1-p_i)(1-p_j)}} \tag{10.84}$$

The regresion of X_i on X_j is linear :

$$E[X_i|X_j = x_j] = (N - x_j)p_i(1-p_j)^{-1}$$

The mode(s) of the distribution (10.83) is near the expected values, is at $x_{1m} \simeq Np_1, \cdots, x_{km} \simeq Np_k$.

The joint factorial moments of this distribution is

$$\mu_{(r_1,\cdots,r_k)} = E[(X_1)_{r_1}\cdots(X_k)_{r_k}] = N^{\Sigma_1^k r_j}p_1^{r_1}\cdots p_k^{r_k} \tag{10.85}$$

The probability generating function is

$$(p_1t_1+\cdots+p_kt_k)^N \tag{10.86}$$

The cumulant generating function is

$$\log_e M(t_1,\cdots,t_k) = N\log_e \sum_{j=1}^{k} p_je^{t_j}. \tag{10.87}$$

10.7. Bivariate Normal Distribution

The two-dimensional random variable (X,Y) is said to follow a bivariate normal distribution with parameters $\mu_X,\mu_Y,\sigma_X,\sigma_Y$, and ρ if their joint pdf is given by

$$f_{X,Y}(x,y) = \frac{1}{2\pi\sigma_X\sigma_Y\sqrt{1-\rho^2}}\times \exp\left[-\frac{1}{2(1-\rho^2)}\left\{\left(\frac{x-\mu_X}{\sigma_X}\right)^2 - 2\rho\left(\frac{x-\mu_X}{\sigma_X}\right)\left(\frac{y-\mu_Y}{\sigma_Y}\right)+\left(\frac{y-\mu_Y}{\sigma_Y}\right)^2\right\}\right] \tag{10.88}$$

$-\infty < x < \infty, -\infty < y < \infty$, when the parameters satisfy $-\infty < \mu_X < \infty, -\infty < \mu_Y < \infty, \sigma_X > 0, \sigma_Y > 0, |\rho| < 1$.

To show that

$$\int_{-\infty}^{\infty}\int_{-\infty}^{\infty} f(x,y)dxdy = 1 \tag{10.89}$$

substitute

$$\frac{x-\mu_X}{\sigma_X} = u, \quad \frac{y-\mu_Y}{\sigma_Y} = v \tag{10.90}$$

The left hand side of (10.89) becomes

$$\frac{1}{2\pi\sqrt{1-\rho^2}}\int_{-\infty}^{\infty}\int_{-\infty}^{\infty} e^{-\frac{1}{2(1-\rho^2)}[u^2-2uv+v^2]}dudv \tag{10.91}$$

$$= \frac{1}{2\pi\sqrt{1-\rho^2}}\int_{-\infty}^{\infty}\int_{-\infty}^{\infty} e^{-\frac{1}{2(1-\rho^2)}[(u-\rho v)^2+(1-\rho^2)v^2]}dudv \tag{10.92}$$

Letting $w = \frac{u-\rho v}{\sqrt{1-\rho^2}}$, (10.92) becomes

$$\left[\frac{1}{\sqrt{2\pi}}\int_{-\infty}^{\infty} e^{-\frac{w^2}{2}}dw\right]\left[\frac{1}{\sqrt{2\pi}}\int_{-\infty}^{\infty} e^{-\frac{v^2}{2}}dv\right] = 1.$$

Note. If $\rho^2 = 1$, the pdf (10.88) is not defined. In this case it is known there exist constants a, b such that $P(Y = aX + b) = 1$. The distribution then is said to be a bivariate degenerate (or singular) normal distribution.

We shall here consider non-degenerate bivariate normal distribution, where $\rho^2 < 1$.

The moment generating function is

$$M_{X,Y}(t_1,t_2) = E(e^{t_1X+t_2Y}) = E(e^{t_1(\mu_X+U\sigma_X)+t_2(\mu_Y+V\sigma_Y)})$$

where U and V are given in (10.90).

$$= \frac{e^{t_1\mu_X+t_2\mu_Y}}{2\pi\sqrt{1-\rho^2}}\int_{-\infty}^{\infty}\int_{-\infty}^{\infty} e^{t_1u\sigma_X+t_2v\sigma_Y-\frac{1}{2(1-\rho^2)}[u^2-2\rho uv+v^2]}dudv \tag{10.93}$$

The exponent in (10.93) is

$$-\frac{1}{2(1-\rho^2)}[u^2-2\rho uv+v^2-2(1-\rho^2)\{t_1u\sigma_X+t_2v\sigma_Y\}]$$

$$= -\frac{1}{2}S^2 - \frac{1}{2}Z^2 + \frac{1}{2}(t_1^2\sigma_X^2 + 2\rho t_1t_2\sigma_X\sigma_Y + t_2^2\sigma_Y^2)$$

where $S = \frac{\{u-\rho v-(1-\rho^2)t_1\sigma_X\}}{\sqrt{1-\rho^2}}$,

$Z = v - \rho t_1\sigma_X - t_2\sigma_Y$.

Hence the integral becomes

$$e^{t_1\mu_X+t_2\mu_Y+\frac{1}{2}(t_1^2\sigma_X^2+2\rho t_1t_2\sigma_X\sigma_Y+t_2^2\sigma_Y^2)}\left[\frac{1}{\sqrt{2\pi}}\int_{-\infty}^{\infty} e^{-\frac{s^2}{2}}ds\right]\left[\frac{1}{\sqrt{2\pi}}\int_{-\infty}^{\infty} e^{-\frac{z^2}{2}}dz\right]$$

$$= \exp\left[t_1\mu_X + t_2\mu_Y + \frac{1}{2}(t_1^2\sigma_X^2 + 2\rho t_1t_2\sigma_X\sigma_Y + t_2^2\sigma_Y^2)\right] \tag{10.93$'$}$$

The mgf of the distribution (10.88) about the point (μ_X, μ_Y) is

$$\begin{aligned} M_{X-\mu_X, Y-\mu_Y} &= e^{-\mu_X t_1 - \mu_Y t_2} M_{X,Y}(t_1, t_2) \\ &= \exp\left[\frac{1}{2}(t_1^2\sigma_X^2 + 2\rho t_1 t_2 \sigma_X \sigma_Y + t_2^2 \sigma_Y^2)\right] \end{aligned} \tag{10.94}$$

From (10.93)′, the following follow :

$$E(X) = \left.\frac{\partial M(t_1, t_2)}{\partial t_1}\right|_{t_1=t_2=0} = \mu_X$$

$$E(X^2) = \left.\frac{\partial^2 M(t_1, t_2)}{\partial t_1^2}\right|_{t_1=t_2=0} = \mu_X^2 + \sigma_X^2$$

$$E(XY) = \left.\frac{\partial^2 M(t_1, t_2)}{\partial t_1 \partial t_2}\right|_{t_1=t_2=0} = \rho\sigma_X\sigma_Y + \mu_X\mu_Y$$

Hence $E(Y) = \mu_Y, V(X) = \sigma_X^2, V(Y) = \sigma_Y^2$, Cov $(X, Y) = \rho\sigma_X\sigma_Y$, Corr. Coeff. $(X, Y) = \rho$.

Writing $\dfrac{(X - \mu_X)}{\sigma_X} = U$ and $\dfrac{(Y - \mu_Y)}{\sigma_Y} = V$ we have the standardised bivariate normal density with parameter ρ,

$$f(u, v) = \frac{1}{2\pi\sqrt{1-\rho^2}} e^{-\frac{1}{2(1-\rho^2)}\{u^2 - 2\rho uv + v^2\}} \quad -\infty < u < \infty, -\infty < v < \infty.$$

The mgf of a standardised bivariate normal density is

$$M_{U,V}(t_1, t_2) = e^{\frac{1}{2}(t_1^2 + 2\rho t_1 t_2 + t_2^2)}$$

For this distribution

$$\begin{array}{lll} \mu_{21} = \mu_{12} = 0, & \mu_{22} = 1 + 2\rho^2, & \\ \mu_{31} = \mu_{13} = 3\rho, & \mu_{32} = \mu_{23} = 0, & \mu_{33} = 3\rho(3 + 2\rho^2) \\ \mu_{41} = \mu_{14} = 0 & \mu_{42} = \mu_{24} = 3(1 + 4\rho^2); & \\ \mu_{51} = \mu_{15} = 15\rho; & & \end{array}$$

All the cumulants of order higher than 2 vanish.

The marginal mgf's of the distribution (10.88) are :

$$M_X(t_1) = M_{X,Y}(t_1, 0) = e^{\mu_X t_1 - \frac{1}{2}t_1^2\sigma_X^2}$$

$$M_Y(t_2) = M_{X,Y}(0, t_2) = e^{\mu_Y t_2 - \frac{1}{2}t_2^2\sigma_Y^2}$$

so that the marginal distribution of X is univariate normal $N(\mu_X, \sigma_X)$ and similarly for Y.

We have already noted that if two random variables are uncorrelated they are not necessarily independent.

• **THEOREM 10.19.** *If (X, Y) have a bivariate normal distribution, then X, Y are independent if and only if they are uncorrelated.*

Proof. If X, Y are uncorrelated i.e. if $\rho = 0$, the bivariate density (10.88) can be written as the product of two univariate normal densities so that X, Y are independent.

Normal variables, like other variables, may however be dependent with zero correlations. Thus X, Y may each have univariate normal density with zero correlation, but the joint density of X, Y may not be a bivariate normal.

• **THEOREM 10.20.** *If (X, Y) have a bivariate normal distribution, the marginal density of X is $N(\mu_X, \sigma_X)$ and of Y is $N(\mu_Y, \sigma_Y)$.*

Proof.

$$f_X(x) = \int_{-\infty}^{\infty} f(x, y)dy$$

substituting $v = \dfrac{y - \mu_Y}{\sigma_Y}$

$$= \frac{1}{2\pi\sigma_X\sqrt{1-\rho^2}} \int_{-\infty}^{\infty} \exp\left[-\frac{1}{2}\left(\frac{x-\mu_X}{\sigma_X}\right)^2 - \frac{1}{2(1-\rho^2)}\left(v - \rho\frac{x-\mu_X}{\sigma_X}\right)^2\right] dv;$$

Substituting $w = \left\{v - \rho\left(\dfrac{x-\mu_X}{\sigma_X}\right)\right\} \Big/ \sqrt{1-\rho^2}$,

$$f_X(x) = \frac{1}{\sigma_X\sqrt{2\pi}} \exp\left[-\frac{1}{2}\left(\frac{x-\mu_X}{\sigma_X}\right)^2\right].$$

Similarly for Y.

• **THEOREM 10.21.** *If (X, Y) has a bivariate normal distribution, then the conditional distribution of Y given $X = x$ is univariate normal $N\left(\mu_X + \rho\frac{\sigma_Y}{\sigma_X}(x - \mu_X), \sigma_Y^2(1-\rho^2)\right)$.*

Similarly the conditional distribution of X given $Y = y$ is univariate normal

$$N\left(\mu_X + \rho\frac{\sigma_Y}{\sigma_X}(y - \mu_Y), \sigma_X^2(1-\rho^2)\right).$$

Proof.

$$f_{Y|X}(y|x) = \frac{f_{X,Y}(x,y)}{f_X(x)} = \frac{1}{\sigma_X\sqrt{2\pi(1-\rho^2)}} e^{-\frac{1}{2\sigma_Y^2(1-\rho^2)}\{y-\mu_Y-\rho\frac{\sigma_Y}{\sigma_X}(x-\mu_X)\}^2} \tag{10.95}$$

obtained by the substitutions as in (10.90) and proceeding similarly as in (10.92). Hence the conditional distribution of Y is univariate normal with parameters as stated above. Similarly for $f_{Y|X}(y|x)$.

COROLLARY 10.4. The regression curve of Y on X is

$$y = \mu_Y + \rho\frac{\sigma_Y}{\sigma_X}(x - \mu_X) \tag{10.96}$$

which is a linear function of x. More specifically, the regression curve of Y on X is a straight line with gradient $\rho\frac{\sigma_Y}{\sigma_X}$ and intercept $\left(\mu_Y - \rho\frac{\sigma_Y}{\sigma_X}\mu_X\right)$.

Similarly the regression curve of X on Y is a straight line (Fig. 10.3)

$$x = \mu_X + \rho\frac{\sigma_X}{\sigma_Y}(y - \mu_Y) \tag{10.97}$$

Note. The variance of the conditional distribution of Y given x (X given y) does not depend on $x(y)$. The conditional distributions are homoscedastic.

The locus of the points in the xy-plane such that $f(x, y)$ is a constant is an ellipse

$$\frac{(x - \mu_X)^2}{\sigma_X^2} - 2\rho\frac{(x - \mu_X)(y - \mu_Y)}{\sigma_X\sigma_Y} + \frac{(y - \mu_Y)^2}{\sigma_Y^2} = k^2 \tag{10.98}$$

with centre (μ_X, μ_Y) [because the discriminant $B^2 - 4AC$ of the quadratic form $Ax^2 + Bxy + Cy^2$ is negative for (10.98)] where k is a constant (independent of x and y). Thus if the probability density surface (10.88) is cut by the plane $z = z_o$ (a sufficiently small possitive quantity), parallel to the xy-plane, the resulting contour curves are similar and concentric ellipses (10.98) with centre (μ_X, μ_Y). The curves (10.98) are called equi-probability contours. k is a decreasing function of z_o. The density $f(x, y)$ is maximum at the centre and is equal to

$$f(\mu_X, \mu_Y) = \frac{1}{2\pi\sigma_X\sigma_Y(1 - \rho^2)}.$$

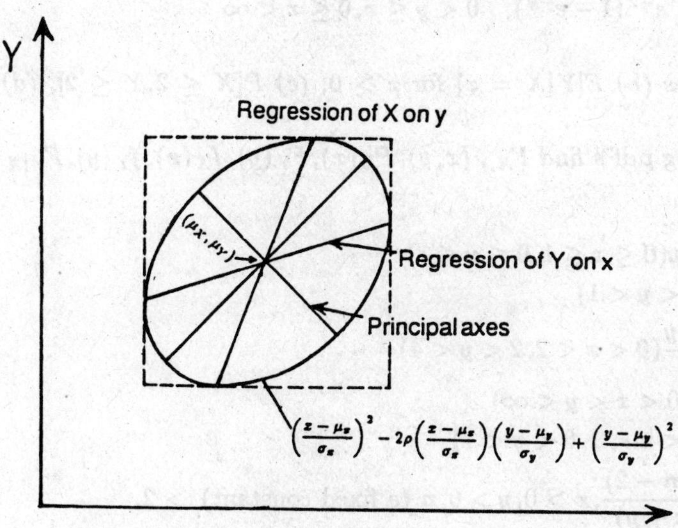

Fig. 10.4 An equiprobability contour of a bivariate normal density

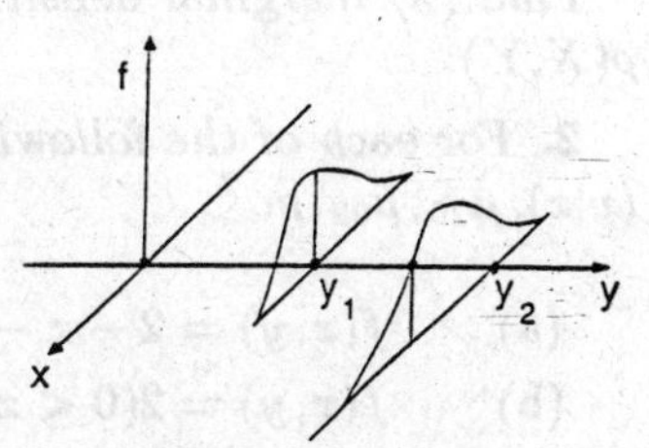

Fig. 10.3

The line of regression of y on x(x on y) passes through the two points where the tangent to the ellipse are vertical (horizontal) [Fig. 10.4].

As the value of $|\rho|$ tends to one the ellipse becomes elongated and both the lines of regression approaches the same diagonal of the circumscribed rectangle.

If $\rho = 0$, the principal axes of each ellipse are parallel to the co-ordinate axes.

Consider a bivariate normal r.v. (X, Y) with $\rho(X, Y)$ not necessary equal to zero. For the transformed variables

$$U = X\cos\theta + Y\sin\theta, \quad V = -X\sin\theta + Y\cos\theta \tag{10.99}$$

$\text{Cov}\,(U, V) = \frac{1}{2}(\sigma_Y^2 - \sigma_X^2)\sin 2\theta + \rho\sigma_X\sigma_Y\cos 2\theta.$

Hence U, V are uncorrelated if

$$2\theta = \tan^{-1}\left(\frac{2\rho\sigma_X\sigma_Y}{\sigma_X^2 - \sigma_Y^2}\right) \tag{10.100}$$

Therefore, by the transformation of co-ordinates from (X, Y) to (U, V) as in (10.99), (10.100), the new axes of co-ordinates become parallel to the principal axes of the ellipse (10.98).

10.8. Exercises

1. *(X, Y) have joint pdf*

$$f(x, y) = \begin{cases} e^{-y}(1 - e^{-x}), & 0 < x \le y, 0 \le y < \infty \\ e^{-x}(1 - e^{-y}), & 0 < y \le x, 0 \le x < \infty \end{cases}$$

Find (a) marginal densities (b) $E[Y|X = x]$ for $x > 0$, (c) $P[X \le 2, Y \le 2]$, (d) $\rho(X, Y)$.

2. *For each of the following pdf's find $F_{X,Y}(x, y), F_X(x), F_Y(y), f_X(x), f_Y(y), F_{Y|X}(y|x), \mu_{20}, \mu_{02}, \rho$.*

(a) $f(x, y) = 2 - x - y (0 \le x \le 1, 0 \le y \le 1)$

(b) $f(x, y) = 2 (0 < x < y < 1)$

(c) $f(x, y) = \dfrac{6 - x - y}{8} (0 < x < 2, 2 < y < 4)$

(d) $f(x, y) = 2e^{-x-y} (0 < x < y < \infty)$

(e) $f(x, y) = x + y (0 < x < 1, 0 < y < 1)$

(f) $f(x, y) = \dfrac{(n-1)(n-2)}{n(1 + x + y)}, x > 0, y > 0, n$ (a fixed constant) > 2.

(g) $f(x, y) = x^2 + \dfrac{xy}{3}, (0 < x < 1, 0 < y < 2)$.

3. *(X, Y) have bivariate cumulative distribution function*

$$F_{X,Y}(x, y) = \begin{cases} 1 - e^{-x} - e^{-y} + e^{-(x+y)}, x, y \ge 0 \\ 0 \textit{ elsewhere.} \end{cases}$$

Find joint pdf and marginal pdf's.

4. *(X, Y) have joint pdf*

$$f(x,y) = \begin{cases} \frac{1}{2} & \text{for } |x|+|y| \le 1 \\ 0 & \text{elsewhere.} \end{cases}$$

Find marginal pdf's and conditional pdf's.

$$\left[f_X(x) = \int\limits_{|x|+|y|\le 1} \frac{1}{2} dy = \int\limits_{|x|-1}^{1-|x|} \frac{1}{2} dy = 1 - |x|, \quad |x| \le 1. \right.$$
$$= 0 \quad \text{(elsewhere).}$$

$$f_X(x|y) = \begin{cases} \frac{1}{2(1-|y|)} & \text{for } |x| \le 1-|y| \text{ if } |y| \le 1 \\ 0 & \text{for } |x| > 1-|y| \text{ if } |y| \le 1 \\ \text{undefined} & \text{for all } x \text{ if } |y| > 1 \end{cases}$$

5. *(X, Y) have joint pdf $f(x, y) = 3x^2y + 3y^2x, 0 \le x \le 1, 0 \le y \le 1; = 0$ elsewhere. Find the marginal pdf's. Find the conditional $P\left(\frac{1}{2} \le Y \le \frac{3}{4} \middle| \frac{1}{3} \le X \le \frac{2}{3}\right)$*

$$\left[f_X(x) = x + \frac{3}{2}x^2, \quad f_Y(y) = y + \frac{3}{2}y^2, \quad \text{Prob.} = \frac{515}{4096} \right]$$

6. *X has a discrete probability distribution $P(X = x) = \frac{x}{3}, x = 1, 2; f_{Y|X}(y|x)$ is binomial with parameters $\left(x, \frac{1}{2}\right), y = 0, 1, \cdots, x$. Find $E(X), V(X), f(x, y), E(Y)$.*

7. *The joint pdf of bivariate gamma variables (X, Y) is*

$$f(x,y) = \frac{\lambda^{r+p}}{\Gamma(r)\,\Gamma(p)} e^{-\lambda x} y^{p-1}(x-y)^{r-1}, \quad 0 < y < x < \infty; \lambda, p, r > 0$$

Find the marginal pdf's and conditional pdf's.

8. *Consider (X, Y) with $Y = 2X$. Show that*

$$F_{X,Y}(x,y) = \begin{cases} F_X(x) & \text{for } x \le \frac{y}{2} \\ F_X\left(\frac{y}{2}\right) = F_Y(y) & \text{for } x > \frac{y}{2} \end{cases}$$

[(a) Suppose $2x \le y$. The events, $\{X \le x\} \Leftrightarrow \{Y \le 2x\}$. Hence

$$P\{X \le x, Y \le y\} = P\{X \le x, Y \le 2x\} + P\{X \le x, 2x < Y \le y\}$$
$$= P\{X \le x, Y \le 2x\} = P\{X \le x\} = F_X(x).$$

(b) Suppose $y < 2x$. The events, $\{Y \le y\} \Leftrightarrow \left\{X \le \frac{y}{2}\right\}$.

$$P\{X \le x, Y \le y\} = P\left\{X \le \frac{y}{2}, Y \le y\right\} + P\left\{\frac{y}{2} < X \le \frac{x}{2}, Y \le y\right\}$$

$$= P\left\{X \le \frac{y}{2}, Y \le y\right\} = P\left\{X \le \frac{y}{2}\right\} = F_X\left(\frac{y}{2}\right)\Bigg]$$

9. *Consider (X, Y) with $Y = g(X)$. Show that if*

(i) $g(t)$ is monotonically increasing

$$F_{X,Y}(x, y) = \begin{cases} F_Y(y) & \text{if } y < g(x) \\ F_X(x) & \text{if } y \ge g(x) \end{cases}$$

(ii) $g(t)$ is monotonically decreasing

$$F_{X,Y}(x, y) = \begin{cases} 0 & \text{if } y < g(x) \\ F_X(x) + F_Y(y) - 1 & \text{if } y \ge g(x) \end{cases}$$

10. *(X, Y) have joint pdf $f(x, y)$.*

(A) Show that for the following pdf's

(a) X, Y are stochastically independent :

(i) $f(x, y) = 6xy^2, \quad 0 < x, y < 1.$

(ii) $f(x, y) = 12xy(1 - y), \quad 0 < x, y < 1.$

(iii) $f(x, y) = 4xy, \quad 0 < x, y < 1.$

(b) X, Y are stochastically dependent :

(iv) $f(x, y) = \frac{1}{2}, \quad |x| + |y| \le 1.$

(v) $f(x, y) = 8xy, \quad 0 \le y \le x < 1.$

(vi) $f(x, y) = 2e^{-(x+y)}, 0 < x < y < \infty.$

(vii) $f(x, y) = \frac{1}{4\pi^2}[1 - \sin(x + y)], \quad -\pi \le x, y \le \pi.$

(viii) $f(x, y) = x + y, \quad 0 < x, y < 1.$

(B) Verify if X, Y are stochastically independent when

(ix) $f(x, y) = p_1^x(1 - p_1)^{1-x} p_2^y(1 - p_2)^{1-y}[1 + \lambda(x - p_1)(y - p_2)],$

$$0 < x, y < 1, 0 < p_1, p_2 < 1, |\lambda| \le 1$$

(x) $f(x, y) = \frac{\sqrt{3}}{2\pi} \exp\{-[(x - 7)^2 + (x - 7)y + y^2]\}.$

11. *Show that the r.v. (X, Y) with joint pdf*

$$f(x, y) = \frac{1}{2\pi\sqrt{1 + y^2}} \exp\left\{-\frac{x^2}{2(1 + y^2)} - \frac{y^2}{2}\right\}$$

have covariance zero and yet X, Y are not independent.

12. *Show that X, Y are dependent r.v.'s. but uncorrelated, when*

(a) (X, Y) is a discrete r.v. with

$$P(X = i, Y = j) = \frac{1}{8}, \quad i, j = 1, -1; \quad P(X = 0, Y = 0) = \frac{1}{2}.$$

(b) (X, Y) are continuous r.v.'s. with $f(x, y) = \frac{1}{2}$ inside the square with corners at points (1, 0), (0, 1), (−1, 0), (0, −1).

(c) $X = \sin 2\pi R, Y = \cos 2\pi R$ when R has a uniform distribution over (0, 1).

13. *Let (X, Y) have joint pdf*

$$f(x, y) = \begin{cases} \dfrac{1 + xy}{4}, & |x| < 1, |y| < 1 \\ 0 & \textit{elsewhere.} \end{cases}$$

Show that X, Y are not independent. The r.v.'s. X^2, Y^2 are, however, independent.

14. *X and θ are independently and uniformly distributed in the intervals $(0, a)$ and $\left(0, \frac{\pi}{2}\right)$ respectively. Show that*

$$P[X < b\cos\theta] = \frac{2b}{\pi a}, \quad b < a.$$

15. *Show that r.v.'s. X and $g(X)$ are not independent when $g(X)$ is a monotonic function of X.*

16. *(X, Y) have joint pdf $f(x, y) = \frac{3}{2\sqrt{x}}, 0 < y < x < 1$.*

Find

$$f_{Y|X}\left(y|x = \frac{1}{2}\right), \quad f_{X|Y}\left(x|y = \frac{1}{2}\right); \quad \left[2, 0 < y < \frac{1}{2}; \frac{1}{(2 - \sqrt{2})\sqrt{x}}, \frac{1}{2} < x < 1\right]$$

17. *A r.v. X has pdf $f(x) = \frac{e^{-\frac{x}{2}}}{\sqrt{2\pi\, x}}, x > 0$.*

Conditional pdf of Y given x is

$$f_{X|Y}(y|x) = \frac{\sqrt{x}}{\sqrt{2\pi}} e^{-\frac{xy^2}{2}}, \quad -\infty < y < \infty$$

Show that the marginal density of Y is

$$f_Y(y) = \frac{1}{\pi(1 + y^2)}, \quad -\infty < y < \infty.$$

18. *(X, Y) have joint pdf $f(x, y) = A\frac{x}{y}, 0 < x < 1, 1 < y < 2; = 0$ otherwise; A, a constant. Evaluate A.*

Find

$$f_{Y|X}(y|x),\quad f_{X|Y}(x|y),\quad \left[\frac{2}{\log_e 2}; 2x, 0 < x < 1; \frac{1}{y\log_e 2}, 1 < y < 2\right]$$

19. *For the bivariate distribution*

$$f(x, y) = 1,\quad |y| < x,\quad 0 < x < 1.$$

find the conditional densities, $E(Y|x), \rho(X, Y)$.

$$\left[f_{X|Y}(x|y) = (1+y)^{-1}, -1 < y \le 0; (1-y)^{-1}, 0 < y < 1\right.$$

$$f_{Y|X}(y|x) = \frac{1}{2x}, 0 < x < 1$$

$$\left.E(Y|x) = 0, \rho = 0\right]$$

20. *Show that for the distribution*

$$f(x, y) = x + y,\quad 0 < x < 1, 0 < y < 1$$

$$\mu'_{r,s} = \frac{1}{(r+2)(s+1)} + \frac{1}{(r+1)(s+2)}, r, s, \text{ being positive}$$

integers; hence $\rho(X, Y) = -\dfrac{1}{11}$.

21. *(X, Y) follows a bivariate normal distribution with $(0, 0, \sigma_X^2, \sigma_Y^2, \rho)$. Show that $U = \dfrac{X}{\sigma_X} - \rho\dfrac{Y}{\sigma_Y}$ and $V = (\sqrt{1-\rho^2})Y/\sigma_Y$ are independent variates.*

22. *Let X_1, X_2 be independent normal with parameters $(\mu_1, \sigma_1), (\mu_2, \sigma_2)$ respectively. Show that the correlation coefficient between U and V where $U = X\cos\theta + Y\sin\theta, V = -X\sin\theta + Y\cos\theta$ is given by*

$$\rho^2 = 1 - \frac{4\sigma_1^2\sigma_2^2}{4\sigma_1^2\sigma_2^2 + (\sigma_1^2 - \sigma_2^2)^2\sin^2 2\theta}.$$

23. *X, Y are standard normal variates with correlation coefficient ρ. Show that correlation coefficient $(X^2, Y^2) = \rho^2$.*

$$\left[\rho(X^2, Y^2) = \frac{E(X^2Y^2) - E(X^2)E(Y^2)}{\sqrt{E(X^4) - E(X^2))^2}\sqrt{E(Y^4) - (E(Y^2))^2}}\right.$$

The mgf $M_{X,Y}(t_1, t_2) = \exp\left[\frac{1}{2}(t_1^2 + 2\rho t_1 t_2 + t_2^2)\right]$. $E(X^2Y^2) =$ coefficient of $\dfrac{t_1^2}{2!}\dfrac{t_2^2}{2!}$ in $M_{X,Y}(t_1, t_2) = 2\rho^2 + 1$. Similarly $E(X^4) = E(Y^4) = 3.\Big]$

24. *Let (X,Y) have a bivariate normal distribution $(\mu_1,\mu_2,\sigma_1^2,\sigma_2^2,\rho)$. Show that $Z = aX + bY + c$ is a univariate normal $N(a\mu_1 + b\mu_2 + c, a^2\sigma_1^2 + 2ab\rho\sigma_1\sigma_2 + b^2\sigma_2^2)$.*

[use mgf of (X,Y) to find mgf of Z]

25. *(X,Y) follow bivariate $N(\mu_x,\mu_y,\sigma_x,\sigma_y,\rho)$-distribution. Show that $U = \frac{X-\mu_x}{\sigma_x} + \frac{Y-\mu_y}{\sigma_y}$ and $V = \frac{X-\mu_x}{\sigma_x} - \frac{Y-\mu_y}{\sigma_y}$ are independently normally distributed with zero means and variances $2(1+\rho)$ and $2(1-\rho)$ respectively.*

26. *Multivariate Chebychev inequality : Let $(X_1,\cdots,X_n)$ be jointly distributed with $E(X_i) = \mu_i, V(X_i) = \sigma_i^2, i = 1,\cdots,n$. Define $A_i = \{|X_i - \mu_i| \le \sqrt{n}\, t\sigma_i\}$. Prove that $P\left(\bigcap_{i=1}^{n} A_i\right) \ge 1 - \frac{1}{t^2}$ for $t > 0$.*

[We have by Chebychev inequality $P(A_i) \ge 1 - \frac{1}{nt^2}$; hence $P(A_i^c) \le \frac{1}{nt^2}$. Now by corollary 2.4

$$P\left(\bigcap_{i=1}^{n} A_i\right) \ge 1 - \sum_{i=1}^{n} P(A_i^c) \ge 1 - \frac{n}{nt^2} = 1 - \frac{1}{t^2}, t > 0 \Bigg]$$

27. *X_1, X_2 are r.v.'s. with means 0, variances 1 and correlation coefficient ρ. Show that*

$$E[\max(X_1^2, X_2^2)] \le 1 + \sqrt{1-\rho^2}.$$

Hence show that for r.v.'s. X, Y that are jointly distributed with means μ_X and μ_Y and variances σ_X^2 and σ_Y^2 respectively and correlation coefficient ρ and for any $t > 0$,

$$P[|X - \mu_X| \ge t\sigma_X \quad \text{or} \quad |Y - \mu_Y| \ge t\sigma_Y] \le \frac{1}{t^2}\{1 + \sqrt{1-\rho^2}\}.$$

[First part : $2 \max (X_1^2, X_2^2) = |X_1^2 - X_2^2| + X_1^2 + X_2^2$

Hence

$$E\{\max(X_1^2, X_2^2)\} = \frac{1}{2}E|X_1^2 - X_2^2| + \frac{1}{2}E(X_1^2 + X_2^2) \qquad (i)$$

Now

$$(E|X_1^2 - X_2^2|)^2 = (E\{|X_1 - X_2||X_1 + X_2|\})^2 \le E\{|X_1 - X_2|\}^2 E\{|X_1 + X_2|^2\}$$
$$= \{2(1-\rho)\}\{2(1+\rho)\} \qquad (ii)$$

(the inequality in (ii) follows from Cauchy-Schwartz inequality)

Hence the result follows from (i).

Second part : Let $\frac{X-\mu_X}{\sigma_X} = Z_1, \frac{Y-\mu_Y}{\sigma_Y} = Z_2$. Hence $E(Z_1) = E(Z_2) = 0, V(Z_1) = V(Z_2) = 1, \rho(Z_1,Z_2) = \rho$.

Let max $(Z_1^2, Z_2^2) = Z$, which is a random variable. By Markov inequality

$$P(Z \ge t^2) \le \frac{E(Z)}{t^2} \le \frac{1+\sqrt{1-\rho^2}}{t^2}$$

i.e.,

$$P[|X-\mu_X| \geq t\sigma_X \quad \text{or} \quad |Y-\mu_Y| \geq t\sigma_Y] \leq \frac{1+\sqrt{1-\rho^2}}{t^2}.\Bigg]$$

28. *For the bivariate normal distribution* $N(0,0,1,1,\rho)$, *show that the moments obey the recurrence relation*

$$\mu_{r,s} = (r+s-1)\rho\mu_{r-1,s-1} + (r-1)(s-1)(1-\rho^2)\mu_{r-2,s-2}.$$

Hence or otherwise show that $\mu_{r,s} = 0$ *if* $r+s$ *is odd;* $\mu_{3,1} = 3\rho, \mu_{22} = 1+2\rho^2$.

[*Consider mgf* $M(t_1,t_2) = \exp\left\{\frac{1}{2}(t_1^2 + 2\rho t_1 t_2 + t_2^2)\right\}$. *Find* $\frac{\partial^2 M}{\partial t_1 \partial t_2}, t_1\frac{\partial M}{\partial t_1} + t_2\frac{\partial M}{\partial t_2}$ *and get relationship betwen these quantities. Identify coefficients.*]

29. (X,Y) *is a discrete r.v. with pmf* $P(X=i, Y=j) = p_{ij}, i,j = 0,1$. *Find the mgf of* X, *mgf of* Y, *mgf of* (X,Y). *Show that* X, Y *are independent if and only if*

$$p_{00}p_{11} - p_{01}p_{10} = 0.$$

30. *Find the mgf of*

(i) $f(x,y) = x+y, 0 < x,y < 1$,

(ii) $f(x,y) = e^{-(x+y)}, 0 < x,y < \infty$,

31. *Determine the mgf of the r.v. with pdf*

$$f(x) = \begin{cases} k_1 & \text{for } a \leq x \leq a+c_1 \\ k_2 & \text{for } b \leq x \leq b+c_2 \\ 0 & \text{otherwise} \end{cases}$$

where a, b, c_1, c_2, k_1, k_2 *are constants such that*

$$c_1 > 0, c_2 > 0, k_1 > 0, k_2 > 0, a + c_1 \leq b, k_1c_1 + k_2c_2 = 1.$$

32. *If* $f_{X,Y}(x,y) = \frac{1}{8}(6-x-y), 0 \leq x \leq 2, 2 \leq y \leq 4$ *calculate (i)* $P(X+Y<3)$, *(ii)* $P\left(X < \frac{3}{2}, Y < \frac{5}{2}\right)$ $\left[\frac{3}{8}, \frac{9}{32}\right]$

33. *A r.v.* X *has an exponential density* $f(x) = \lambda e^{-\lambda x}, x > 0, \lambda > 0$. *Find (a)* $E(Y), V(Y)$ *where* $Y = e^{-X}$. *(b)* $E(Z), V(Z)$ *when* $Z = e^X$ *and examine the conditions for their existence.*

[(a) $\frac{\lambda}{\lambda+1}, \frac{\lambda}{(\lambda+1)(\lambda+2)}$ (b) $\frac{\lambda}{\lambda-1}$ (provided $\lambda > 1$), $\frac{\lambda}{\lambda-2} - \left(\frac{\lambda}{\lambda-1}\right)^2$ (provided $\lambda > 2$).]

34. *X has a Poisson distribution with parameter λ. Conditional distribution of Y given x is binomial with parameter (x, p). Show that the marginal distribution of Y is Poisson with parameter λp.*

$$\left[f(x,y) = h(x)g(y|x) = \left\{e^{-\lambda}\frac{\lambda^x}{x!}\right\}\left\{\binom{x}{y}p^y q^{x-y}\right\}\right.$$

$$x = 0, 1, 2, \cdots, y = 0, 1, \cdots, x, \lambda > 0, p > 0$$

Let $x - y = z, z = 0, 1, 2, \cdots$

$$f(z,y) = e^{-\lambda}\frac{\lambda^{y+z}}{(y+z)!}\binom{y+z}{y}p^y q^z$$

Marginal distribution of y is

$$\left. g(y) = \sum_{z=0}^{\infty} f(z,y) = \frac{(\lambda p)^y}{y!}e^{-\lambda}\sum_{z=0}^{\infty}\frac{(\lambda q)^z}{z!} = e^{-\lambda p}\frac{(\lambda p)^y}{y!}, \quad y = 0, 1, \cdots\right]$$

35. *$X_1, \cdots, X_k$ are k independent Poisson variates with parameters $\lambda, \cdots, \lambda_k$ respectively. Show that the conditional distribution of $X_1, \cdots, X_k$ given $\sum_{i=1}^{k} X_i = x$ is multinomial.*

$$P\left(X_1 = x_1, \cdots, X_k = x_k \middle| \sum_{i=1}^{k} X_i = x\right)$$

$$= \frac{P\left(X_1 = x_1, \cdots, X_k = x_k; \sum_1^k X_i = x\right)}{P\left(\sum_1^k X_i = x\right)}$$

$$= \frac{\left(e^{-\lambda_1}\frac{\lambda_1^{x_1}}{x_1!}\right)\cdots\left(e^{-\lambda_k}\frac{\lambda_k^{x_k}}{x_k!}\right)}{e^{-\lambda}\frac{\lambda^x}{x!}}, \lambda = \sum_{i=1}^{k}\lambda_i$$

(by the reproductive property of Poisson distribution)

$$= \frac{x!}{x_1!\cdots x_k!}\left(\frac{\lambda_1}{\lambda}\right)^{x_1}\cdots\left(\frac{\lambda_k}{\lambda}\right)^{x_k}$$

which is multinomial with parameters x and $\frac{\lambda_i}{\lambda}, i = 1, \cdots, k-1$.

36. *Show that for positive values of the parameters, $p_0^a\{1 - p_1t_1 - p_2t_2\}^{-a}$ is the pgf of (X, Y) such that marginal distributions of X, Y and $X + Y$ have negative binomial distributions.*

37. *(X, Y) have joint pdf $f(x, y) = 2, 0 < x < y < 1$. Find $E(X|y), V(X|y)$,*

$$P\left(X < \frac{1}{2}\middle|Y = \frac{3}{4}\right), P\left(X < \frac{1}{2}\right)\left[\frac{y}{2}, \frac{y^2}{12}, \frac{2}{3}, \frac{3}{4}\right]$$

38. *X, Y are jointly distributed discrete r.v.'s. with*

$$p_{x,y} = \frac{2}{n(n+1)}(0), \quad x = 1, \cdots, n; y = 1, \cdots, x \quad \textit{(otherwise)}.$$

Show that X, Y are not independent. Find regression of Y on X and of X on Y.

$$\left[\text{marginal densities : } f_X(x) = \frac{2x}{n(n+1)}, f_Y(y) = \frac{2}{n+1}.\right.$$

$$E(X|y) = \frac{n+1}{2}, \quad E(Y|x) = \frac{x+1}{2}.$$

$$\left.\text{Hence regression of } Y \text{ on } x \text{ is } y = \frac{x+1}{2}\right]$$

39. *X, Y have joint pdf $f(x, y) = 1, 0 < x < 1, |y| < x$. Show that on the set of positive probability density, the graph of $E(Y|x)$ is a straight line, whereas that of $E(X|y)$ is not a straight line.*

40. *Show that the angle θ between the regression lines (of Y on X and of X on Y) is*

$$\tan^{-1}\left(\frac{1-\rho^2}{\rho}\frac{\sigma_X\sigma_Y}{\sigma_X^2+\sigma_Y^2}\right).$$

What can be concluded from this about the relative location of the two lines of regression for ρ close to 0 and for ρ close to 1 ?

41. *(X, Y) has a pdf $f(x, y) = k \cdot e^{-(ax+by)}, 0 \le y \le x+1, a(>0), b(>0)$ are given constants. Find the normalising constant k; find the regression curve of Y on X. Sketch this regression curve for $a = 2, b = 1$.*

42. *X, Y, Z have distribution*

$$dF = Ax^{l-1}y^{m-1}z^{n-1}, x, y, z \ge 0, x+y+z \le 1.$$

Show that $A = \dfrac{\Gamma(l+m+n+1)}{\Gamma(l)\,\Gamma(m)\,\Gamma(n)}$.

43. *The r.v. (X_1, X_2, X_3) has the trivariate normal probability density*

$$f_{X_1,X_2,X_3}(x_1, x_2, x_3) = k \exp\left[-\frac{1}{2}(2X_1^2 + X_2^2 + 4X_3^2 - X_1X_2 - 2X_1X_3)\right].$$

Write the matrix of coefficients of the quadratic from in the expression and compute k. Obtain (a) marginal pdf of (X_1, X_3), (b) the conditional pdf of (X_1, X_3) given X_2.

11 PROBABILITY DISTRIBUTION OF FUNCTIONS OF RANDOM VARIABLES

11.1. Introduction

In this chapter we consider the problem of finding the joint probability distribution of several functions $Y_1 = g_1(X_1, \cdots, X_n), \cdots, Y_n = g_n(X_1, \cdots, X_n)$ of random variables $(X_1, \cdots, X_n)$ starting from the joint distribution of $(X_1, \cdots, X_n)$. Section 11.2 considers the function $Y = g(X)$ of a single r.v. X. The probability integral transformation is considered in section 11.3. 11.4 considers functions $Y_1 = g_1(X_1, X_2), Y_2 = g_2(X_1, X_2)$ of two random variables (X_1, X_2) — first dealing with some simple univariate functions, $X+Y, X-Y, XY, X/Y$ (section 11.4.1) and then considering the technique of transformation of variables (section 11.4.2). 11.5 considers functions of n random variables, 11.6 the distribution of maximum and minimum of several random variables. Since the moment generating function, when it exists, uniquely determines the distribution of random variables, such functions can be used to find the joint distribution of $Y_1, \cdots, Y_n$ starting from the distribution of $(X_1, \cdots, X_n)$. The technique has been discussed in section 11.7.

11.2. Functions of One Random Variable

Let X be a r.v. and $g(X)$ a (borel measurable) function of X. $Y = g(X)$ is a r.v. (vide theorem 5.9). We want to find the *df* and pdf of Y.

$$F_Y(y) = P[Y \leq y] = P[X \in I_y] = \int_{I_y} dF_X(x)$$

where $I_y = \{x : g(x) \leq y\}$ and $X \in C$ denotes the event $\{w : X(w) \in C\}$. $x \in I_y$ if and only if $g(x) \leq y$.

If X is a discrete r.v.,

$$P[Y = y] = P[g(X) = y] = \sum_x P\{X(w) = x : g(x) = y\}$$

provided g has a finite number of inverses for each y.

❑ **EXAMPLE 11.1.** $Y = aX + b$,

(i) If $a > 0$, $\quad ax + b \leq y \quad$ if and only if $\quad x \leq \dfrac{y-b}{a}$.

(ii) If $a < 0$, $\quad ax + b \leq y \quad$ if and only if $\quad x \geq \dfrac{y-b}{a}$.

Hence, in case (i)

$$F_Y(y) = F_X\left(\frac{y-b}{a}\right)$$

in case (ii)

$$F_Y(y) = 1 - F_X\left(\frac{y-b}{a} - 0\right).$$

If $a = 0$,

$$F_Y(y) = \begin{cases} 0, & y < b \\ 1, & y \geq b \end{cases}$$

❑ **EXAMPLE 11.2.** $Y = X^2$.

$$F_Y(y) = 0 \quad \text{for } y < 0$$

$$F_Y(y) = P(X^2 \leq y) = P(-\sqrt{y} \leq X \leq \sqrt{y}) = F_X(\sqrt{y}) - F_X(-\sqrt{y} - 0), \quad y \geq 0.$$

If X is absolutely continuous, X^2 is absolutely continuous and differentiation gives

$$f_Y(y) = \begin{cases} \dfrac{1}{2\sqrt{y}}[f_X(\sqrt{y}) + f_X(-\sqrt{y})] & (y \geq 0) \\ 0 & (y < 0) \end{cases}$$

If X is discrete, pmf of Y is

$$f_Y(y) = f_X(-\sqrt{y}) + f_X(\sqrt{y}).$$

In particular if X is a standard normal deviate

$$F_Y(y) = \Phi(\sqrt{y}) - \Phi(-\sqrt{y}).$$

❑ **EXAMPLE 11.3.** *Let X be a $N(\mu, \sigma)$-variable and $Y = g(X) = |X - \mu|$.*

$$\begin{aligned} F_Y(y) &= P(|X-\mu| \le y) = P(\mu - y \le X \le \mu + y) \\ &= \frac{1}{\sigma\sqrt{2\pi}} \int_{\mu-y}^{\mu+y} e^{-(x-\mu)^2/2\sigma^2} dx \\ &= \frac{2}{\sigma\sqrt{2\pi}} \int_0^y e^{-t^2/2\sigma^2} dt, \quad y > 0. \end{aligned}$$

Hence

$$f_Y(y) = \begin{cases} 0, & y \le 0 \\ \dfrac{\sqrt{2}}{\sigma\sqrt{\pi}} e^{-y^2/2\sigma^2}, & y > 0 \end{cases}$$

❑ **EXAMPLE 11.4.** *Let $Y = g(X) = aX^2 + bX + c, a \neq 0$.*

(i) Let $a > 0$. $g(x) < y$ iff $ax^2 + bx + (c - y) \le 0$ i.e. iff $\alpha \le x \le \beta$ where

$$\alpha = \frac{-b - \sqrt{b^2 - 4a(c-y)}}{2a}, \quad \beta = \frac{-b + \sqrt{b^2 - 4a(c-y)}}{2a}.$$

Hence $F_Y(y) = P(\alpha \le X \le \beta) = F_X(\alpha) - F_X(\beta - 0)$.

(ii) Similarly if $a < 0$,

$$F_Y(y) = 1 - F_X(\alpha - 0) + F_X(\beta).$$

• **THEOREM 11.1.** *Let X be a r.v. of continuous type with pdf $f_X(x)$. Let $g(x)$ be a continuous function of x and the equation $y = g(x)$ have a finite number of solutions $x_1, \cdots, x_n$ for every y. Assume also $F_X(x)$ is differentiable at all these points. Then*

$$f_Y(y) = \sum_{i=1}^{n} \frac{f_X(x_i)}{|g'(x_i)|} \tag{11.1}$$

where $g'(x_i) = \left.\dfrac{dg(x)}{dx}\right|_{x_i}$

If, in particular, there is only one solution of the equation $y = g(x)$, $f_Y(x) = \dfrac{f_X(x)}{g'(x)}$.

Proof. Since $y = g(x)$ whenever $x = x_1, x_2, \cdots, x_n$,

$$\begin{aligned} f_Y(y)dy &= P\{y < Y < y + dy\} \\ &= P\{x_1 < X < x_1 + dx_1\} + \cdots + P\{x_n < X < x_n + dx_n\} \quad (11.2) \\ &= f(x_1)|dx_1| + \cdots + f(x_n)|dx_n|, \end{aligned}$$

when in some of these terms in (11.2), where y is a decreasing function of x around x_i, dx_i will be negative (and the inequality will be $x_i + dx_i < X < x_i$).

Now

$$g'(x_i)dx_i = dy, \quad i = 1, \cdots, n.$$

Hence

$$f_Y(y) = \sum_{i=1}^{n} \frac{f_X(x_i)}{|g'(x_i)|}.$$

Note : The theorem is valid if the equation $y = g(x)$ has only a countable number of solutions $x_1, x_2, \cdots$.

❑ **EXAMPLE 11.5.** *If* $Y = aX + b, a \neq 0, f_Y(y) = \frac{1}{|a|} f_X\left(\frac{y-b}{a}\right)$.

In particular if X is $N(\mu, \sigma)$ and $Y = a + bX, b \neq 0$,

$$f_Y(y) = \frac{1}{|b|\sigma\sqrt{2\pi}} \exp\left[-\frac{1}{2\sigma^2}\left\{\frac{1}{b}(y-a) - \mu\right\}^2\right]$$

$$= \frac{1}{|b|\sigma\sqrt{2\pi}} \exp\left[-\frac{1}{2b^2\sigma^2}\{y - (a + b\mu)\}^2\right], \quad -\infty < y < \infty$$

Hence Y is normally distributed with mean $a + b\mu$ and variance $b^2\sigma^2$.

❑ **EXAMPLE 11.6.** *(i) Let* $g(x) = |x|$. *Here the equation* $y = g(x)$ *has no solution if* $y < 0$ *and has two solutions*

$$x_1 = x, x_2 = -x \quad \text{for } y > 0.$$

Hence

$$f_Y(y) = f_X(x) + f_X(-x), \quad y > 0.$$
$$f_Y(0) = f_X(0)$$

Again

$$F_Y(y) = \begin{cases} P\{|X| \leq y\} = F_X(y) - F_X(-y-0), & y \geq 0 \\ 0, & y < 0 \end{cases}$$

❑ **EXAMPLE 11.7.** $g(x) = x(0)$ *if* $x > 0$ *(otherwise).*

If $y > 0$,

$$P[Y \leq y] = P[g(X) \leq y] = P[g(X) = 0] + P[0 < g(X) \leq y]$$
$$= P[X \leq 0] + P[0 < X \leq y] = F_X(y).$$

If $y < 0$, $\quad F_Y(y) = 0$.

If $y = 0$, $\quad F_Y(0) = P(Y = 0) = P(X \leq 0) = F_X(0)$.

Fig (11.1)

Thus $F_Y(y)$ is discontinuous with discontinuity at $y = 0$.

$$f_Y(y) = \begin{cases} f_X(y) & \text{for } y > 0 \\ F_X(0) & \text{for } y = 0 \\ 0 & \text{for } y < 0 \end{cases}$$

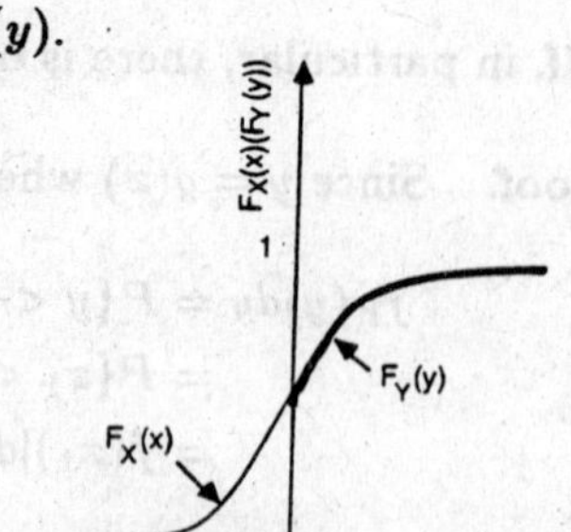

Fig. 11.1

❑ **EXAMPLE 11.8.** $y = a\cos(x+\theta), \quad a > 0$

If $|y| < a$, the equation $y = a\cos(x+\theta)$ has infinitely many solutions $x_n, n = \cdots -1, 0, 1, \cdots$

$$|g'(x_n)| = a\sin(x_n+\theta) = \sqrt{a^2-y^2}$$

Hence

$$f_Y(y) = \begin{cases} \dfrac{1}{\sqrt{a^2-y^2}} \displaystyle\sum_{n=-\infty}^{\infty} f_X(x_n) & |y| < a \\ 0, & |y| > a \end{cases} \tag{11.3}$$

Note that $f_Y(\pm a) = \infty$.

In particular if X is uniformly distributed over $(-\pi, \pi), f_X(x) = \dfrac{1}{2\pi}(0)$ for $|x| \le \pi$ (otherwise). The equation $y = a\cos(x+\theta)$, has just two solutions in $(-\pi, \pi), f_X(x) = \dfrac{1}{2\pi}$ at both these solutions. Hence from (11.3) we obtain

$$f_Y(y) = \frac{2}{\sqrt{a^2-y^2}} \cdot \frac{1}{2\pi} \quad \text{for } |y| < a. \tag{11.4}$$

The distribution function obtained by integrating (11.4) is

$$F_Y(y) = \begin{cases} 0, & y < -a \\ \dfrac{1}{\pi} arc\cos\dfrac{y}{a} - 1, & |y| < a \\ 1, & y > a \end{cases}$$

Note that $f_Y(\pm a) = \infty$; however $P(Y = \pm a) = 0$.

❑ **EXAMPLE 11.9.** $Y = a\tan X, \quad a > 0$.

The equation $y = a\tan x$ has infinitely many solutions

$$x_n = arc\tan\frac{y}{a} \qquad n = \cdots -1, 0, 1, \cdots$$

$$g'(x) = \frac{a}{\cos^2 x} = \frac{{}^2 + y^2}{a}$$

$$f_Y(y) = \frac{a}{a^2+y^2} \sum_{n=-\infty}^{\infty} f_X(x_n)$$

If in particular X is uniformly distributed over $(-\pi, \pi)$ then $y = a\tan x$, has only two solutions in $(-\pi, \pi)$ and $f_Y(y) = \dfrac{2a}{a^2+y^2} \cdot \dfrac{1}{2\pi} = \dfrac{a}{\pi(a^2+y^2)}$, a Cauchy distribution.

❑ **EXAMPLE 11.10.**

$$g(x) = k \cdot e^{-\beta x}, x \ge 0, \beta > 0, k > 0.$$
$$= 0, \qquad x < 0 \quad \text{(Fig. 11.2)}$$

The equation $y = g(x)$ has no solution for $y < 0$ and $y > k$. For $0 < y < k$,

$$x = -\frac{1}{\beta} ln \frac{y}{ak}$$

$$g'(x) = -\beta y$$

$$f_Y(y) = \begin{cases} \frac{1}{\beta y} f_X\left(-\frac{1}{\beta} ln \frac{y}{k}\right), & 0 < y < k \\ 0, & \text{otherwise} \end{cases}$$

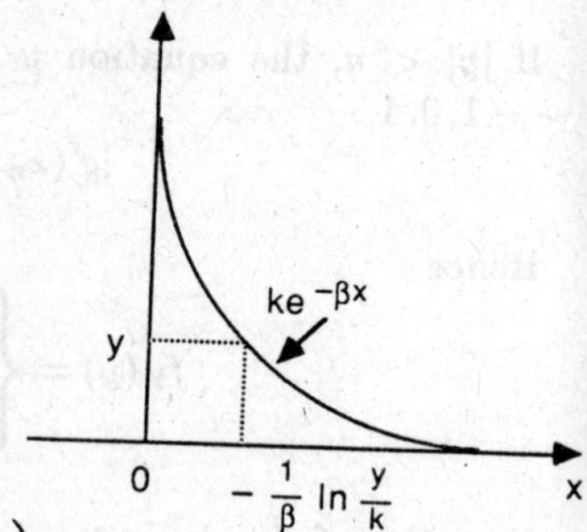

Fig. 11.2

For $0 < y < k, g(x) \leq y$ for $x \leq 0$ and $x > -\frac{1}{\beta} \log\left(\frac{y}{k}\right)$

$$F_Y(y) = P[X \leq 0] + P\left[X > -\frac{1}{\beta} \log \frac{y}{k}\right] = F_X(0) + 1 - F_X\left(-\frac{1}{\beta} \log \frac{y}{k}\right), \quad 0 < y < k$$

In particular if X is rectangular on (a, b), $f_X(x) = (b-a)^{-1}, a < x < b$,

$$f_Y(y) = \frac{1}{\beta y}, \quad ke^{-b\beta} < y < ke^{-a\beta}$$

and is zero elsewhere.

❑ **EXAMPLE 11.11.** *X is a. $N(\mu, \sigma)$ variable and $Y = bX^2, b > 0$*

$$f_Y(y) = \frac{1}{2a\sqrt{\frac{y}{b}}}\left[f_X\left(\sqrt{\frac{y}{b}}\right) + f_X\left(-\sqrt{\frac{y}{b}}\right)\right]$$

$$= \frac{1}{2\sigma\sqrt{2\pi by}}\left[\exp\left\{-\frac{1}{2\sigma^2}\left(\sqrt{\frac{y}{b}} - \mu\right)^2\right\} + \exp\left\{-\frac{1}{2\sigma^2}\left(-\sqrt{\frac{y}{b}} - \mu\right)^2\right\}\right]$$

$$= \frac{1}{2\sigma\sqrt{2\pi by}} e^{-\frac{1}{2\sigma^2}(\mu^2 + \frac{y}{b})}\left[e^{\frac{\mu}{\sigma^2}\sqrt{\frac{y}{b}}} + e^{-\frac{\mu}{\sigma^2}\sqrt{\frac{y}{b}}}\right]$$

$$= \frac{e^{-\frac{1}{2\sigma^2}(\mu^2 + \frac{y}{b})}}{\sigma\sqrt{2\pi b}} y^{-\frac{1}{2}} \cosh\left[\frac{\mu}{\sigma^2}\sqrt{\frac{y}{b}}\right], \quad 0 < y < \infty \tag{11.5}$$

In particular if $\mu = 0$,

$$f_Y(y) = \frac{y^{\frac{-1}{2}} e^{\frac{-y}{2b\sigma^2}}}{\Gamma\left(\frac{1}{2}\right)\sqrt{2b\sigma^2}}, \quad y > 0$$

so that Y has a gamma distribution with parameters $\left(\frac{1}{2b\sigma^2}, \frac{1}{2}\right)$. If $b = 1, \sigma^2 = 1, Y$ follows a χ^2-distribution with one degree of freedom. If $\mu \neq 0, b = 1, \sigma^2 = 1$, the resulting distribution (11.5) is called a non-central χ^2-distribution with one d.f.

11.3. Probability Integral Transformation

Consider a r.v. X with d.f. $F_X(x), 0 \leq F_X(x) \leq 1$. We define an inverse function F_X^{-1} on $0 \leq y \leq 1$ by setting $F_X^{-1}(y)$ equal to the smallest value of x for which $F_X(x) \geq y$.

• **THEOREM 11.2.** *If X is a r.v. with continuous distribution function $F_X(x)$, then the r.v. $Y = F_X(x)$ has a rectangular distribution over $(0, 1)$.*

Conversely if Y be a rectangular variable over (0, 1) then $X = F_X^{-1}(Y)$ has probability distribution function F_X.

Proof. The d.f. of Y is $F_Y(y) = P[Y \leq y] = P[F_X(x) \leq y] = P[X \leq F_X^{-1}(y)] = F_X(F_X^{-1}(y)) = y, 0 \leq y \leq 1$. Hence Y follows a uniform distribution over (0, 1).

Conversely, suppose Y has a uniform distribution over (0, 1). The d.f. of $X, P[X \leq x] = P[F_X^{-1}(Y) \leq x] = P[Y \leq F_X(x)] = F_X(x)$, so that X has d.f. $F_X(x)$.

The transformation $Y = F_X(x)$ is called probability integral transformation.

The theorem 11.2 is of fundamental importance and states that every continuous probability distribution can be transformed into a normalised uniform distribution. Theoretically, therefore, every continuous distribution can be transformed into every other distribution by applying the above-mentioned transformations. The transformation $Y = F_X(x)$ is very useful in drawing a random sample from the distribution of X. Since $0 \leq Y \leq 1$, if one draws a random number y between [0, 1]. and calculates $x = F_X^{-1}(y)$ then x is a random sample from the distribution of X.

❑ **EXAMPLE 11.12.** *$X_1, \cdots, X_n$ have each continuous uniform distribution*

$$dF = dx, \quad 0 \leq x \leq 1.$$

Show that

$$Z = -2\sum_{i=1}^{n} \log_e X_i \quad \text{has a } \chi^2_{(2n)} \text{ distribution.}$$

Consider transformed variable

$$Y = -2\log_e X; \quad 0 < Y < \infty.$$

Hence

$$x = e^{\frac{-y}{2}}, \quad dx = -\frac{1}{2}e^{\frac{-y}{2}}\,dy.$$

The pdf of Y is

$$f_Y(y) = \frac{1}{2}e^{\frac{-y}{2}}, \quad 0 < y < \infty$$

so that Y follows a $\chi^2_{(2)}$ distribution. Hence $Z = \sum_{i=1}^{n} Y_i$ follows $\chi^2_{(2n)}$-distribution by the reproductive property of this distribution.

❑ **EXAMPLE 11.13.** *A variable X has a uniform distribution over [0, 1]. Find a function of X say Y, with probability element*

$$f_Y(y)dy = e^{-y}dy, \quad 0 \leq y \leq \infty.$$

It is known that the d.f. of Y follows a uniform distribution over [0, 1]. Now

$$F_Y(y) = \int_0^y e^{-t}dt = 1 - e^{-y}.$$

Hence $1 - e^{-Y}$ is uniform over [0, 1].

Therefore, $X = 1 - e^{-Y}$ with probability one, when $Y = \log_e\left(\frac{1}{1-X}\right)$

❑ **EXAMPLE 11.14.** *$X_1, \cdots, X_n$ are random samples from a uniform distribution over (0, 1). Find the pdf of $u = (X_1, \cdots, X_n)^{1/n}$.*

It is known that $Z = -2\sum_{i=1}^{n} \log_e X_i$ has a χ^2-distribution with $2n$ degrees of freedom.

$$f_Z(z)dz = \frac{1}{2^n \Gamma(n)} e^{-\frac{1}{2}z} z^{n-1}dz, \quad 0 \le z < \infty$$

Make the transformation

$$u = e^{\frac{-z}{2n}}; \quad Z = 2n \log_e\left(\frac{1}{u}\right)$$

$$\frac{du}{dz} = -\frac{1}{2n}e^{\frac{-z}{2n}} = -\frac{u}{2n}; \quad 0 < u < 1$$

$$f_U(u)du = \frac{1}{2^n \Gamma(n)} e^{-\frac{1}{2}[2n \log_e \frac{1}{u}]} \left(2n \log_e \frac{1}{u}\right)^{n-1} \frac{2n}{u} du$$

$$= \frac{n^n u^{n-1}}{\Gamma(n)} \left(\log_e \frac{1}{u}\right)^{n-1} du, \quad 0 < u < 1$$

11.4. Functions of Two Random Variables

11.4.1. Using Distribution Functions

Let (X, Y) be two jointly continuous random variables with pdf $f_{X,Y}(x, y)$. We first consider the problem of finding probability distributions of some simple functins $U = X + Y, V = X - Y, W = XY, Z = X/Y$ by using the procedure of finding the marginal distributions.

• **THEOREM 11.3.** *Let (X, Y) be jointly distributed continuous random variables with pdf $f_{X,Y}(x, y)$. The pdf of $U = X + Y, V = X - Y$ are respectively given by*

(i) $$f_U(u) = \int_{-\infty}^{\infty} f_{X,Y}(x, u - x)dx = \int_{-\infty}^{\infty} f_{X,Y}(u - y, y)dy \tag{11.6}$$

(ii) $$f_V(v) = \int_{-\infty}^{\infty} f_{X,Y}(x, x - v)dx = \int_{-\infty}^{\infty} f_{X,Y}(v + y, y)dy \tag{11.7}$$

Proof. The d.f. of $U = X + Y$ is

$$F_U(u) = P(X + Y \leq u)$$

$$= \iint_{x+y\leq u} f_{X,Y}(x, y)dxdy = \int_{-\infty}^{\infty}\left[\int_{-\infty}^{u-x} f_{X,Y}(x, y)dy\right]dx$$

$$= \int_{-\infty}^{\infty}\left[\int_{-\infty}^{u} f_{X,Y}(x, z - x)dz\right]dx \quad \text{(Putting } y = z - x\text{)}$$

$$= \int_{-\infty}^{u}\left[\int_{-\infty}^{\infty} f_{X,Y}(x, z - x)dx\right]dz = \int_{-\infty}^{u} f_U(z)dz \tag{11.8}$$

Differentiating both sides we get the pdf of U. The change from a double integral to a repeated integral, as well as the change of order of integration are justified since $f(x, y)$ is abolutely integrable over the entire xy-plane. The second expression in (11.6) can be derived similarly. (11.7) is proved in an analogous manner.

COROLLERY 11.1. *If X, Y are independent continuous random variables, the pdf of $U = X + Y$ is*

$$f_U(u) = \int_{-\infty}^{\infty} f_X(x)f_Y(u - x)dx = \int_{-\infty}^{\infty} f_X(u - y)f_Y(y)dy \tag{11.9}$$

The density $f_U(.)$ is called, here, the convolution of the densities $f_X(.), f_Y(.)$.

❑ **EXAMPLE 11.15.** *X, Y are independent uniform variables over (0, 1). Find the pdf of (i) $U = X + Y$, (ii) $V = X - Y$ (iii) $W = |X - Y|$.*

$$\text{(i)} \quad f_U(u) = \int_{-\infty}^{\infty} f_X(x)f_Y(u - x)dx, \quad 0 \leq u \leq 2.$$

When $0 \leq u \leq 1, 0 \leq x \leq u$

$$f_U(u) = \int_0^u dx = u$$

When $1 \leq u \leq 2, u - 1 \leq x \leq 1$

$$f_U(u) = \int_{u-1}^{1} dx = 2 - u.$$

Hence

$$f_U(u) = \begin{cases} u, & 0 \leq u \leq 1 \\ 2 - u, & 1 \leq u \leq 2 \end{cases}$$

This is known as the triangular distribution.

$$\text{(ii)} \quad f_V(v) = \int_{-\infty}^{\infty} f_X(x)f_Y(x - v)dx, \quad -1 \leq v \leq 1.$$

When $-1 \leq v \leq 0, 0 \leq x \leq 1 + v$.

$$f_V(v) = \int_0^{1+v} dx = 1 + v.$$

When $0 \le v \le 1$, $v \le x \le 1$

$$f_V(v) = \int_v^1 dx = 1 - v.$$

Hence

$$f_V(v) = \begin{cases} 1+v, & -1 \le v \le 0 \\ 1-v, & 0 \le v \le 1 \end{cases} \tag{i}$$

as in fig. 11.3.

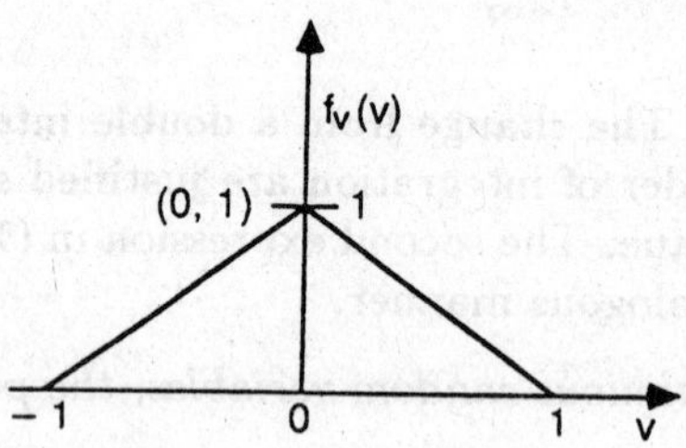

Fig. 11.3

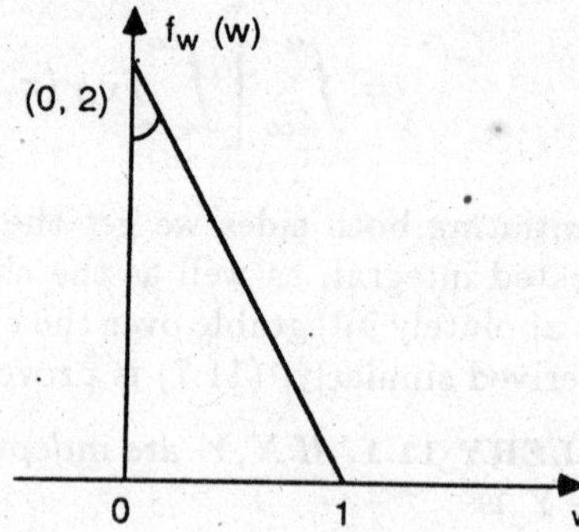

Fig. 11.4

(iii) Let $w = |x-y|; 0 \le w \le 1$. Solving the equations $w = |v|$, $w = +v$ or $-v$, $\left|\dfrac{dw}{dv}\right| = 1$ at each solution.

Hence from (i),

$$f_W(w) = 2(1-w), \quad 0 \le w \le 1,$$

as in fig. 11.4.

❑ **EXAMPLE 11.16.** *(X, Y) has a bivariate normal density with parameters $(\mu_1, \mu_2, \sigma_1, \sigma_2, \rho)$. Find the pdf of $Z = X + Y$.*

$$f(x,y) = k\exp\left[-\frac{1}{2(1-\rho^2)}\left\{\left(\frac{x-\mu_1}{\sigma_1}\right)^2 - 2\rho\left(\frac{x-\mu_1}{\sigma_1}\right)\left(\frac{y-\mu_2}{\sigma_2}\right) + \left(\frac{y-\mu_2}{\sigma_2}\right)^2\right\}\right]$$

$$= k\exp\left[-\frac{Q(x,y)}{2}\right] \quad \text{(say)}.$$

Writing $x + y = Z, \mu_1 + \mu_2 = \mu$,

$$A_{11} = [\sigma_1^2(1-\rho^2)]^{-1}, \quad A_{12} = -\rho[\sigma_1\sigma_2(1-\rho^2)]^{-1}, \quad A_{22} = [\sigma_2^2(1-\rho^2)]^{-1},$$

$$\begin{aligned} Q(x,y) &= A_{11}(x-\mu_1)^2 + 2A_{12}(x-\mu_1)(y-\mu_2) + A_{22}(y-\mu_2)^2 \\ &= A_{11}[(z-y)-(\mu-\mu_2)]^2 + 2A_{12}[(z-y)-(\mu-\mu_2)](y-\mu_2) + A_{22}(y-\mu_2)^2 \\ &= \left(A_{11} - \frac{(A_{12}-A_{11})^2}{A_{11}-2A_{12}+A_{22}}\right)(z-\mu)^2 + \Bigg[(A_{11}-2A_{12}+A_{22})^{\frac{1}{2}}(y-\mu_2) \\ &\quad + \frac{A_{12}-A_{11}}{(A_{11}-2A_{12}+A_{22})^{\frac{1}{2}}}(z-\mu)\Bigg]^2 \end{aligned} \tag{i}$$

Writing the coefficient of $(z-\mu)^2$ in (i) as a,

$$b = \sqrt{A_{11} - 2A_{12} + A_{22}}, \quad c = \frac{A_{12} - A_{11}}{b}$$

and noting that $a > 0$, pdf of z is

$$f_Z(z) = ke^{-\frac{a}{2}(z-\mu)^2} \int_{-\infty}^{\infty} \exp\left\{ -\frac{1}{2}[b(y-\mu_2) + c(z-\mu)]^2 dy \right\} = \frac{k\sqrt{2\pi}}{b} e^{-\frac{a}{2}(z-\mu)^2}$$

so that z is $N\left(\mu, \frac{1}{\sqrt{a}}\right)$ when $a = (\sigma_1^2 + \sigma_2^2 + 2\rho\sigma_1\sigma_2)^{-1}$. In particular, when $\rho = 0$, z is $N(\mu, \sqrt{\sigma_1^2 + \sigma_2^2})$.

- **THEOREM 11.4.** *Let X, Y be jointly distributed continuous random variables with pdf $f_{X,Y}(x,y)$. The pdf of $W = XY$ and $Z = \frac{X}{Y}$ are respectively,*

$$\text{(i)} \quad f_W(w) = \int_{-\infty}^{\infty} \frac{1}{|x|} f_{X,Y}\left(x, \frac{w}{x}\right) dx = \int_{-\infty}^{\infty} \frac{1}{|y|} f_{X,Y}\left(\frac{w}{y}, y\right) dy \tag{11.10}$$

$$\text{(ii)} \quad f_Z(z) = \int_{-\infty}^{\infty} |y| f_{X,Y}(yz, y) dy, \qquad \text{provided } P(Y=0) = 0. \tag{11.11}$$

We omit the proof.

In particular if $f_{X,Y}(x,y) = f_{X,Y}(-x,-y)$, (11.11) reduces to

$$f_Z(z) = 2\int_0^{\infty} y f_{X,Y}(yz, y) dy \tag{11.12}$$

❑ **EXAMPLE 11.17.** *(X, Y) follows bivariate normal with $(0, 0, \sigma_1^2, \sigma_2^2, \rho)$. Find the pdf of $Z = \frac{X}{Y}$.*

By (11.12)

$$f_Z(z) = \frac{1}{\pi\sigma_1\sigma_2\sqrt{1-\rho^2}} \int_0^{\infty} y e^{-\frac{y^2}{2(1-\rho^2)}\left[\frac{z^2}{\sigma_1^2} - \frac{2\rho z}{\sigma_1\sigma_2} + \frac{1}{\sigma_2^2}\right]} dy = \frac{\sqrt{1-\rho^2}}{\pi\sigma_1\sigma_2\left[\frac{z^2}{\sigma_1^2} - \frac{2\rho z}{\sigma_1\sigma_2} + \frac{1}{\sigma_2^2}\right]}$$

$$= \frac{\sigma_1\sigma_2\sqrt{1-\rho^2}}{\pi\left[\sigma_2^2\left(z - \rho\frac{\sigma_1}{\sigma_2}\right)^2 + \sigma_1^2(1-\rho^2)\right]}$$

so that Z is a Cauchy variable with parameters $\lambda = \frac{\sigma_1}{\sigma_2}\sqrt{1-\rho^2}$ and $\theta = \rho\frac{\sigma_1}{\sigma_2}$.

In particular if $\sigma_1 = \sigma_2 = 1$, $\rho = 0$,

$$f_Z(z) = \frac{1}{\pi(z^2+1)}, \quad -\infty < z < \infty.$$

❑ **EXAMPLE 11.18.** *X, Y are independent variables, each uniform over (0, 1). Find pdf of (i) W = XY, (ii) Z = X/Y.*

(i) $$f_W(w) = \int_{-\infty}^{\infty} \frac{1}{|x|} f_X(x) f_Y\left(\frac{w}{x}\right) dx = \int_w^1 \frac{1}{x} dx$$

$$\left(\because \quad x = \frac{w}{y} \text{ and hence } w < x < 1\right) = -\log_e w, \qquad 0 < w < 1.$$

(ii) $$f_Z(z) = \int_{-\infty}^{\infty} |y| f_X(yz) f_Y(y) dy.$$

When $0 < z < 1, 0 < y < 1$.

$$f_Z(z) = \int_0^1 y\, dy = \frac{1}{z}$$

When $1 \le z < \infty, 0 < y < \frac{1}{z}$

$$f_Z(z) = \int_0^{\frac{1}{z}} y\, dy = \frac{1}{2z^2}$$

Hence

$$f_Z(z) = \begin{cases} \frac{1}{2}, & 0 < z < 1. \\ \frac{1}{2z^2}, & 1 \le z < \infty. \end{cases}$$

We now consider the problem of finding the distribution of general functions of two random variables by using the method of transformation of variables.

11.4.2. Transformation of Variables

Consider random variables (X_1, X_2) having joint density $f_{X_1,X_2}(x_1, x_2)$. Let $Y_1 = g_1(X_1, X_2), Y_2 = g_2(X_1, X_2)$ be two functions of (X_1, X_2). We want to find the joint distribution of (Y_1, Y_2).

The joint d.f. of (Y_1, Y_2) is

$$F_{Y_1,Y_2}(y_1, y_2) = P(Y_1 \le y_1, \quad Y_2 \le y_2) = \iint_{R_{y_1 y_2}} f_{X_1,X_2}(x_1, x_2) dx_1 dx_2$$

where $R_{y_1 y_2} = \{(x_1, x_2) : Y_1 \le y_1, \quad Y_2 \le y_2\}$.

Define

$$J = \begin{vmatrix} \frac{\partial x_1}{\partial y_1} & \frac{\partial x_2}{\partial y_1} \\ \frac{\partial x_1}{\partial y_2} & \frac{\partial x_2}{\partial y_2} \end{vmatrix}$$

as the Jacobian of transformation.

To find the joint pdf of (Y_1, Y_2), solve the simultaneous equations

$$g_1(x_1, x_2) = y_1, \quad g_2(x_1, x_2) = y_2 \tag{11.13}$$

giving (x_1, x_2) in terms of (y_1, y_2). Let

$$x_{1i} = g_{1i}^{-1}(y_1, y_2), \quad x_{2i} = g_{2i}^{-1}(y_1, y_2) \qquad i = 1, \cdots, m$$

be the m sets of solutions, all real. Calculate the Jacobians,

$$J_i = \begin{vmatrix} \dfrac{\partial g_{1i}^{-1}}{\partial y_1} & \dfrac{\partial g_{2i}^{-1}}{\partial y_1} \\ \dfrac{\partial g_{1i}^{-1}}{\partial y_2} & \dfrac{\partial g_{2i}^{-1}}{\partial y_2} \end{vmatrix}$$

The joint pdf of (Y_1, Y_2) is

$$f_{Y_1,Y_2}(y_1, y_2) = \sum_{i=1}^{m} f_{X_1,X_2}(x_{1i}, x_{2i})|J_i| \tag{11.14}$$

We have assumed that the first partial derivatives of g_{1i}^{-1}, g_{2i}^{-1} are continuous on the domain of (Y_1, Y_2) and $|J_i| \neq 0$ on the domain of $(Y_1, Y_2), i = 1, \cdots, m$.

If for certain (y_1, y_2), the equations (11.13) have no real solution, $f(y_1, y_2) = 0$. The proof is omitted.

If $(x_{1i}, x_{2i}) = (x_1, x_2)$ for all $i = 1, \cdots, m$,

$$f_{Y_1,Y_2}(y_1, y_2) = f(x_1, x_2)|J|.$$

If only distribution of Y_1 is desired, a variable $Y_2 = g_2(X_1, X_2)$ is suitably chosen, joint distribution of (Y_1, Y_2) is obtained and distribution of Y_1 is found out by integrating over Y_2.

❑ **EXAMPLE 11.19.** *X_1, X_2 are independent gamma variables with parameters (α, p_1), (α, p_2) respectively. Show that $U = X_1 + X_2, V = X_1/(X_1 + X_2)$ are independently distributed, U following a gamma distribution with parameters $(\alpha, p_1 + p_2), V$ a beta distribution with parameters (p_1, p_2). Also find the distribution of $W = \dfrac{X_1}{X_2}$.*

The probability element of (X_1, X_2) is

$$dF = \frac{\alpha^{p_1+p_2}}{\Gamma(p_1)\Gamma(p_2)} e^{-\alpha(x_1+x_2)} x_1^{p_1-1} x_2^{p_2-1} dx_1 dx_2, \quad 0 < x_1, x_2 < \infty$$

Use transformation

$$\left.\begin{aligned} u &= x_1 + x_2 \\ v &= \frac{x_1}{x_1 + x_2} \end{aligned}\right\} \text{ i.e., } \quad \begin{aligned} x_1 &= uv & 0 \leq u < \infty \\ x_2 &= u(1 - v). & 0 \leq v < 1 \end{aligned}$$

$|J| = u$

Joint probability element of (U, V) is

$$dF = \frac{\alpha^{p_1+p_2}}{\Gamma(p_1)\Gamma(p_2)} e^{-\alpha u}(uv)^{p_1-1}[u(1-v)]^{p_2-1} u\,du\,dv$$
$$= \left\{\frac{\alpha^{p_1+p_2}}{\Gamma(p_1+p_2)} u^{p_1+p_2-1} e^{-\alpha u} du\right\}\left\{\frac{1}{B(p_1,p_2)} v^{p_1-1}(1-v)^{p_2-1} dv\right\}$$
$$= f_U(u) f_V(v).$$

Again $w = \frac{v}{1-v}$; hence $v = \frac{w}{1+w}$; $dv = \frac{dw}{(1+w)^2}$.

The pdf of W is obtained from the pdf of V as

$$f_W(w) = \frac{1}{B(p_1,p_2)}\left[\frac{w}{1+w}\right]^{p_1-1}\frac{1}{(1+w)^{p_2+1}} = \frac{1}{B(p_1,p_2)}\frac{w^{p_1-1}}{(1+w)^{p_1+p_2}}, \quad 0 \le w < \infty$$

so that W follows a beta distribution of the second kind with parameters p_1, p_2.

[Alternatively, put $X_1 = r\cos^2\theta, X_2 = r\sin^2\theta; |J| = 2r\cos\theta\sin\theta, X_1 + X_2 = r, \frac{X_1}{(X_1+X_2)} = \cos^2\theta$; show that r and $\cos^2\theta$ are independently distributed.]

❑ **EXAMPLE 11.20.** *Let X, Y be independent beta variates with parameters (a, b), (c, d) respectively such that $a = c + d$. Show that XY is a beta variate with parameters $(c, b + d)$.*

$$f_{X,Y}(x,y) = kx^{a-1}(1-x)^{b-1}y^{c-1}(1-y)^{d-1} \quad \text{where} \quad k = [B(a,b)B(c,d)]^{-1}.$$

Let $u = x, z = xy$; hence $x = u, y = \frac{z}{u}.0 < u, z < 1.|J| = \frac{1}{|u|}$

The joint density of U and Z is

$$f_{U,Z}(u,z) = ku^{a-c-d}(1-u)^{b-1}z^{c-1}(u-z)^{d-1}$$
$$= k(1-u)^{b-1}z^{c-1}(u-z)^{d-1} \quad (\text{since } a = c + d)$$

The marginal density of z is

$$f_z(z) = kz^{c-1}\int_0^1 (1-u)^{b-1}(u-z)^{d-1}du$$
$$= kz^{c-1}(1-z)^{b+d-1}\int_0^1 t^{b-1}(1-t)^{d-1}dt \quad \left[\text{where } t = \frac{1-u}{1-z}\right]$$
$$= \frac{B(b,d)}{B(a,b)B(c,d)}z^{c-1}(1-z)^{b+d-1} = \frac{1}{B(c,b+d)}z^{c-1}(1-z)^{b+d-1}$$

In general product of Beta variates with parameters $(a_1, b_1), \cdots (a_k, b_k)$ such that $a_i = a_{i+1} + b_{i+1}, i = 1, \cdots, k-1$ is distributed as a Beta variate with parameters $\left(a_k, \sum_{i=1}^{k} b_i\right)$.

❑ **EXAMPLE 11.21.** *Let X, Y be independent, each $N(0,\sigma)$-variable. Find the joint pdf of $Z = \sqrt{X^2+Y^2}, W = \frac{X}{Y}$.*

Solving the equations

$$z = \sqrt{x^2+y^2}, \quad w = \frac{x}{y}$$

we have two solutions

$$\left(x_1 = \frac{zw}{\sqrt{1+w^2}}, \quad y_1 = \frac{z}{\sqrt{1+w^2}}\right), \quad (-x_1, -y_1)$$

for any w. Here

$$\begin{vmatrix} \frac{\partial z}{\partial x} & \frac{\partial z}{\partial y} \\ \frac{\partial w}{\partial x} & \frac{\partial w}{\partial y} \end{vmatrix} = \begin{vmatrix} \frac{x}{\sqrt{x^2+y^2}} & \frac{y}{\sqrt{x^2+y^2}} \\ \frac{1}{y} & \frac{-x}{y^2} \end{vmatrix} = \frac{1+w^2}{-z}; \quad \text{hence } |J_i| = \frac{|z|}{1+w^2},$$

for each point $(x_i, y_i), i = 1, 2$.

$$f_{X,Y}(x,y) = \frac{1}{2\pi\sigma^2} e^{-\frac{1}{2\sigma^2}(x^2+y^2)}$$

Since $f(x_1, y_1) = f(x_2, y_2)$

$$f_{Z,W}(z,w) = \frac{2}{2\pi\sigma^2} e^{\frac{-z^2}{2\sigma^2}} \left(\frac{z}{1+w^2}\right) = \left\{\frac{z}{\sigma^2} e^{\frac{-z^2}{2\sigma^2}}\right\} \left\{\frac{1}{\pi(1+w^2)}\right\}$$

which shows that (Z, W) are independently distributed, Z having a Rayleigh distribution and W a Cauchy distribution.

Alternative Solution : Consider polar transformation from (x, y) to (z, θ) : $x = z\cos\theta$, $y = z\sin\theta$, $z > 0, 0 < \theta < 2\pi$, $w = \cot\theta$, $J = z$.

The joint probability element of (Z, θ)

$$dF(z,\theta) = \frac{z}{2\pi\sigma^2} e^{\frac{-z^2}{2\sigma^2}} dz d\theta.$$

Hence pdf of Z is

$$f_Z(z) = \frac{z}{\sigma^2} e^{-z^2/2\sigma^2}; \quad \text{pdf of } \theta \text{ is } f(\theta) = \frac{1}{2\pi}, \quad 0 < \theta < 2\pi \tag{11.15}$$

For each value of w there are two values of θ in $(0, 2\pi)$ satisfying $\cot\theta = w$.

$$\frac{dw}{d\theta} = \operatorname{cosec}^2\theta = (1+w^2)$$

Hence from (11.15), pdf of W is

$$f_W(w) = \frac{1}{\pi(1+w^2)}, \quad -\infty < w < \infty.$$

11.5. Functions of n Random Variables

Let $X_1, \cdots, X_n$ be jointly continuous random variables with joint pdf $f_{X_1,\cdots,X_n}(x_1, \cdots, x_n)$ and let $\mathcal{X} = \{(x_1, \cdots, x_n) : f(x_1, \cdots, x_n) > 0\}$. Let $Y_1 = g_1(X_1, \cdots, X_n), \cdots, Y_n = g_n(X_1, \cdots, X_n)$. Let $\mathcal{Y}$ = domain of $(Y_1, \cdots, Y_n)$.

Suppose $\mathcal{X}$ can be decomposed into m disjoint subsets $\mathcal{X}_1, \cdots, \mathcal{X}_m$ such that $(x_1, \cdots, x_n) \to (y_1, \cdots, y_n)$ is a one-one tranformation of $\mathcal{X}_i$ onto $\mathcal{Y}$. Let $x_1 = g_{1i}^{-1}(y_1, \cdots, y_n), \cdots, x_n = g_{ni}^{-1}(y_1, \cdots, y_n)$ denote the inverse transformations of $\mathcal{Y}$ onto $\mathcal{X}_i, i = 1, \cdots, m$. Define

$$J_i = \begin{vmatrix} \dfrac{\partial g_{1i}^{-1}}{\partial y_1} & \cdots & \dfrac{\partial g_{ni}^{-1}}{\partial y_1} \\ \vdots & & \\ \dfrac{\partial g_{1i}^{-1}}{\partial y_n} & \cdots & \dfrac{\partial g_{ni}^{-1}}{\partial y_n} \end{vmatrix}$$

We assume all the partial derivatives exist and are continuous over $\mathcal{Y}$ and $J_i \neq 0, i = 1, \cdots, m$. Then

$$f_{Y_1,\cdots,Y_n}(y_1, \cdots, y_n) = \sum_{i=1}^{m} f_{X_1,\cdots,X_n}(x_{1i}, \cdots, x_{ni})|J_i|,$$

for $(y_1, \cdots, y_n) \in \mathcal{Y}$.

If joint distribution of $Y_1, \cdots, Y_k (k < n)$ are desired, additional $(n-k)$ variables $Y_{k+1} = g_{k+1}(X_1, \cdots, X_n), \cdots, Y_n = g_n(X_1, \cdots, X_n)$ are suitably chosen, joint pdf of $(Y_1, \cdots, Y_n)$ are obtained and then distribution of $(Y_1, \cdots, Y_k)$ is found by integrating out $(Y_{k+1}, \cdots, Y_n)$.

❑ **EXAMPLE 11.22.** *Let X_1, X_2, X_3 be three independently and identically distributed random variables, each with pdf $f(x) = e^{-x} (x > 0)$. Show that the variables.*

$$Y_1 = X_1 + X_2 + X_3, \quad Y_2 = \frac{X_1 + X_2}{X_1 + X_2 + X_3}, \quad Y_3 = \frac{X_1}{X_1 + X_2}$$

are independently distributed.

Using the transformation

$$y_1 = x_1 + x_2 + x_3, \qquad y_2 = \frac{x_1 + x_2}{x_1 + x_2 + x_3}, \qquad y_3 = \frac{x_1}{x_1 + x_2},$$

$$x_1 = y_1 y_2 y_3, \qquad x_2 = y_1 y_2 (1 - y_3), \qquad x_3 = y_1 (1 - y_2)$$

$$|J| = y_1^2 y_2.$$

$$f_{X_1,X_2,X_3}(x_1, x_2, x_3) = e^{-y_1}.$$

Hence

$$f_{Y_1,Y_2,Y_3}(y_1, y_2, y_3) = e^{-y_1} y_1^2 y_2$$

which shows that Y_1, Y_2, Y_3 are independently distributed.

❑ **EXAMPLE 11.23.** *Let (X, Y, Z) have joint pdf*

$$f(x, y, z) = 6(1 + x + y + z)^{-4}, \quad x, y, z > 0.$$

Find the pdf of $X + Y + Z$.

Use the transformation : $u = x + y + z, uv = y + z, uvw = z$.

$J = u^2 v; 0 < u < \infty, 0 < v < 1, 0 < w < 1$.

$$f(u, v, w) = \frac{6u^2 v}{(1 + u)^4}.$$

Hence integrating over v and w,

$$f_U(u) = \frac{Cu^2}{(1 + u)^4}, \quad 0 < u < \infty$$

where C is a suitable constant.

❑ **EXAMPLE 11.24.** *Let $X_i, i = 1, \cdots, n$ be n independently distributed normal variables with parameters (μ_i, σ_i). Find the pdf of $Y = a + \sum_{i=1}^{n} b_i X_i$ where $a, b_1, \cdots, b_n$ are arbitrary constants.*

The probability element of the joint distribution of $(X_1, \cdots, X_n)$ is

$$dF = \frac{1}{(2\pi)^{\frac{n}{2}} \sigma_1 \cdots \sigma_n} \exp\left\{ -\frac{1}{2} \sum_{i=1}^{n} \left(\frac{x_i - \mu_i}{\sigma_i^2} \right)^2 \right\} \tag{11.16}$$

We make the following transformation from $(x_1, \cdots, x_n)$ to $(y_1, \cdots, y_n)$

$$\left.\begin{aligned} y_1 &= \frac{b_1\sigma_1\left(\frac{x_1 - \mu_1}{\sigma_1}\right) + \cdots + b_n\sigma_n\left(\frac{x_n - \mu_n}{\sigma_n}\right)}{\sqrt{\sum_1^n b_i \sigma_i^2}} \\ y_2 &= l_{21}\left(\frac{x_1 - \mu_1}{\sigma_1}\right) + \cdots + l_{2n}\left(\frac{x_n - \mu_n}{\sigma_n}\right) \\ &\cdots\cdots\cdots\cdots\cdots\cdots \\ y_n &= l_{n1}\left(\frac{x_1 - \mu_1}{\sigma_1}\right) + \cdots + l_{nn}\left(\frac{x_n - \mu_n}{\sigma_n}\right) \end{aligned}\right\} \tag{11.17}$$

where $(l_{i1}, \cdots l_{in}), i = 2, \cdots, n$, are mutually orthogonal vectors of unit length and each is also orthogonal to the vector

$$\left(\frac{b_1\sigma_1}{\sqrt{\sum b_i^2 \sigma_i^2}}, \cdots, \frac{b_n\sigma_n}{\sqrt{\sum b_i^2 \sigma_i^2}} \right).$$

The transformation (11.17) can be written as

$$\begin{bmatrix} y_1 \\ \vdots \\ y_n \end{bmatrix} = L \begin{bmatrix} \dfrac{(x_1 - \mu_1)}{\sigma_1} \\ \vdots \\ \dfrac{(x_n - \mu_n)}{\sigma_n} \end{bmatrix} \quad \text{i.e., } \boldsymbol{Y} = L\boldsymbol{Z} \quad \text{(say)}$$

where L is an orthogonal matrix, $LL' = L'L = I = \text{Diag}\,(1, \cdots, 1)$.

Hence

$$\sum_1^n y_i^2 = \boldsymbol{Y}'\boldsymbol{Y} = \boldsymbol{Z}'L'L\boldsymbol{Z} = \boldsymbol{Z}'\boldsymbol{Z} = \sum_1^n \frac{(x_i - \mu_i)^2}{\sigma_i^2}$$

$$J\left(\frac{x_1 \cdots x_n}{y_1 \cdots y_n}\right) = \sigma_1 \cdots \sigma_n |L| = \pm \sigma_1 \cdots \sigma_n.$$

Hence pdf of $y_1, \cdots, y_n$ is

$$\frac{1}{(2\pi)^n} \exp\left\{-\frac{1}{2}\sum_1^n y_i^2\right\}, \quad -\infty < y_i < \infty, \quad i = 1, \cdots, n$$

which shows that $Y_1, \cdots, Y_n$ are independent, each $N(0, 1)$. In particular, Y_1 is $N(0, 1)$.

Now

$$Y = a + \sum b_i X_i = a + \sum b_i \mu_i + Y_1 \sqrt{\sum_1^n b_i^2 \sigma_i^2}$$

$$E(Y) = a + \sum b_i \mu_i, \quad V(Y) = \sum b_i^2 \sigma_i^2.$$

Hence Y is $N(a + \sum b_i \mu_i, \sum b_i^2 \sigma_i^2)$ since if Z is normal (μ, σ), any linear function $\alpha Z + \beta, (\alpha \neq 0)$ itself follows a normal distribution $N(\alpha\mu + \beta, \alpha^2\sigma^2)$ (vide example 7.15).

[The results in this example were proved in theorem 9.5 by using characteristic function].

11.6. Distributions of Maximum and Minimum

Let $X_1, \cdots, X_n$ be r.v.'s each defined on the probability space $(\mathcal{S}, \mathcal{A}, P)$. Let

$$Y_1 = \max(X_1, \cdots, X_n), \quad Y_2 = \min(X_1, \cdots, X_n).$$

Clearly Y_1, Y_2 are r.v.'s, because

$$\{w : Y_2(w) \leq z\} = \{w : X_1(w) \geq z, \cdots, X_n(w) \geq z\} = \bigcap_{i=1}^n \{w : X_i(w) \geq z\} \in \mathcal{A}.$$

Thus Y_2 is a $\mathcal{A}$-measurable function and hence is a r.v. Similarly Y_1 is a r.v.

• **THEOREM 11.5.** *If $X_1, \cdots, X_n$ are independent r.v.'s, X_i having d.f. $F_{X_i}(x)$, the d.f. of Y_1, Y_2 respectively are*

(i) $F_{Y_1}(x) = \prod_{i=1}^{n} F_{X_i}(x)$

(ii) $F_{Y_2}(x) = 1 - \prod_{i=1}^{n}[1 - F_{X_i}(x)]$

Proof.

$$\begin{aligned} F_{Y_1}(x) &= P(Y_1 \le x) = P(X_1 \le x, \cdots, X_n \le x) \\ &= P(X_1 \le x) \cdots P(X_n \le x) \quad \text{[by independence]} \\ &= \prod_{i=1}^{n} F_{X_i}(x) \end{aligned}$$

$$\begin{aligned} F_{Y_2}(x) &= P(Y_2 \le x) = 1 - P(Y_2 > x) = 1 - P(X_1 > x, \cdots, X_n > x) \\ &= 1 - P(X_1 > x) \cdots P(X_n > x) \quad \text{[by independence]} \\ &= 1 - \prod_{i=1}^{n}[1 - F_{X_i}(x)] \end{aligned}$$

COROLLERY 11.2 If $X_1, \cdots, X_n$ are iid variables each having d.f. $F_X(x)$,

$$F_{Y_1}(x) = (F_X(x))^n; \quad F_{Y_2}(x) = 1 - (1 - F_X(x))^n.$$

COROLLERY 11.3. If, further, $X_1, \cdots, X_n$ are continuous with common pdf f, the pdf of Y_1, Y_2 are respectively given by

(i) $f_{Y_1}(x) = nF^{n-1}(x)f(x)$

(ii) $f_{Y_2}(x) = n[1 - F(x)]^{n-1}f(x)$.

❑ **EXAMPLE 11.25.** *Let X, Y be independent r.v.'s with negative exponential distributions having respective parameters λ_1, λ_2. Find $E[\max(X, Y)]$.*

Let $Z = \max(X, Y)$. The d.f. of Z is

$$\begin{aligned} F_Z(z) &= F_X(z)F_Y(z) = (1 - e^{-\lambda_1 z})(1 - e^{-\lambda_2 z}) \\ f_Z(z) &= \frac{d}{dz}(F_Z(z)) = \frac{d}{dz}[1 - e^{-\lambda_1 z} - e^{-\lambda_2 z} + e^{-(\lambda_1+\lambda_2)z}] \\ &= \frac{e^{-\lambda_1 z}}{\lambda_1} + \frac{e^{-\lambda_2 z}}{\lambda_2} + \frac{e^{-(\lambda_1+\lambda_2)z}}{-(\lambda_1+\lambda_2)} = \frac{e^{-\lambda_1 z}}{\lambda_1} + \frac{e^{-\lambda_2 z}}{\lambda_2} - \frac{e^{-(\lambda_1+\lambda_2)z}}{(\lambda_1+\lambda_2)}. \\ E(Z) &= \int_0^\infty z f_Z(z) dz \end{aligned}$$

Now

$$\int_0^\infty z\frac{e^{-\lambda_1 z}}{\lambda_1} dz = \int_0^\infty \frac{e^{-\lambda_1 z}(\lambda_1 z)^{2-1}}{\lambda_1^3} d(\lambda_1 z) = \frac{\Gamma(2)}{\lambda_1^3} = \frac{1}{\lambda_1^3}.$$

Thus

$$E(Z) = \frac{1}{\lambda_1^3} + \frac{1}{\lambda_2^3} - \frac{1}{(\lambda_1+\lambda_2)^3}$$

❑ **EXAMPLE 11.26.** *Show that a necessary and sufficient condition for independent r.v.'s $X_1, \cdots, X_n$ to be geometrically distributed with parameter p is that $Y = \min(X_1, \cdots, X_n)$ has a geometric distribution with parameter $(1-q^n), q = 1-p$.*

Suppose X has a geometric distribution with parameter p. The d.f. is $F_X(x) = P(X \leq x) = 1 - q^{x+1}$ Hence d.f. of Y is

$$F_Y(x) = 1 - [1 - F_X(x)]^n = 1 - q^{n(x+1)} \tag{11.18}$$

so that Y is geometric with parameter $(1-q^n)$. Conversity, suppose Y has d.f. (11.18). Now

$$F_Y(x) = 1 - [1 - F_X(x)]^n$$

Hence

$$1 - F_X(x) = q^{x+1} \quad \text{when} \quad F_X(x) = 1 - q^{x+1}$$

so that each X is geometric with parameter p.

• **THEOREM 11.6.** *If $X_1, \cdots, X_n$ are each independent r.v.'s, the joint d.f. of Y_1, Y_2 is*

$$F_{Y_1,Y_2}(y_1, y_2) = \begin{cases} \prod_{i=1}^{n} F_{X_i}(y_1) - \prod_{i=1}^{n} [F_{X_i}(y_1) - F_{X_i}(y_2)], & y_1 > y_2 \\ \prod_{i=1}^{n} F_{X_i}(y_1), & y_1 \leq y_2 \end{cases}$$

Proof.

$$\left.\begin{aligned} F_{Y_1,Y_2}(y_1, y_2) &= P[Y_1 \leq y_1, Y_2 \leq y_2] = P[Y_1 \leq y_1] - P[Y_1 \leq y_1, Y_2 \geq y_2] \\ &= P[X_1 \leq y_1, \cdots, X_n \leq y_1] - P[y_2 \leq X_1, \cdots, X_n \leq y_1] \\ &= \prod_{i=1}^{n} F_{X_i}(y_1) - \prod_{i=1}^{n} [F_{X_i}(y_1) - F_{X_i}(y_2)] \\ & \qquad \text{(by independence).} \end{aligned}\right\} \tag{11.19}$$

If $y_1 \leq y_2$, the last term in the right hand side of (11.19) vanishes. Hence the result.

COROLLERY 11.4. If $X_1, \cdots, X_n$ are iid with common d.f. F then

$$F_{Y_1,Y_2}(y_1, y_2) = \begin{cases} F^n(y_1) - [F(y_1) - F(y_2)]^n, & y_1 \leq y_2 \\ F^n(y_1), & y_1 \leq y_2 \end{cases}$$

COROLLERY 11.5. If $X_1, \cdots, X_n$ are continuous with common density f, the joint pdf of Y_1, Y_2 is

$$f_{Y_1,Y_2}(y_1, y_2) = \begin{cases} n(n-1)[F(y_1) - F(y_2)]^{n-2} f(y_1) f(y_2), & y_1 > y_2 \\ 0, & y_1 \leq y_2 \end{cases}$$

❑ **EXAMPLE 11.27.** $X_1, \cdots, X_n$ *are independent exponential variables with the same parameter λ. Find the joint distribution of Y_1, Y_2.*

$$F_{Y_1,Y_2}(y_1, y_2) = \begin{cases} (1 - e^{-\lambda y_1})^n - [e^{-\lambda y_2} - e^{-\lambda y_1}]^n, & y_1 > y_2 \\ (1 - e^{-\lambda y_1})^n, & y_1 \leq y_2 \end{cases}$$

The joint pdf is

$$f_{Y_1,Y_2}(y_1, y_2) = \begin{cases} n(n-1)\lambda^2[e^{-\lambda y_2} - e^{-\lambda y_1}]^{n-2}e^{-\lambda(y_1+y_2)}, & y_1 > y_2 \\ 0, & y_1 \leq y_2 \end{cases}$$

11.7. Use of Moment Generating Functions

Since moment generating function, when it exists, uniquely determines the distribution function, one may use the mgf to determine the distribution. Thus to find the joint distribution of $Y_1 = g_1(X_1, \cdots, X_n), \cdots, Y_n = g_n(X_1, \cdots, X_n)$ one may find the joint mgf of $(Y_1, \cdots, Y_n)$ with the help of the joint pdf of $(X_1, \cdots, X_n)$. If the mgf so obtained can be identified to be the mgf of a known distribution ξ, then ξ would be the joint distribution of $(Y_1, \cdots, Y_n)$. However, since only mgf of some one-dimensional random variables and some two-dimensional random variables are readily recognisable, the use of mgf to obtain the distribution will be limited in use to univariate functions or bivariate functions of random variables only.

Mostly by using the property that the mgf of a sum of n independent random variables is the product of their mgf's we have been able to prove several properties (including the reproductive property) of binomial (theorem 8.1), poisson (example 7.16), geometric, negative binomial (theorem 8.4), exponential, gamma (theorem 9.2), cauchy (theorem 9.4) and normal variables (theorem 9.5) and hence have been able to identify the distributions of the relevant functions of these variables. Here we consider the mgf of arbitrary functions of random variables, restricting our discussion to the case of jointly continuous two-dimensional variables only.

Let (X_1, X_2) be two continuous variables having joint pdf $f_{X_1,X_2}(x_1, x_2)$. The joint moment generating function of $Y_1 = g_1(X_1, X_2), Y_2 = g_2(X_1, X_2)$ is

$$M_{Y_1Y_2}(t_1, t_2) = E(e^{t_1Y_1+t_2Y_2}) = \int_{-\infty}^{\infty}\int_{-\infty}^{\infty} e^{t_1g_1(x_1,x_2)+t_2g_2(x_1,x_2)} f(x, y)dxdy \tag{11.20}$$

The case of discrete random-variables is treated analogously.

❑ **EXAMPLE 11.28.** *Let X, Y be independent standard normal variables. Find the mgf of XY.*

$$M_{XY}(t) = E(e^{txy}) = \frac{1}{2\pi}\int_{-\infty}^{\infty}\int_{-\infty}^{\infty} e^{txy - \frac{x^2+y^2}{2}} dxdy$$

$$= \frac{1}{2\pi}\int_{-\infty}^{\infty} e^{-\frac{y^2}{2}(1-t^2)} \left[\int_{-\infty}^{\infty} e^{-\frac{1}{2}(x-ty)^2} dx\right] dy = \frac{1}{\sqrt{1-t^2}}.$$

❑ **EXAMPLE 11.29.** *(X, Y) follows a bivariate normal distribution with parameters $(\mu_1, \mu_2, \sigma_1, \sigma_2, \rho)$. Show that $Z = aX + bY + c$ where a, b, c are constants, is $N(a\mu_1 + b\mu_2 + c, \sqrt{a\sigma_1^2 + b\sigma_2^2 + 2ab\rho\sigma_1\sigma_2})$.*

Moment generating function of Z is

$$M_Z(t) = \int_{-\infty}^{\infty}\int_{-\infty}^{\infty} e^{t(ax+by+c)} f_{X,Y}(x,y)dxdy \tag{11.21}$$

Now it is known

$$M_{X,Y}(t) = E(e^{t_1X+t_2Y}) = e^{\mu_1t_1+\mu_2t_1+\frac{1}{2}(\sigma_1^2t_1^2+\sigma_2^2t_2^2+2\rho\sigma_1\sigma_2t_1t_2)} \tag{11.22}$$

Putting $t_1 = at, t_2 = bt$ in (11.22), we obtain

$$M_Z(t) = \exp\left[c + at\mu_1 + bt\mu_2 + \frac{t^2}{2}(a^2\sigma_1^2 + b^2\sigma_2^2 + 2\rho\sigma_1\sigma_2ab)\right]$$

which is the mgf of a $N(a\mu_1+b\mu_2+c, a^2\sigma_1^2+b^2\sigma_2^2+2\rho ab\sigma_1\sigma_2)$. Hence the result follows by the uniqueness property of the mgf.

11.8. Exercises

1. *X is a r.v. with d.f. $F_X(x)$. Let*

$$g(X) = \begin{cases} x, & |x| \geq c \\ 0, & |x| < c \end{cases}$$

Show that d.f. of $Y = g(x)$ is $F_Y(y) =$

(i) $F_X(y)$ for $y \geq c$, (ii) $F_X(c)$ for $0 \leq y < c$. (iii) $F_X(-c)$ for $-c \leq y < 0$, (iv) $F_X(y)$ for $y < -c$.

2. ***X is uniformly distributed over $(-a, a)$. Find the distribution of Y =***

(i) $\frac{1}{X^2}$, (ii) $\sqrt{|X|}$, (iii) $\cos X$, (iv) e^X, (v) $bX^2 + cX + d$.

3. ***X has a uniform distribution over (0,1). Find the probability distribution of Y =***

(i) X^2, (ii) $\frac{1}{X}$, (iii) $\sqrt{X}$, (iv) $\sin(\pi X)$, (v) $-2\log_e X$, (vi) $\tan \pi X$, (vii) aX^2+bX+c, (viii) e^X, (ix) $\frac{X}{(1+X)}$.

4. ***Let X be a r.v. with pdf***

$$f(x) = \begin{cases} \lambda e^{-\lambda x}, & x > 0 \\ 0, & x \leq 0 \end{cases}, \quad \lambda > 0$$

Find the pdf of Y where Y is (i) $\sin X$, (ii) $\cos X$, (iii) $\tan X$.

5. ***X has a uniform distribution over (0,1). Find the distribution of $\frac{1}{X}$.***

6. ***X, Y are independent, each $N(0, \sigma)$ variable. Find the pdf of $w = \tan^{-1}\left(\frac{x}{y}\right)$.***

$$\left[f(w) = \frac{1}{\pi}, |w| < \frac{\pi}{2}\right]$$

7. *X has a pdf $f_X(x) = Ce^{-|x|}$. Find the constant C and then find the distribution of $Y = X^2$.*

8. *X had pdf*

$$f_X(x) = \begin{cases} 0 & \text{for } X \le -k \\ \frac{1}{k^2}(k+X) & \text{for } -k \le X \le 0 \\ \frac{1}{k^2}(k-X) & \text{for } 0 \le X \le k \\ 0 & \text{for } k \le X \end{cases}$$

Find the pdf of $Y = e^X$

9. *Let X be a r.v. with pdf $f(x) = \frac{2x}{\pi^2}(0), 0 < x < \pi$ (otherwise). Find the d.f. and pdf of $Y = sinX$.*

10. *Let X have a binomial distribution with parameters (n, p). Find the pdf of $Y = (X - np)^2$.*

11. *X, Y are independent standard normal variates. Find the pdf of $T = |X - Y|$.*

12. *Let X_1, X_2 be independent $\chi^2_{(2)}$-variables. Show that $Y = \frac{1}{2}(X_1 - X_2)$ has pdf*

$$g_Y(y) = \frac{1}{2}e^{-|y|}, \quad -\infty < y < \infty.$$

13. *X_1, X_2 are random samples from rectangular population*

$$dF = dx, \quad 0 \le x \le 1.$$

Find the pdf of standard deviation $S = \frac{|X_1 - X_2|}{2}$ [4(1–2s)]

14. *Let X, Y, Z be independent random variables with the same geometric distribution $\{q^k p\}$. Find*

(a) $P\{X = Y\}$, (b) $P\{X \ge 2Y\}$, (c) $P\{X + Y \le Z\}$

$$\left[(a)\ \frac{p}{1+q},\ (b)\ \frac{1}{1+q+q^2},\ (c)\ \frac{1}{(1+q)^2}\right]$$

15. *The r.v.'s. X, Y are uniformly distributed over the circle C of radius 1 so that*

$$f_{X,Y}(x, y) = \frac{1}{\pi}(0) \quad \text{if} \quad (x, y) \in C \quad \text{(otherwise)}.$$

Find the pdf of $R = \pm\sqrt{X^2 + Y^2}$.

[Hints : Use polar transformation; $g(r) = 2r$]

16. *X, Y have joint density*

$$f_{X,Y}(x, y) = \frac{1}{4}, \quad -1 \le x, y \le 1.$$

Find (i) $P(X+Y \leq t)$ where $0 < t \leq 1$, (ii) $P(X^2+Y^2 \leq r^2)$, (iii) the d.f. and pdf of $R^2(= X^2+Y^2)$.

$$\left[\text{(i) } \frac{t^2}{4}, \text{ (ii) } r^2, \text{ (iii) } dF = \frac{\pi}{4}d(r^2)\right]$$

17. *X has a Rayleigh distribution* $f(x) = \frac{x}{\sigma^2}e^{-\frac{x^2}{2\sigma^2}}, x > 0$. *Find pdf of* $Y = e^{-x^2}$.

$$\left[\frac{1}{2y\sigma^2}y^{\frac{1}{2\sigma^2}}, 0 < y < 1\right]$$

18. *X has a Cauchy distribution* $f(x) = \frac{1}{\pi(1+x^2)}, -\infty < x < \infty$. *Find pdf of* $Y = \frac{1}{X}$.

$$\left[\frac{1}{\pi(1+y^2)}, -\infty < y < \infty\right]$$

19. *X had pdf,* $f_1(x) = \sqrt{\frac{2}{\pi}}e^{-\frac{x^2}{2}}, 0 \leq x < \infty$, *Y has pdf* $f_2(y) = 2y^3e^{-y}, 0 < y < \infty$. *Find distribution of* $U = \frac{X}{Y}$.

$$\left[\frac{6}{(u^2+2)^{5/2}}, \quad 0 < u < \infty\right]$$

20. *X, Y are each independent binomial variables with parameter* (n, p). *Find the pmf of (i)* $X+Y$, *(ii)* $X-Y$, *(iii)* $U = \frac{X}{Y+1}$.

21. *Let X have a binomial distribution with parameters* (n, p). *Find the pmf of* $Y = X^2$.

$$\left[\frac{n!}{(\sqrt{y})!(n-\sqrt{y})!}p^{\sqrt{y}}q^{n-\sqrt{y}}, p+q = 1\right]$$

22. *Two independent variables X, Y are each uniformly distributed over the range* $(-a, a)$. *Show that the sum* $Z = X+Y$ *has a triangular density.*

$$f(z) = \begin{cases} \frac{2a+z}{4a^2} & (-2a \leq z \leq 0) \\ \frac{2a-z}{4a^2} & (0 \leq z \leq 2a) \end{cases} \qquad \text{(i)}$$

Show that the mgf of Z is $\left(\frac{1}{at}\sinh at\right)^2$.

23. *(X, Y) are continuous random variables with joint pdf*

$$f_{X,Y}(x, y) = \begin{cases} \frac{1}{4a^2}[1+xy(x^2-y^2)], & |x|, |y| \leq a, a > 0 \\ 0 & \text{elsewhere.} \end{cases}$$

Show that the pdf of $Z = X+Y$ is triangular with pdf given in (i).

24. *Show that the pdf of $Z = X_1 + X_2$ when X_1, X_2 are independent, each rectangular over (a, b) is,*

$$f_Z(z) = \begin{cases} \dfrac{z-2a}{(b-a)^2}, & 2a \leq z \leq a+b \\[2ex] \dfrac{2b-z}{(b-a)^2}, & a+b \leq z \leq 2b \end{cases}$$

25. *X is a continous r.v. with pdf $f_X(x)$. Another r.v. Y, distributed independently of X has a uniform distribution over $(-a, a)$. Show that pdf of $Z = X + Y$ is*

$$f_Z(z) = \frac{1}{2a}[F_X(z+a) - F_X(z-a)]$$

26. *X, Y are independent normal, each $N(0, \sigma^2)$-variable. Show that*

(i) $Z = +\sqrt{X^2+Y^2}$ and $W = arc\tan\dfrac{X}{Y}$ are independently distributed, Z having a Rayleigh distribution (vide exercise 17) and W a uniform distribution over $\left(-\dfrac{\pi}{2}, \dfrac{\pi}{2}\right)$.

(ii) Z and $V = \dfrac{X}{Y}$ are independently distributed, Z having a Rayleigh distribution and V a Cauchy distribution (Note that if X, Y are normal, but their expected values are not zero, Z and V are not independent).

27. *Let X be a non-negative r.v. of the continuous type with pdf $f_X(x)$. Show that pdf of $Y = X^m, m > 0$ is*

$$f_Y(y) = \begin{cases} \dfrac{1}{m} y^{\frac{1}{m}-1} f(y^{\frac{1}{m}}), & y > 0 \\[2ex] 0, & y \leq 0. \end{cases}$$

28. *Let X be a standard normal variate. Show that pdf of $Y = X^{2m}$ is*

$$f_Y(y) = \begin{cases} \dfrac{1}{\sqrt{2\pi}\, m y^{1-\frac{1}{2}m}} . \exp\left\{-\dfrac{y^{\frac{1}{m}}}{2}\right\}, & y > 0, \\[2ex] 0 & y \leq 0 \end{cases}$$

29. *Let $X_1, \cdots, X_n$ be iid rectangular variables over $(0,1)$. Let $X_{(1)} = \min\{X_1, \cdots, X_n\}$, $X_{(n)} = \max\{X_1, \cdots, X_n\}$.*

(a) Find the d.f. and pdf of (i) $X_{(1)}$, (ii) $X_{(n)}$.

(b) Find the joint d.f. and joint pdf of $X_{(1)}, X_{(n)}$.

(c) Find the d.f. and pdf of the range, $X_{(n)} - X_{(1)}$.

(d) Find $E[X_{(1)}], \sigma^2_{X_{(1)}}, E[X_{(n)}], \sigma^2_{X_{(n)}}, \rho_{X_{(1)}, X_{(n)}}$.

(e) Find the conditional pdf of $X_{(1)}$ given $X_{(n)}$ and the conditional mean and variance of $X_{(1)}$, given $X_{(n)}$.

30. *(Continuation). Suppose $X_1, \cdots, X_n$ are iid rectangular variables over (a, b). Extend the results of the previous problem to this probability density function.*

31. *Life of an electric bulb manufactured by a certain company is assumed to be exponentially distributed with mean 200 hours. Bulbs, nine in number are bought. What is the chance that no bulb will get extinct before 100 hours ? Find the average life of the bulb that lasts the longest.*

32. *X, Y, Z have joint pdf*

$$g(x, y, z) = \frac{1}{\sqrt{xyz}} f(x + y + z), \quad 0 < x, y, z < \infty$$

Derive the distribution of (i) $U = X + Y + Z$, (ii) $Z = \dfrac{Y}{X}$, (iii) $T = \dfrac{Z}{X + Y}$.

[Hints : (i) transformation : $u = x + y + z, uv = y + z, uvw = z$.

(ii) and (iii) : transformation : $u = x + y + z, uv = x + y, uvw = x$.]

33. *Find the distribution of $\bar{X} = \dfrac{1}{n}\sum_{i=1}^{n} X_i$, where $X_1, \cdots, X_n$ are independently and identically distributed r.v.'s with common pdf $f(x) = \dfrac{1}{\theta}e^{-\frac{x}{\theta}}$, $0 < x < \infty, \theta > 0$.*

$$\left[f(\bar{x}) = \frac{n}{\theta}\exp\left(-\frac{n\bar{x}}{\theta}\right)\right]$$

34. *Let X, Y be independently distributed with a common pdf $f(x) = \dfrac{1}{\lambda}e^{-\frac{x}{\lambda}}, x > 0, \lambda > 0$. Define $U = X + Y, V = \dfrac{X}{(X + Y)}, W = \dfrac{X}{Y}$. Check if (i) U is independent of V (ii) U is independent of W.*

35. *X_1, X_2 are independent standard normal deviates. Let $Y = \min(X_1, X_2), Z = \max(X_1, X_2)$. Find $E(Y), E(Z), V(Y), V(Z)$.*

36. *X is a $N(0, 1)$ r.v. Find the moment generating function of $Y = X^2$.*

$$\left[M_Y(t) = E(e^{tX^2}) = \frac{1}{\sqrt{1 - 2t}}, \quad t < \frac{1}{2}\right]$$

37. *(X, Y) have joint pdf*

$$f(x, y) = \frac{y}{(1 + x)^4}\exp\left(-\frac{y}{1 + x}\right), \quad x, y > 0$$

Show that joint density of U and V where $U = \dfrac{Y}{1 + X}, V = \dfrac{1}{1 + X}$ is $dF = ue^{-u}, 0 \le u \le \infty, 0 \le v \le 1$.

38. *X_1, X_2, X_3 are each independent standard normal deviates. By using polar transformations,*

$$x_1 = r\cos\theta_1\cos\theta_2, x_2 = r\cos\theta_1\sin\theta_2, x_3 = r\sin\theta_1,$$

$$\left(0 \le r < \infty, -\frac{\pi}{2} < \theta_1 < \frac{\pi}{2}, 0 < \theta_2 < 2\pi\right)$$

show that the variates corresponding to r, θ_1, θ_2 are independently distributed and distribution of r is

$$dF = \sqrt{\frac{2}{\pi}} e^{-\frac{r^2}{2}} r^2 dr, \quad r \geq 0.$$

39. *X_1, X_2, X_3 are independent random samples from the distribution $f(x) = \frac{x}{\lambda^2} e^{-\frac{x}{\lambda}}$, $x > 0, \lambda > 0$. find the joint pdf of $Y_1 = X_1 + X_2 + X_3, Y_2 = X_2, Y_3 = X_3$ and the marginal pdf of Y_1.* $\left[\frac{y_1^5}{120\lambda^6} e^{-\frac{y_1}{\lambda}}\right]$

40. *Let X_1, X_2 be independent random variables, each rectangular over $(0, 1)$. Show that*

$$Y = \sqrt{-2 \log X_1} \cos 2\pi X_2 \quad \text{and} \quad Z = \sqrt{-2 \log X_1} \sin 2\pi X_2$$

are independent random variables, each $N(0, 1)$.

41. *(X, Y) have joint pdf*

$$f_{X,Y}(x, y) = \frac{e^{-y}}{\lceil(k_1) \lceil(k_2)} x^{k_1-1} (y - x)^{k_2-1}, \quad 0 < x < y < \infty$$

which is zero otherwise. Show that marginal distribution of X and Y are gamma distributions $G(k_1)$ and $G(k_1+k_2)$ respectively. Find the correlation coefficient between (X, Y) $\left[\sqrt{\frac{k_1}{k_1 + k_2}}\right]$

42. *X, Y are independent random variables with pdf*

$$f_X(x) = \alpha e^{-\alpha x}, x \geq 0, \quad f_Y(y) = \beta e^{-\beta y}, y \geq 0.$$

Show that pdf of $Z = X + Y$ is

$$f_Z(z) = \begin{cases} \frac{\alpha\beta}{\beta - \alpha}(e^{-\alpha z} - e^{-\beta z}), & \beta \neq \alpha \\ \alpha^2 z e^{-\alpha z}, & \beta = \alpha \end{cases}$$

43. *X, Y are independent random variables, each with Reyleigh density*

$$f_X(x) = \frac{x}{\alpha^2} e^{-\frac{x^2}{2\alpha^2}}, \quad x \geq 0$$

$$f_Y(y) = \frac{y}{\beta^2} e^{-\frac{y^2}{2\beta^2}}, \quad y \geq 0$$

Show that the pdf and d.f. of $Z = \frac{X}{Y}$ are, respectively.

$$f_Z(z) = \frac{2\alpha^2}{\beta^2} \cdot \frac{z}{\left(z^2 + \frac{\alpha^2}{\beta^2}\right)^2}, \quad z \geq 0.$$

$$F_Z(z) = \frac{z^2}{z^2 + \frac{\alpha^2}{\beta^2}}, \quad z > 0.$$

44. *Let X, Y, Z be mutually independent random variables with the pdf*

$$f(u) = \begin{cases} \alpha e^{-\alpha u}, & 0 \le u < \infty, \alpha > 0 \\ 0 & \text{otherwise} \end{cases}$$

Find the pdf of (i) $X+Y+Z$, (ii) $\dfrac{(X+Y)}{Z}$, (iii) $\dfrac{(X+Y)}{X}$.

45. *Consider the one-one transformation* $(x_1, \cdots, x_n) \to (r, \theta_1, \cdots, \theta_{n-1})$ *such that*

$$\begin{aligned}
x_1 &= r\cos\theta_1, & 0 \le r < \infty \\
x_2 &= r\sin\theta_1\cos\theta_2, & -\frac{\pi}{2} < \theta_i < \frac{\pi}{2}, i = 1, \cdots, n-2 \\
x_3 &= r\sin\theta_1\sin\theta_2\cos\theta_3, & 0 < \theta_{n-1} < 2\pi \\
&\cdots\cdots\cdots\cdots\cdots\cdots \\
x_{n-1} &= r\sin\theta_1\sin\theta_2\sin\theta_3\cdots\sin\theta_{n-2}\cos\theta_{n-1} \\
x_n &= r\sin\theta_1\sin\theta_2\sin\theta_3\cdots\sin\theta_{n-2}\sin\theta_{n-1}
\end{aligned}$$

Find the Jacobian of the transformation.

$$\left[|J| = r^{n-1}(\sin\theta_1)^{n-2}(\sin\theta_2)^{n-3}\cdots\sin\theta_{n-2}\right]$$

12 CONVERGENCE OF A SEQUENCE OF RANDOM VARIABLES

12.1. Introduction

In this chapter we consider a sequence of random variables $\{X_n\}, n = 1, 2, \cdots$, all defined on the same probability space $(\mathcal{S}, \mathcal{A}, P)$ and study their different modes of convergence. For each point w we have, therefore, an infinite number of values $\{X_n(w)\}$. Section 12.2 discusses convergence of a sequence $\{X_n\}$ to a r.v. X in distribution. Sections 12.3 through 12.5 deals with respectively convergence in probability, almost sure convergence, mean-square convergence and gradation of different modes of convergence. Weak laws of large numbers are considered in 12.6. The section 12.7 deals with the celebrated Central Limit Theorem — which discusses the situations when a sequence of random sums converges in distribution to a normal law.

We have not considered in this treatise convergence almost surely in details. There are situations when a random sum does not converge to a normal distribution. Results on such topics have not been considered. Convergence of sequences of random vectors are also not considered.

12.2. Convergence of a Sequence of Distribution Functions

Let $\{X_n\}$ be a sequence of r.v.'s all defined on the probability space $(\mathcal{S}, \mathcal{A}, P)$, with $F_n(x)$ as the d.f. of $X_n, n = 1, 2, \cdots$.

Definition 12.1. Let F_n be a sequence of d.f.'s. If there exists a d.f. F such that

$$F_n(x) \to F(x)$$

at every point x where $F(x)$ is continuous, then F_n is said to converge weakly to F and we write $F_n \xrightarrow{w} F$.

If $\{X_n\}$ be a sequence of r.v.'s with $\{F_n\}$ as the corresponding sequence of d.f.'s and if there exists a r.v. X with d.f. F such that $F_n \xrightarrow{w} F$ then we say X_n converges in law to X and write $X_n \xrightarrow{L} X$.

Note 12.1 : If the sequence $\{F_n\}$ converges to F, it is not necessary that F will be a d.f. always.

Note 12.2 : As the definition 12.1 implies, it is not necessary that $\{F_n(x)\}$ would converge to $F(x)$ at every point of discontinuity of $F(x)$, even if $F_n \xrightarrow{w} F$.

❑ **EXAMPLE 12.1.** *Consider* $\{X_n\}$ *where* $P(X_n = n) = 1, n = 1, 2, \cdots$.

Here

$$F_n(x) = \begin{cases} 0, & x < n \\ 1, & x \geq n \end{cases}$$

$$\lim_{n\to\infty} F_n(x) = 0 \quad (-\infty < x < \infty).$$

Thus $\{F_n\}$ does not converge to a d.f.

❑ **EXAMPLE 12.2.** *Let* $\{X_n\}$ *be a sequence of independent r.v.'s each uniform over* $(0, \theta)$. *Let* $X_{(n)} = \max_{1\leq i\leq n} X_i$. *The d.f. of* $X_{(n)}$ *is*

$$F_{(n)}(x) = \begin{cases} 0, & x \leq 0 \\ \left(\dfrac{x}{\theta}\right)^n, & 0 < x < \theta \\ 1, & x \geq \theta. \end{cases}$$

Consider $Y_n = \dfrac{X_{(n)}}{n}$. The d.f. of Y_n is

$$F_n(x) = P(Y_n \leq x) = P(X_{(n)} \leq nx) = \begin{cases} 0, & x \leq 0 \\ \left(\dfrac{x}{\theta}\right)^n, & 0 < x < \dfrac{\theta}{n} \\ 1, & x \geq \dfrac{\theta}{n}. \end{cases}$$

Hence

$$\lim_{n\to\infty} F_n(x) = F(x) = \begin{cases} 0, & x < 0 \\ 1, & x \geq 0 \end{cases}$$

which is the d.f. of a degenerate r.v. X with $P(X = 0) = 1$. Hence $F_n \xrightarrow{w} F$ and $X_n \xrightarrow{L} X$ Note that $x = 0$ is a point of discontinuity of $F(x)$, $F_n(0) = 0$ for all n and $F(0) = 1$. so that F_n does not converge to F at the point $x = 0$.

[A r.v. X is said to be degenerate if it has a degenerate distribution $F(x)$ i.e., if there exists a real number 'c' such that

$$F(x) = \begin{cases} 0, & x < c \\ 1, & x \geq c. \end{cases}$$

Here $P(X = c) = 1$].

Let the sequence of d.f.'s $F_n(x) \xrightarrow{w} F$, the d.f. of a r.v. X. Then

$$\lim_{n\to\infty} P(a < X_n \leq b) = F(b) - F(a), \tag{12.1}$$

for any two points of continuity a, b of $F(x), a < b$. Thus

$$\lim_{n\to\infty} F_n(t) = F(t) \tag{12.2}$$

for any t at which $F(x)$ is continuous.

Note 12.3 : Whenever we say that a standardised variate $\dfrac{\{t_n - E(t_n)\}}{\sqrt{V(tn)}} = Y_n \to \tau$ as $n \to \infty$, where τ is a $N(0,1)$ variate, we actually say, $Y_n \xrightarrow{L} \tau$ i.e., $F_n(y) \xrightarrow{w} \Phi(y)$ for all $y, -\infty < y < \infty$ (since $\Phi(y)$ has no point of discontinuity).

The following theorems hold.

- **THEOREM 12.1.** *(a) If $X_n (n = 1, 2, \cdots), X$ are continuous r.v.'s, with density functions $f_n(x), f(x)$ respectively, then $X_n \xrightarrow{L} X$ iff*

$$f_n(x) \to f(x) \qquad \textit{almost everywhere as } n \to \infty.$$

(b) If $X_n(n = 1, 2, \cdots), X$ are discrete r.v.'s with pmf's, $P(X_n = j) = p_n(j), P(X = j) = p(j), j = 0, 1, 2, \cdots$, then $X_n \xrightarrow{L} X$ iff

$$p_n(j) \to p(j) \qquad \text{for all } j = 0, 1, 2, \cdots$$

- **THEOREM 12.2.** *Let $\{X_n\}$ be a sequence of r.v.'s such that $X_n \xrightarrow{L} X$. Then for any constant 'a',*

$$a + X_n \xrightarrow{L} a + x, \quad aX_n \xrightarrow{L} aX \quad (a \neq 0).$$

- **THEOREM 12.3.** *Let $\{X_n\}, \{Y_n\}$ be two sequences of r.v.'s with $X_n \xrightarrow{L} X, Y_n \xrightarrow{L} c$ where X is a r.v. and 'c' is a constent; then (i) $X_n + Y_n \xrightarrow{L} X + c$, (ii) $X_n Y_n \xrightarrow{L} cX$, (iii) $\dfrac{X_n}{Y_n} \xrightarrow{L} \dfrac{X}{c}$ $(c \neq 0)$.*

It will often be tedious to see if $F_n \xrightarrow{w} F$ by calculating the limit of $F_n(x)$ at every continuity point of $F(x)$. The following theorem stated without proof shows that this difficulty can be overcome by part by the use of characteristic function. The proof can be seen in the treatise by Lukacs.

• **THEOREM 12.4.** *(Levy-Cramer). Let $\{X_n\}(n = 1, 2, \cdots)$ be a sequence of r.v.'s and let $F_n(x), \Psi_n(t)$ be respectively the d.f. and characteristic function of X_n. The sequence $\{F_n(x)\}$ tends to the d.f. $F(x)$ as $n \to \infty$ iff the sequence $\{\Psi_n(t)\}$ converges uniformly to a function $\Psi(t)$ for all t, where $\Psi(t)$ is continuous at $t = 0$. Then the limiting function $\Psi(t)$ is the c.f. of the limiting d.f. $F(x)$.*

A weaker version of the above theorem stated in terms of moment generating function is as follows (Rohatgi, p. 277)

• **THEOREM 12.5.** *Let $\{X_n\}$ be a sequence of r.v.'s and let $F_n(x)$ and $M_n(t)$ be respectively the d.f. and moment generating function of $F(x)$, where $M_n(t)$ exists for $|t| < T$ for every n. If there exists a d.f. F with corresponding mgf $M(t)$ which exists for $|t| \le t_1 < T$ such that $M_n(t) \to M(t)$ as $n \to \infty$ for every $|t| \le t_1$, then $F_n(x)$ converges to the d.f. $F(x)$.*

❑ **EXAMPLE 12.3.** *Consider $\{X_n\}$ of example 12.1, where*

$$F_n(x) = \begin{cases} 0, & x < n \\ 1, & x \ge n \end{cases}$$

Here $\Psi_n(t) = e^{itn} = \cos nt + i \sin nt$. For $t = \frac{\pi}{2}$, $\Psi_n(t)$ takes the values $1, i, -i, -1$ only and $\Psi_n(t)$ does not converge. Hence $F_n(x)$ does not converge.

❑ **EXAMPLE 12.4.** *Let $\{X_n\}$ be a sequence of independent Cauchy variates. Consider the sequence $Y_n = \frac{S_n}{n^2}$ where $S_n = \sum_{i=1}^{n} X_i$. The c.f. of $Y_n = \Psi_n(t) = e^{-|t|/n}$. Hence*

$$\lim_{n\to\infty} \Psi_n(t) = \Psi(t) \quad \text{(say)} = \begin{cases} 1 & \text{if } t = 0 \\ 0 & \text{if } t \ne 0 \end{cases}$$

$\Psi(t)$ is not continuous at $t = 0$ and cannot be a characteristic function. Therefore, $\frac{S_n}{n^2}$ does not converge in law.

❑ **EXAMPLE 12.5.** *Let $\{X_n\}$ be a sequence of independent r.v.'s such that*

$$P\left(X_k = \frac{k}{n}\right) = \frac{1}{n}, \quad k = 1, 2, \cdots, n.$$

The c.f. of X_n is

$$\Psi_n(t) = E(e^{itX_n}) = \frac{1}{n}\sum_{1}^{n} e^{itk/n} = \frac{e^{it/n}}{n}\left(\frac{1 - e^{it}}{1 - e^{it/n}}\right)$$

Now $\displaystyle\lim_{n\to\infty} \frac{n(1 - e^{it/n})}{e^{it/n}} = -it.$

Hence $\displaystyle\lim_{n\to\infty} \Psi_n(t) = \frac{e^{it} - 1}{it} = \Psi(t)$, which is the c.f. of the uniform density on [0, 1]. Hence X_n converges in law to a uniform distribution.

We now apply theorem 12.4 to show that for many distributions, limiting distribution is a standard normal distribution.

❑ **EXAMPLE 12.6.** *Let $\{X_n\}$ be a sequence of binomial variables,*

$$P(X_n = r) = \binom{n}{r} p^r q^{n-r}, \quad 0 < p < 1, \quad q = 1 - p, \quad r = 0, 1, \cdots, n$$

$E(X_n) = np, V(X_n) = npq$. Consider the sequence of standardised binomial variables $Y_n = \dfrac{(X_n - np)}{\sqrt{npq}}$.

The c.f. of Y_n is

$$\begin{aligned}\Psi_n(t) &= E(e^{it(X_n - np)/\sqrt{npq}}) \\ &= e^{-it\sqrt{np/q}}(pe^{it/\sqrt{npq}} + q)^n\end{aligned}$$

$$\begin{aligned}\log_e \Psi_n(t) &= -it\sqrt{np/q} + n\log_e(pe^{it/\sqrt{npq}} + q) \\ &= -it\sqrt{np/q} + n\log_e\left[1 + p\frac{it}{\sqrt{npq}} - \frac{pt^2}{2npq} + o\left(\frac{t^2}{n}\right)\right] \\ &\qquad \left(\because \quad \log_e(1+x) = x - \frac{x^2}{2} + \frac{x^3}{3} - \cdots, |x| < 1\right) \\ &= -it\sqrt{np/q} + n\left[\frac{ipt}{\sqrt{npq}} - \frac{pt^2}{2npq} + \frac{p^2t^2}{2npq} + 0\left(\frac{t^3}{n^{3/2}}\right)\right] \\ &= -\frac{t^2}{2} + 0\left(\frac{t^3}{\sqrt{n}}\right)\end{aligned}$$

[A function f_n is at most of order g_n, as $n \to \infty$, written as $f_n = 0(g_n)$ if $|f_n| < cg_n$ for some constant c, for all $n \geq n_o$.

f_n is of smaller order than g_n as $n \to \infty$, written as $f_n = o(g_n)$ if $\lim_{n\to\infty} [f_n/g_n] = 0$].

Thus for any finite $t, \log_e \Psi_n(t)$ converges to $(-t^2/2)$ uniformly and hence $\Psi_n(t) \to e^{-t^2/2}$ uniformly in t. The latter being the c.f. of a standard normal deviate, it follows by theorem 12.4 that the d.f. of Y_n converges to that of a standard normal variable.

We have thus proved De-Moivre Laplace theorem (theorem 9.6). The same theorem is stated below in a slightly alternative form.

• **THEOREM 12.6.** *Let $\{X_n\}$ be a sequence of binomial variables with parameters $(n, p), Y_n = \frac{(X_n - np)}{\sqrt{npq}}$ and $F_n(Y)$ be the d.f. of Y_n. If $0 < p < 1$, the relation*

$$\lim_{n\to\infty} F_n(Y) = \frac{1}{\sqrt{2\pi}} \int_{-\infty}^{y} e^{-\frac{z^2}{2}} dz \tag{12.3}$$

holds for every value of y (since the d.f. of a normal variable has no point of discontinuity, convergence considered in (12.3) holds for every value of y).

The r.v. X_n is said to have an asymptotically normal distribution $N(np, \sqrt{npq})$ (vide definition 12.5).

COROLLARY 12.1. *The sequence* $\left\{\frac{X_n}{n}\right\}$ *is asymptotically* $N\left(p, \sqrt{\frac{pq}{n}}\right)$.

For a yet-another proof see example 12.14.

❑ **EXAMPLE 12.7.** *Let $\{X_n\}$ be a sequence of binomial variables with parameters (n, p). Let $n \to \infty$ in such a way that np remains constant at λ.*

The c.f. of X_n is

$$\Psi_n(t) = (q + pe^{it})^n = \left(1 - \frac{\lambda}{n} + \frac{\lambda}{n}e^{it}\right)^n = \left(1 + \frac{\lambda}{n}(e^{it} - 1)\right)^n \to \exp[\lambda(e^{it} - 1)] \quad \text{(i)}$$

as $n \to \infty$. (i) is the c.f. of a Poisson distribution with parameter λ. Hence the binomial d.f. approaches a poission d.f. (an alternative proof was given in lemma 8.3).

❑ **EXAMPLE 12.8.** *Let $\{X_n\}$ be a sequence of poisson variables with parameter λ. Consider the standardised variable $Y_n = \frac{(X_n - \lambda)}{\sqrt{\lambda}}$. The c.f. of Y_n is*

$$\Psi_n(t) = e^{-it\sqrt{\lambda}} \exp[\lambda(e^{it\sqrt{\lambda}} - 1)]$$

$$\log_e \Psi_n(t) = -it\sqrt{\lambda} + \lambda(e^{it/\sqrt{\lambda}} - 1) = -it\sqrt{\lambda} + \lambda\left(\frac{it}{\sqrt{\lambda}} - \frac{t^2}{2\lambda} - \frac{it^3}{6\lambda^{\frac{3}{2}}} + \cdots\right)$$

$$= -\frac{t^2}{2} - \frac{it^3}{6\lambda^{\frac{1}{2}}} + \cdots$$

Hence $\Psi_n(t) \to \exp(-t^2/2)$ as $\lambda \to \infty$. Thus Y_n tends in distribution to a $N(0,1)$-variable as $\lambda \to \infty$.

❑ **EXAMPLE 12.9.** *Let $\{X_n\}$ be a sequence of independently and identically distributed $\chi^2_{(1)}$ r.v.'s. Let*

$$Y_n = \frac{(S_n - n)}{\sqrt{2n}}, \qquad \text{where} \quad S_n = \sum_{i=1}^{n} X_i.$$

The mgf of $Y_n = M_n(t) = E(e^{tY_n})$

$$= \left(1 - \sqrt{\frac{2}{n}}t\right)^{-\frac{n}{2}} \exp\left[\left(-t\sqrt{\frac{n}{2}}\right)\right], \quad |t| < \sqrt{\frac{n}{2}}$$

$$\log_e M_n(t) = -t\sqrt{\frac{n}{2}} - \frac{n}{2}\log_e\left(1 - \sqrt{\frac{2}{n}}t\right)$$

$$= -t\sqrt{\frac{n}{2}} + \frac{n}{2}\left[\sqrt{\frac{2}{n}}t + \frac{1}{2}\left(\sqrt{\frac{2}{n}}t\right)^2 + \cdots\right] = \frac{t^2}{2} + 0\left(\frac{1}{n}\right)$$

Hence $M_n(t) \to e^{\frac{t^2}{2}}$ and $Y_n \xrightarrow{L} \tau$ where τ is a $N(0,1)$-variable.

12.3. Convergence in Probability

Definition 12.2. A sequence of r.v.'s $\{X_n\}$ is said to converge in probability to a r.v. X if, for every given $\epsilon(>0)$ and $\delta(>0)$, there exists a N such that

$$P\{X_n - X| > \epsilon\} < \delta \quad \text{for all } n \geq N \quad \text{i.e.} \quad \lim_{n\to\infty} P\{|X_n - X| > \epsilon\} = 0 \tag{12.4}$$

for every $\epsilon(>0)$. This is written as $X_n \xrightarrow{P} X$ or $p\lim X_n = X$. $\{X_n\}$ is said to converge stochastically to X.

In particular if the r.v. X is a constant i.e., $P(X = \mu) = 1, X_n \xrightarrow{P} \mu$ if $P\{|X_n - \mu| < \epsilon\} \to 1$ for every fixed $\epsilon(>0)$. In this case, it is said that the Weak Law of Large Numbers (WLLN) holds for $\{X_n\}$.

❑ **EXAMPLE 12.10.** *Consider the probability space $(\mathcal{S}, \mathcal{A}, P)$ where $\mathcal{S} = [0,1]$ and P is such that $P[a \leq w \leq b] = (b-a)^{-1}, 0 \leq a \leq b \leq 1$. Define $\{X_n\}$ such that $X_n(w) = 1(0)$ if $w \in \left[0, \frac{1}{n}\right]$ (otherwise). Then* $P(X_n = 1) = \frac{1}{n}, P(X_n = 0) = 1 - \frac{1}{n}$.

$$P(|X_n| \leq \epsilon) = \begin{cases} P(X_n = 0) = 1 - \dfrac{1}{n} & \text{if } \epsilon < 1 \\ 1 & \text{if } \epsilon \geq 1 \end{cases}$$

Thus $P(|X_n| \geq \epsilon) \to 1$ as $n \to \infty$ for every $\epsilon(>0)$. Hence $X_n \xrightarrow{P} X$ where X is a degenerate r.v. with $P(X = 0) = 1$.

The following results hold :

• **THEOREM 12.7.**

(i) $X_n \xrightarrow{P} X \quad \Rightarrow \quad X_n - X \xrightarrow{P} 0.$

(ii) $X_n \xrightarrow{P} X \quad \Rightarrow \quad X_n - X_m \xrightarrow{P} X \quad$ as $n, m \to \infty$

• **THEOREM 12.8.** *Let $\{X_n\}, \{Y_n\}$ be two sequences of r.v.'s with $X_n \xrightarrow{P} X, Y_n \xrightarrow{P} Y$. Then*

(i) $cX_n \xrightarrow{P} cX$ ('c', a real constant)

(ii) $X_n + Y_n \xrightarrow{P} X + Y$

(iii) $X_n Y_n \xrightarrow{P} XY$

(iv) $\dfrac{X_n}{Y_n} \xrightarrow{P} \dfrac{X}{Y}$ if $P[Y_n = 0] = 0 \forall n$ and $P[Y = 0] = 0$.

• **THEOREM 12.9.** *If $g(x)$ is a continous real valued function, and $X_n \xrightarrow{P} X$ then $g(X_n) \xrightarrow{P} g(X)$.*

• **THEOREM 12.10.** *(Cramer p. 254). Let $\{X_n\}, \{Y_n\}$ be two sequences of r.v.'s with $X_n \xrightarrow{L} X, Y_n \xrightarrow{P} c$ where X is a r.v. and 'c' (> 0) a constant. Then*

$$\text{(i)} \quad X_n + Y_n \xrightarrow{L} X - c, \quad \text{(ii)} \quad X_n Y_n \xrightarrow{L} \frac{X}{c} \quad (c \neq 0), \quad \text{(iii)} \quad \frac{X_n}{Y_n} \xrightarrow{L} cX$$

12.4. Almost Sure Convergence

Definition 12.3. The sequence $\{X_n\}$ converges almost surely to a r.v. X if

$$P\left\{\lim_{n\to\infty} X_n = X\right\} = 1 \tag{12.5}$$

i.e. if

$$P\left\{w : \lim_{n\to\infty} X_n(w) = X(w)\right\} = 1$$

This is denoted as a.s. $\lim X_n = X$ or $X_n \xrightarrow{as} X$. When $X_n \xrightarrow{as} c$, a constant, we say that the strong law of large numbers (SLLN) holds for the sequence $\{X_n\}$. To verify $X_n \xrightarrow{P} X$ one has to calculate the probability, $P\{|X_n - X| < \epsilon\} = p_n(\epsilon)$ and show that $p_n(\epsilon) \to 1$ as $n \to \infty$. To verify $X_n \xrightarrow{as} X$ one has to show that the set of points w for which $\lim_{n\to\infty} X_n(w) = X(w)$ has probability one.

This concept will not be discussed in details in this treatise.

12.5. Convergence in rth mean

Definition 12.4. Let $\{X_n\}$ and X be r.v.'s having moments of rth order. Then $|X_n - X|$ has also moments of rth order. $\{X_n\}$ is said to converge in rth mean to X if

$$E|X_n - X|^r \to 0 \quad \text{as } n \to \infty, \quad r > 0. \tag{12.6}$$

This is also called L_r-convergence, $X_n \xrightarrow{Lr} X$. For $r = 2$, this is called mean square (or L_2) convergence. $X_n \xrightarrow{ms} X$ if $E|X_n - X|^2 \to 0$.

Lemma 12.1. $X_n \xrightarrow{r} X \Rightarrow E|X_n|^r \to E|X|^r$

❑ **EXAMPLE 12.11.** *Consider $\{X_n\}$ where*

$$P(X_n = n) = \frac{1}{n^r}, \quad P(X_n = 0) = 1 - \frac{1}{n^r}, \quad r > 0.$$

Then

$$P\{|X_n| > \epsilon\} = \begin{cases} 0 & \text{if } \epsilon \geq n \\ P(X_n = n) = \dfrac{1}{n^r} & \text{if } \epsilon < n \end{cases}$$

Hence $X_n \xrightarrow{P} 0$. Again $E|X_n|^r = 1$ so that $X_n \not\xrightarrow{L_r} 0$. If $r = 3, E|X_n|^2 = \frac{1}{n}$ so that $X_n \xrightarrow{L_2} Y$ where $P(Y = 0) = 1$.

Lemma 12.2. Let X, Y be two r.v.'s with d.f.'s $F(x), G(y)$ respectively. Let ϵ, δ be two positive constants such that

$$P(|X - Y| \leq \epsilon) \geq 1 - \delta \tag{12.7}$$

Then $|F(x) - G(x)| \leq [F(x + \epsilon) - F(x - \epsilon)] + \delta$.

Proof. Let A be the event $[|X - Y| \leq \epsilon]$, B the event $[Y \leq x]$.

Then $P[|X - Y| \leq \epsilon, Y \leq x] \geq P[Y \leq x] - \delta \qquad [\because \; P(AB) \geq P(B) + P(A) - 1$

Hence, $P[X \leq x + \epsilon] \geq P[Y \leq x] - \delta \qquad (12.8)$

Similarly considering the event A and $\bar{B}$,

$$P[X > x - \epsilon] \geq P[Y > x] - \delta \tag{12.9}$$

From (12.8) and (12.9) it follows that

$$F(x - \epsilon) - \delta \leq G(x) \leq F(x + \epsilon) + \delta.$$

Since $F(x - \epsilon) \leq F(x) \leq F(x + \epsilon)$,

$$F(x - \epsilon) - F(x + \epsilon) - \delta \leq F(x) - G(x) \leq F(x + \epsilon) - F(x - \epsilon) + \delta.$$

The following theorem shows the gradation of different types of convergence.

• **THEOREM 12.11.** *Let $\{X_n\}$ be a sequence of r.v.'s and X another r.v. Then*

(a) $[X_n \xrightarrow{Lr} X] \Rightarrow [X_n \xrightarrow{Ls} X; 0 \leq s \leq r] \Rightarrow [X_n \xrightarrow{P} X] \Rightarrow [X_n \xrightarrow{L} X]$

(b) $[X_n \xrightarrow{as} X] \Rightarrow [X_n \xrightarrow{P} X]$

Proof. (a) (i) Lr-convergence implies Ls-convergence $(r \geq s)$:

$$[X_n \xrightarrow{L_r} X] \quad \Rightarrow \quad E|X_n - X|^r \to 0.$$

Similarly $[X_n \xrightarrow{Ls} X] \Rightarrow E|X_n - X|^s \to 0$.

Now $(E|X_n - X|^r)^{\frac{1}{r}} \geq (E|X_n - X|^s)^{\frac{1}{s}}, r \geq s > 0$ (by theorem 6.11). (12.10)

Hence if $X_n \xrightarrow{Lr} X$, the lhs of (12.10) $\to 0$ and hence the rhs of (12.10) $\to 0$, which in turn implies $X_n \xrightarrow{Ls} X$.

(ii) Lr convergence implies P-convergence.

By Markov's inequality (6.45)

$$P\{|X_n - X| > \epsilon\} \leq \frac{E|X_n - X|^r}{\epsilon^r}, r > 0,$$

because $|X_n - X|^r$ is a positive increasing function of $|X_n - X|$; but $E|X_n - X|^r \to 0$, hence $P\{|X_n - X| > \epsilon\} \to 0$ for every ϵ.

(iii) P-convergence implies convergence in law.

Proof. Suppose $X_n \xrightarrow{P} X$ so that

$$P(|X_n - X| \leq \epsilon) \geq 1 - \delta$$

for every $\epsilon(> 0), \delta(> 0)$ and n sufficiently large.

By lemma 12.2

$$|F(x) - F_n(x)| \leq F(x + \epsilon) - F(x - \epsilon) + \delta \tag{12.11}$$

If x is a continuity point of F, the rhs of (12.11) can be made arbitrarily small by choosing ϵ and δ sufficiently small. Hence $F_n(x) \xrightarrow{w} F(x)$.

(b) $X_n \xrightarrow{as} X$ means $P\{X_n \to X\} = 1$ i.e., $P\{|X_n - X| < \epsilon$ for $n = n_o + 1, n_o + 2, \cdots, n_o + r\} > 1 - \eta$ for all r. Putting $r = 1$, we get $P\{|X_n - X| < \epsilon\} > 1 - \eta$ for $n \geq n_o$, which is convergence in probability.

Note 12.4. The converses in theorem 12.11 are not true.

Note 12.5. Although convergence in rth mean is stronger than convergence in probability, it is neither implied by nor implies convergnece a.s.

❑ **EXAMPLE 12.12.** *Show that if $X_n \xrightarrow{P} a$, where 'a' is a constant, then $[X_n \xrightarrow{P} a] \Leftrightarrow [X_n \xrightarrow{L} a]$.*

(i) $[X_n \xrightarrow{P} a] \Rightarrow [X_n \xrightarrow{L} a]$ follows by theorem 12.11. This can also be seen as follows :

$$[X_n \xrightarrow{P} a] \Rightarrow P\{|X_n - a| \leq \epsilon\} \to 0$$

for any $\epsilon(> 0)$.

i.e., $P\{X_n \leq a - \epsilon, X_n \geq a + \epsilon\} \to 0$

i.e., $P\{X_n \leq a - \epsilon\} P\{X_n \geq a + \epsilon\} \to 0$ (the events being independent)

i.e., $F_n(a - \epsilon) \to 0, 1 - F_n(a + \epsilon - 0) \to 0,$ (i)

F_n being d.f. of X_n. Since ϵ is arbitrary, taking limit of (i) as $n \to \infty$

$$\lim F_n(x) = F(x) \quad \text{(say)} = \begin{cases} 0, & x < a \\ 1, & x \geq a \end{cases} \tag{ii}$$

i.e., $X_n \xrightarrow{L} X$ where $P(X = a) = 1$.

(ii) Let $X_n \xrightarrow{L} a$ i.e., limiting distribution of X_n is $F(x)$ in (ii).

$$P\{|X_n - a| \geq \epsilon\} = P\{X_n \leq a - \epsilon\}\{X_n \geq a + \epsilon\} = F_n(a - \epsilon)\{1 - F_n(a + \epsilon - 0)\} \quad \text{(iii)}$$

Taking limit of both sides of (iii).

$$\lim P\{|X_n - a| > \epsilon\} = F(a - \epsilon)(1 - F(a + \epsilon - 0)) = 0(1 - 0) = 0.$$

❑ **EXAMPLE 12.13.** *$\{X_n\}$ is a sequence of r.v.'s with $P(X_n = 0) = 1 - \frac{1}{n}$, $P(X_n = n^{2/r}) = \frac{1}{n}$. Show that P-convergence does not imply L_r-convergence.*

$P\{|X_n| < \epsilon\} = P\{X_n = 0\} = 1 - \frac{1}{n} > 1 - \delta$ for $n > \frac{1}{\delta}$. Hence $X_n \xrightarrow{P} 0$. Now $E|X_n|^r = n \to \infty$. Hence $X_n \not\xrightarrow{L_r} 0$.

12.6. Weak Laws of Large Numbers

We have already defined when a sequence of variables $\{X_n\}$ is said to satisfy the weak law of large numbers (WLLN) and strong law of large numbers (SLLN). These are the situations when $X_n \xrightarrow{P} c$ or $X_n \xrightarrow{as} c$ ('c' a constant) respectively. We will consider in details the situations when a sequence of random sums, $S_n = \sum_{i=1}^{n} X_i$ satisfies the weak laws of large numbers. The SLLN will not be considered.

We recall Tchebychev's inequality : If X is a r.v. with $E(X) = \mu$, $V(X) = \sigma^2$,

$$P\{|X - \mu| \geq \epsilon\sigma\} \leq \frac{1}{\epsilon^2} \tag{12.12}$$

for any arbitrary real $\epsilon(> 0)$.

(a) Bernoulli's WLLN : Consider n Bernoulli trials each with probability p of success. Let $X_i = 1(0)$ if the ith trial results in success (failure). $E(\bar{X}_n) = p, V(\bar{X}_n) = \frac{pq}{n}$ where $\bar{X}_n = \frac{1}{n}\sum_{i=1}^{n} X_i, q = 1 - p$. Applying (12.12) to the sequence of r.v.'s $\{\bar{X}_n\}$,

$$P\{|\bar{X}_n - p| \leq \epsilon\} \geq 1 - \frac{pq}{n\epsilon^2} > 1 - \frac{1}{n\epsilon^2}$$

$$\text{i.e.} \quad \lim_{n \to \infty} P\{\bar{X}_n - p| \leq \epsilon\} = 1 \tag{12.13}$$

for every $\epsilon) > 0$). Hence $\bar{X}_n \xrightarrow{P} p$ (i.e., $\bar{X}_n \xrightarrow{L} X$ where X is a degenerate r.v. at $x = p$). This is known as Bernoulli's LLN.

(b) Tchebychev's WLLN : Let $\{X_n\}$ be a sequence of uncorrelated r.v.'s for which $E(X_i) = \mu_i, V(X_i) = \sigma_i^2$. Then $E(\bar{X}_n) = \frac{1}{n}\sum_{i=1}^{n} \mu_i = \bar{\mu}_n$, $V(\bar{X}_n) = \frac{1}{n^2}\sum_{i=1}^{n} \sigma_i^2$.

If the condition

$$\frac{1}{n^2}\sum_{i=1}^{n}\sigma_i^2 \to 0 \quad \text{as } n \to \infty \tag{12.14}$$

is satisfied then $\bar{X}_n \xrightarrow{P} \bar{\mu}_n$.

Proof. Follows immediately by applying (12.12) to $\{\bar{X}_n\}$, when

$$P\{|\bar{X}_n - \bar{\mu}_n| > \epsilon\} \le \sum_{1}^{n}\sigma_i^2 \Big/ n^2\epsilon^2$$

for every $\epsilon(> 0)$. Assumption (12.14) is immediately satisfied if $V(X_i) = \sigma^2(< \infty)$ for all i.

We note that here $E|\bar{X}_n - \bar{\mu}_n|^2 \to 0$ (by assumption (12.14)) and hence $\bar{X}_n \xrightarrow{ms} \bar{\mu}_n$.

The condition (12.14) is a sufficient (not necessary) condition for $\{S_n\}$ to satisfy the WLLN.

(c) Poisson's WLLN : Consider Poisson's scheme of sampling : n independent Bernoulli trials are performed, the ith trial with probability of success as p_i. Let $X_i = 1(0)$ if the ith trial results in success (failure). $E(\bar{X}_n) = \frac{\Sigma p_i}{n} = \bar{p}$, $V(\bar{X}_n) = \frac{1}{n^2}\sum_{i=1}^{n} p_i q_i \le \frac{1}{4n}$ (since $p_i q_i \le \frac{1}{4}\forall i$) $\to 0$ as $n \to \infty$. Hence by inequality (12.12), $\bar{X}_n \xrightarrow{P} \bar{p}$.

(d) Khintchine's WLLN : Assume $\{X_n\}$ to be a sequence of independently and identically distributed r.v.'s with $E(X_i) = \mu \forall i$. Then the sequence $\bar{X}_n \xrightarrow{P} \mu$.

Proof. Let $\Psi(t)$ be the c.f. of $X_i (i = 1, \cdots, n)$. The c.f. of $\bar{X}_n$ is

$$\Psi_n(t) = \left[\Psi\left(\frac{t}{n}\right)\right]^n.$$

Expanding $\Psi(t)$ in Maclaurin series about $t = 0$,

$$\Psi(t) = 1 + i\mu t + o(t)$$

Hence

$$\Psi_n(t) = \left[1 + \frac{\mu i t}{n} + o\left(\frac{t}{n}\right)\right]^n, \quad \log_e \Psi_n(t) = n\left[\frac{\mu i t}{n} + o\left(\frac{t}{n}\right)\right] = \mu i t + n o\left(\frac{t}{n}\right).$$

Hence $\log_e \Psi_n(t) \to \mu i t$ and $\Psi_n(t) \to \exp(\mu i t)$, as $n \to \infty$. Now $\exp(\mu i t)$ is the c.f. of a degenerate variable Y with $P(Y = \mu) = 1$. Hence by theorem 12.4 $\bar{X}_n \xrightarrow{P} \mu$.

(e) Markoff's WLLN : Let $\{X_n\}$ be a sequence of variables not necessarily uncorrelated. Then $\bar{X}_n \xrightarrow{P} \bar{\mu}_n$ if $V(\bar{X}_n) \to 0$ as $n \to \infty$.

12.7. Central Limit Theorems

Let $\{X_n\}$ be a sequence of independent random variables. In laws of large numbers we considered when a sequence of random sums $S_n = \sum_{i=1}^{n} X_i$ converges to $E(S_n)$ either in probability or almost certainly. Here we consider the situations when $\sum_{i=1}^{n} X_i \xrightarrow{L} \tau$ where τ is a standard normal variable. If the sequence $S_n \xrightarrow{L} \tau$, S_n is said to follow the central limit law or normal convergence. In this section we consider different Central Limit Theorems (CLT).

Definition 12.5. A variable Y follows (in the limit) asymptotically normal distribution with mean 'a' and variable 'b' if $\frac{(Y-a)}{\sqrt{b}}$ follows in the limit, as $n \to \infty$, a $N(0,1)$-distribution i.e., if $\frac{(Y-a)}{\sqrt{b}} \xrightarrow{L} \tau$. Here the distribution of Y depends on the parameter n. The quantities a and b may or may not depend on n.

- **THEOREM 12.12.** *Lindberg-Levy Central Limit Theorem : Let $\{X_n\}$ be a sequence of independently and identically distributed random variables, with $E(X_i) = \mu, V(X_i) = \sigma^2 < \infty, i = 1, 2, \cdots$. Then the sum $S_n = \sum_{i=1}^{n} X_i$ is asymptotically normal $N(n\mu, n\sigma^2)$.*

Proof. Consider the standardised variable

$$Z_n = \frac{1}{\sqrt{n}\sigma} \sum_{i=1}^{n} (X_i - \mu)$$

The r.v.'s $Y_i = (X_i - \mu)$ are iid with $E(Y_i) = 0, V(Y_1) = \sigma_i^2, (i = 1, \cdots, n)$.

Let $\Psi_Y(t)$ be the characertistic function of Y_i. The c.f. of Z_n is

$$\Psi_n(t) = E[e^{it(S_n - n\mu)/\sigma\sqrt{n}}] = \left[\Psi_Y\left(\frac{t}{\sigma\sqrt{n}}\right)\right]^n$$

Developing $\Psi_Y(t)$ in Maclaurin's series about $t = 0$

$$\Psi_Y(t) = \Psi_Y(0) + t\Psi'_Y(0) + \frac{t^2}{2!}\Psi''_Y(0) + o(t^2) = 1 - \frac{t^2}{2!}\sigma^2 + o(t^2)$$

Hence

$$\Psi_n(t) = \left[1 - \frac{t^2}{2n} + o\left(\frac{t^2}{n}\right)\right]^n$$

$$\log_e \Psi_n(t) = n \log_e \left[1 + \left\{\frac{-t^2}{2n} + o\left(\frac{t^2}{n}\right)\right\}\right]$$

$$= n\left[-\frac{t^2}{2n} + o\left(\frac{t^2}{n}\right)\right] = -\frac{t^2}{2} + no\left(\frac{t^2}{n}\right).$$

Hence $\log_e \Psi_n(t) = -\frac{t^2}{2}$ (since $\lim_{n\to\infty} no\left(\frac{t^2}{n}\right) = 0$ for every fixed t) and $\Psi_n(t) \to e^{-\frac{t^2}{2}}$ the c.f. of τ, a standard normal deviate. Therefore, by the continuity theorem, $Z_n \xrightarrow{L} \tau$ and hence S_n is asymptotically $N(n\mu, \sqrt{n}\sigma)$.

COROLLARY 12.2 Under the conditions of theorem 12.12, $\bar{X}_n$ is asymptotically $N\left(\mu, \frac{\sigma}{\sqrt{n}}\right)$.

❑ **EXAMPLE 12.14.** *Consider binomial r.v.'s X_n with parameters (n,p). Now $X_n = Y_1+\cdots+Y_n$ where Y_i denotes the result of the ith bernoulli trial, $Y_i = 1(0)$ if a success (failure), $Y_1, \cdots, Y_n$ being independent with $E(Y_i) = p, V(Y_i) = p(1-p)$. Hence by applying theorem 12.12, X_n is asymptotically normal $(np, \sqrt{npq})$.*

❑ **EXAMPLE 12.15.** *Let $\{X_n\}$ be a sequence of iid standard Cauchy variables. Since the Cauchy variable has no finite mean, theorem 12.12 does not apply.*

The c.f. of each X is $\Psi_X(t) = e^{-|t|}$. The c.f. of $\frac{S_n}{n} = \bar{X}_n$ is $\Psi_n(t) = e^{-|t|}$. Hence $\bar{X}_n$ has a Cauchy distribution.

In case the independent variables $X_1, \cdots, X_n$ are not identically distributed, their sum need not converge to a normal distribution even if all the variables have finite standard deviations.

The following theorem due to Liapounov gives a sufficient condition that a sum of independent r.v.'s should have an asymptotically normal distribution.

• **THEOREM 12.13.** *(Liapounov). Let $\{X_n\}$ be a sequence of mutually independent random variables such that for some $\delta(>0), E\{|X_k - \mu_k|^{2+\delta}\}$ exists for every $k = 1, 2, \cdots$. Then if the condition*

$$\lim_{n\to\infty} \frac{1}{\left(\sum_1^n \sigma_k^2\right)^{1+\frac{\delta}{2}}} \sum_{k=1}^n E\left\{|X_k - \mu_k|^{2+\delta}\right\} = 0 \qquad (12.15)$$

is satisfied, then $\frac{(S_n - \mu_{(n)})}{\sigma_{(n)}}$ is asymptotically normal with mean 0 and variance 1, where $\mu_{(n)} = \sum_{k=1}^n \mu_k, \sigma_{(n)}^2 = \sum_{k=1}^n \sigma_k^2$.

Proof is omitted.

The Liapounov theorem gives only a sufficient condition for $\{S_n\}$ to follow central limit law. The Lindeberg-Feller theorem stated below (without proof) gives a necessary and sufficient condition for the sum S_n to have an asymptotically normal distribution. This is the Central Limit Theorem in the most general form.

• **THEOREM 12.14.** *(Lindeberg-Feller). Let $\{X_n\}$ be a sequence of independent r.v.'s with the cumulative d.f. of X_k as $F_k(x), E(X_k) = \mu_k, V(X_k) = \sigma_k^2 < \infty$. Then*

(i) S_n is asymptotically $N(\mu_{(n)}, \sigma_{(n)})$

(ii) $\lim_{n\to\infty} \max_{1\le k\le n} \frac{\sigma_k}{\sigma_{(n)}} = 0$

if and only if

$$A_n(\epsilon) = \frac{1}{\sigma_{(n)}^2}\sum_{k=1}^{n}\int_{|x_k-\mu_k|\ge\epsilon\sigma_{(n)}} (x-\mu_k)^2 dF_k(x) \to 0 \tag{12.16}$$

as $n \to \infty$, holds for every $\epsilon(> 0)$.

COROLLARY 12.3 If X_k's are iid random variables each with finite variance σ^2, condition (12.16) is always satisfied; because,

$$A_n(\epsilon) = \frac{1}{n\sigma^2}\sum_{k=1}^{n}\int_{|x_k-\mu_k|\ge\epsilon\sigma\sqrt{n}} (x-\mu)^2 dF(x) = \frac{1}{\sigma^2}\int_{y^2\ge n\epsilon^2\sigma^2} y^2 dF(y) \to 0$$

as $n\epsilon^2\sigma^2 \to \infty$ as $n \to \infty$.

COROLLARY 12.4. If Liapounov's condition (12.15) is satisfied, then Lindeberg-Feller condition (12.16) is also satisfied. For

$$\begin{aligned} A_n(\epsilon) &= \frac{1}{\sigma_{(n)}^2}\sum_{k=1}^{n}\int_{|x-\mu_k|\ge\epsilon\sigma_{(n)}} (x-\mu_k)^2 dF_k(x) \\ &\le \frac{1}{\epsilon^\delta\sigma_{(n)}^{2+\delta}}\sum_{k=1}^{n}\int_{|x-\mu_k|\ge\epsilon\sigma_{(n)}} |x-\mu_k|^{2+\delta} dF_k(x) \\ &\le \frac{1}{\epsilon^\delta\sigma_{(n)}^{2+\delta}}\sum_{k=1}^{n} E|X_k-\mu_k|^{2+\delta} \to 0 \quad \text{as } n\to\infty, \end{aligned}$$

by (12.15) and hence (12.16) holds.

Note 12.6. The condition (12.16) can also be stated as follows. Define truncated random variables,

$$U_k = \begin{cases} X_k - \mu_k & \text{if } |X_k-\mu_k| \ge \epsilon\sigma_{(n)} \\ 0 & \text{if } |X_k-\mu_k| < \epsilon\sigma_{(n)} \end{cases} \tag{12.17}$$

Then (12.16) reduces to the condition

$$B_n(\epsilon) = \frac{1}{\sigma_{(n)}^2}\sum_{k=1}^{n} E(U_k^2) \to 0, \quad \text{as } n\to\infty \tag{12.18}$$

The CLT has been studied in the case when X_i's are not all mutually independent random variables. The CLT has been studied for the case of p-variate random vectors $(X_1, \cdots, X_p)$. We state below the p-dimensional form of the central limit theorem due to Lindeberg and Levy.

Let $\boldsymbol{X} = (X_1, \cdots, X_p)$ be a p-dimensional random vector with d.f. $F_{\boldsymbol{X}}(x_1, \cdots, x_p)$ and c.f. $\Psi_{\boldsymbol{X}}(t_1, \cdots, t_p)$. Let $\mu_k = E(X_k)$, $k = 1, \cdots, p$.

Assume that $\boldsymbol{X}$ has a finite dispersion matrix

$$(\sigma_{jk}) = \begin{bmatrix} \sigma_{11} & \cdots & \sigma_{1p} \\ \sigma_{p1} & \cdots & \sigma_{pp} \end{bmatrix}$$

$\sigma_{jk} = \text{cov}(X_j, X_k)$; Let us draw n independent samples from the population and assume sample observations are

$$\begin{bmatrix} x_{11} & \cdots & x_{1n} \\ \vdots & & \vdots \\ x_{p1} & \cdots & x_{pn} \end{bmatrix},$$

x_{jh} being the hth observation of the variable X_j, $\bar{x}_j = \dfrac{1}{n}\sum_{h=1}^{n} x_{jh}$. The random vector

$$\begin{bmatrix} \sqrt{n} & (\bar{x}_1 - \mu_1) \\ & \vdots \\ \sqrt{n} & (\bar{x}_p - \mu_p) \end{bmatrix}$$

is asymptotically normally distributed in the p-variate normal form with a mean vector $(,\cdots,0)'$ and dispersion matrix (σ_{jk}).

❑ **EXAMPLE 12.16.** *Examine if the law of large numbers and CLT hold for the sequence of mutually independent random variables $\{X_k\}$, where*

$$P[X_k = k] = P[X_k = -k] = \frac{1}{2\sqrt{k}}, \quad P[X_k = 0] = 1 - \frac{1}{\sqrt{k}}$$

Taking $\delta = 1$ in (12.15), CLT holds if $\lim\limits_{n\to\infty} \dfrac{\rho}{\sigma} = 0$ where $\rho^3 = \sum\limits_{k=1}^{n} E|X_k - \mu|^3$ and $\sigma = \sqrt{\sum\limits_{k=1}^{n} \sigma_k^2}$. Here $E(X_k) = 0, E|X_k^3| = k^{\frac{5}{2}}$,

$$\lim_{n\to\infty} \frac{\rho}{\sigma} = \lim_{n\to\infty} \frac{\left(\sum\limits_{k=1}^{n} k^{\frac{5}{2}}\right)^{\frac{1}{3}}}{\left(\sum\limits_{k=1}^{n} k^{\frac{3}{2}}\right)^{\frac{1}{2}}} = \lim_{n\to\infty} \frac{\left(\int_0^n x^{\frac{5}{2}}\,dx\right)^{\frac{1}{3}}}{\left(\int_0^n x^{\frac{3}{2}}\,dx\right)^{\frac{1}{2}}} = \lim_{n\to\infty} \frac{A}{\sqrt{n}}$$

$\left(\text{where } A = (\sqrt{5})/(\sqrt[3]{7})(\sqrt[6]{2})\right) = 0.$

Hence CLT holds. Here $\lim\limits_{n\to\infty} \dfrac{V(S_n)}{n^2} = \dfrac{2}{5}\sqrt{n}$ does not tend to zero as $n \to \infty$. Hence WLLN does not hold.

❑ **EXAMPLE 12.17.** *For the sequence of r.v.'s $\{X_n\}$ with $P\{X_k = k^\lambda\} = P\{X_k = -k^\lambda\} = \dfrac{1}{2}, \lambda > 0$, determine if the WLLN holds, central limit law holds.*

$$E(X_k) = 0, \quad \sigma_{(n)}^2 = \sum_{k=1}^{n} k^{2\lambda} \sim \int_0^n x^{2\lambda}\,dx = \frac{n^{2\lambda+1}}{2\lambda+1}.$$

Hence $\sigma^2_{(n)}/n^2 = n^{2\lambda-1}/(2\lambda+1) \to 0$ as $n \to \infty$ if $\lambda < \frac{1}{2}$. Hence WLLN holds for $\lambda < \frac{1}{2}$.

To verify if CLT holds consider the truncated variables U_k defined in (12.17)

$|X_k| = k^\lambda \le n^\lambda$. Now $n^\lambda \le \epsilon\sigma_{(n)} \Rightarrow n \ge \dfrac{2\lambda+1}{\epsilon^2} = n_o(\epsilon)$ (say).

Hence

$$U_k = \begin{cases} X_k & \text{if } n \le n_o(\epsilon) \\ 0 & \text{if } n > n_o(\epsilon) \end{cases}$$

$B_n(\epsilon) = \sum_1^{n_o(\epsilon)} k^{2\lambda} \Big/ \sum_1^n k^{2\lambda} \to 0$ as $n \to \infty$ for every $\epsilon(>0)$ and every $\lambda(>0)$. Hence *CLT* holds for every $\lambda > 0$.

S_n is likely to be of the order of $n^{\lambda+\frac{1}{2}}$ ($\because$ S_n lies within $E(S_n) \pm 3\sigma_{(n)}$ with high probability). Hence for $\lambda \ge \frac{1}{2}$, S_n does not converge in probability to zero.

❑ **EXAMPLE 12.18.** *Show that if the variables X_k are uniformly bounded then (i) $S_n = \sum_{k=1}^n X_k$ satisfies WLLN (ii) S_n obeys the central limit property, provided $\sigma_{(n)} \to \infty$*

(i) $V(S_n) = \sum_1^n \sigma_k^2 \le 2nA^2$ where $|X_k| < A \forall k$. Hence $\dfrac{V(S_n)}{n^2} \le \dfrac{2A^2}{n} \to 0$ as $n \to \infty$. Therefore WLLN holds.

(ii) Suppose $\sigma^2_{(n)} \to \infty$. Here $|X_k| < A\forall k$. Hence, the truncated variables in (12.17) are :

$$\begin{aligned} U_k &= 0 \text{ if } n \text{ is so large that } \sigma_{(n)} > \frac{2A}{\epsilon} \text{ i.e., for } n \ge n_o(\epsilon) \quad \text{(say)} \\ &= X_k - \mu_k \quad \text{otherwise,} \end{aligned}$$

because $|X_k - \mu_k| \le |X_k| + |\mu_k| \le 2A$ and hence $|X_k - \mu_k| \le \epsilon\sigma_{(n)}$ holds if $2A \le \epsilon\sigma_{(n)}$. Therefore, for $n \ge n_o(\epsilon)$, $P\{(U_k = 0), k = 1, \cdots, n\} = 1$. Hence $\lim_{n\to\infty} B_n(\epsilon) = 0$ for every ϵ and the central limit law holds.

The condition $\sigma^2_{(n)} \to \infty$ is a necessary condition for CLT to hold. If $\sigma^2_{(n)} \not\to \infty$, there exists a constant B such that $\sigma^2_{(n)} \uparrow B^2$. For any fixed m, we can find an ϵ such that $P\{|X_m - \mu_m| > \epsilon\sigma_{(n)}\} > P\{|X_m - \mu_m| > \epsilon B\} > 0$. For any $n \ge m$

$$\begin{aligned} A_n(\epsilon) &= \frac{1}{\sigma^2_{(n)}} \sum_{k=1}^n \int_{|x-\mu_k| \ge \epsilon\sigma_{(n)}} (x-\mu_k)^2 dF_k(x) \\ &\ge \epsilon^2 \sum_{k=1}^n P\{|X_k - \mu_k| \ge \epsilon\sigma_{(n)}\} \ge \epsilon^2 P\{|X_m - \mu_m| \ge \epsilon B\} > 0 \end{aligned}$$

and hence condition (12.16) does not hold.

❑ **EXAMPLE 12.19.** *Using the central limit theorem, show that*

$$\lim_{n\to\infty} e^{-n}\sum_{k=0}^{n}\frac{n^k}{k!}=\frac{1}{2}.$$

Let $\{X_n\}$ be a sequence of Poisson variables each with parameter $\lambda = 1$. Then $\sum_{i=1}^{n} X_i$ is Poisson with parameter n. Again by CLT, $\sum_{i=1}^{n} X_i$ is asymptotically $N(n, \sqrt{n})$. Therefore,

$$e^{-n}\sum_{k=0}^{n}\frac{n^k}{k!} = P\left(\sum_{i=1}^{n} X_i - n \leq 0\right) = \frac{1}{\sqrt{2\pi}}\int_{-\infty}^{0} e^{-\frac{z^2}{2}}dz = \frac{1}{2}.$$

12.8. Exercises

1. *Let X_n a sequence of independent random variables, each uniform over (0, 1). Let $Y_n = \frac{\min_{1\leq i\leq n} X_i}{n}$. Show that $Y_n \xrightarrow{L} Z$ where Z is an exponential* r.v. *with mean 1.*

2. *Let X be a $N(0,1)$-variable. Define the sequence of* r.v.'s *$\{X_n\}$ such that $X_{2m+1} = X(m = 0, 1, \cdots), X_{2m} = -X(m = 1, 2, \cdots)$.* Show that $X_n \xrightarrow{L} X$; however, X_n does not converge in probability to X.

3. *Let $\{F_n\}$ be a sequence of d.f.'s such that*

$$F_n(x) = \begin{cases} 0, & x < 0 \\ 1 - \frac{1}{n}, & 0 \leq x < n \\ 1, & n \leq x \end{cases}$$

Show that $F_n \xrightarrow{w} F$ where

$$F(x) = \begin{cases} 0, & x < 0 \\ 1, & x \geq 0 \end{cases}$$

4. *Let $F_n(x)$ be a sequence of negative binomial distributions with parameters (r_n, q_n) with $r_n \to \infty, q_n \to 0$ in such a way that $r_n q_n \to \lambda$. Show that $F_n(x)$ converges in distribution to Poisson distribution.*

5. *Let $\{X_n\}$ be a sequence of iid variables, each uniform over $(0, \theta)$. Consider the variables $Y_n, Z_n = nY_n$ where $Y_n = \min_{i=1,\cdots,n} X_i$. Find the limiting distributions of Y_n and Z_n.*

6. *Let $\{X_n\}$ be a sequence of independent r.v.'s and pmf of X_n is $P(X_n = -n-4) = \frac{1}{n+4}, P(X_n = -1) = 1 - \frac{4}{n+4}, P(X_n = n+4) = \frac{3}{n+4}$. Show that $p \lim_{n\to\infty} X_n = -1$.*

7. *Let $g(y)$ be a continuous function and let $\{X_n\}$ be a sequence of r.v.'s such that $X_n \xrightarrow{P} X$. Show that $g(X_n) \xrightarrow{P} g(X)$.*

8. *Let $\{F_n\}$ be a sequence of d.f.'s such that*

$$F_n(x) = \begin{cases} 0, & x < \theta + \frac{1}{n} \\ 1, & x \geq \theta + \frac{1}{n} \end{cases}$$

Show that the corresponding sequence of random variables $X_n \xrightarrow{L} X$ where X is a r.v. degenerate at $X = \theta$.

9. *Show that the sequence of characteristic functions $\Psi_n(t) = \frac{(\sin nt)}{nt}$ converges to a limit function $\Psi(t)$. However, $F_n(x)$ does not converge to a limiting distribution. Explain why.*

10. *$\{X_n\}$ is a sequence of independent r.v.'s. Determine if they obey WLLN and/or central limit law.*

(i) $P\{X_k = 2^k\} = P\{X_k = -2^k\} = \frac{1}{2}$.

(ii) $P\{X_k = 2^k\} = P\{X_k = -2^k\} = \frac{1}{2^{2k+1}}$

$P\{X_k = 0\} = 1 - \frac{1}{2^{2k}}$

(iii) $P\{X_k = 2^k\} = P\{X_k = -2^k\} = \frac{1}{2^{k+1}}$

$P\{X_k = 1\} = P\{X_k = -1\} = \frac{1}{2}(1 - 2^{-k})$

(iv) $P\{X_k = 2^k\} = P\{X_k = -2^k\} = \frac{1}{2^{k+1}}$,

$P\{X_k = 1\} = P\{X_k = -1\} = \frac{1}{2}\left(1 - \frac{1}{2^k}\right)$

(v) $P\left\{X_k = \frac{1}{2^k}\right\} = P\left\{X_k = -\frac{1}{2^k}\right\} = \frac{1}{2}$

(vi) $P\{X_k = 2^{-k}\} = P\{X_k = -2^{-k}\} = \frac{1}{2^{k+1}}$

$P\{X_k = 1\} = P\{X_k = -1\} = \frac{1}{2}(1 - 2^{-k})$

11. *Let* $\{X_k\}$ *be iid with* $P(X_k = 1) = P(X_k = -1) = \dfrac{1}{2}$. Define $Y_k = \sum_{j=1}^{k} \dfrac{X_j}{2j}$. Find limiting distribution of Y_k as $k \to \infty$.

12. *Let* $\{X_n\}$ *be a sequence of independently distributed* $N\left(0, \dfrac{\sigma}{\sqrt{n}}\right)$*-variables. Show that* $X_n \xrightarrow{L} X$ *where* X *is a degenerate r.v. at* $x = 0$.

REFERENCE

1. *Birnbaum, Z.W.* (1962) Introduction to Probabilities and Mathematical Statistics, Harper and Bros.
2. *Bremud, Pierre* (1988) Introduction to Probabilitic Modelling, Springer Verlag
3. *Cramer, H.* (1950) Mathematical Methods of Statistics, Princeton University Press, 1946 and Asia Publishing House, 1950.
4. *Dwass, M.* (1970) Probability and Statistics, an undergraduate course, W.A. Benjamin, Inc.
5. *Feller, W.* (1968) An Introduction to Probability Theory and its Applications, Vol. 1 (3rd edition), Wiley Eastern.
6. *Fisz, M.* (1963) Probability Theory and Mathematical Statistics, John Wiley and S. Chand.
7. *Fry, T. C.* (1928) Probability and its Engineering uses, Van Nostrand.
8. *Giri, N. C.* (1974) Introduction to Probability and Statistics Part 1 : Probability, Marcel Dekker, Inc. New York.
9. *Gnedenko, B. V.* (1962) Theory of Probability, Chesla Publishing Co. and Mir Publishers.
10. *Goldberg, S.* (1960) Probability : An Introduction, Prentice-Hall, Inc. N. J.
11. *Goon, A. M., Gupta, M. K.* and *Dasgupta B.* An Outline of Statistical Theory, Vol. 1 World Press, 1977.
12. *Halmos, P. R.* Measure Theory, D. Van Nostrand, 1950 and East-West Press, 1964.
13. *Harris, B.* (1966) Theory of Probability, Addison-Wesley

14. *Hastings, N. A. J.* and *Peacock, J. B.* (1975) Statistical DistributionsLongom-Butterworths.

15. *Hodges, J. L.* (Jr.) and *Lehmann, E. L.* (1965) Elements of Probability, Holden Day, Inc.

16. *Hoel, P. G., Port, S. C., Stone, C. J.* (1971) Introduction to Probability Theory, Houghton Mifflin Company.

17. *Hogg, R. V.* and *Craig, A. T.* (1965) Introduction to Mathematical Statistics, Macmillan and Amerind.

18. *Kapur, J. N.* and *Saxena, H. C.* (1961) Mathematical Statistics, S. Chand.

19. *Kendal, M. G.* and *Stuart, A.* (1958) Advanced Theory at Statistics, Vol. I

20. *Keynes, J. M.* (1921) A Treatise of Probability, Macmillan.

21. *Kolmogorov, A. N.* (1956) Foundations of Theory of Probability, Chesla Publishing Co.

22. *Johnson, N. L.* and *Kotz, S. M.* (1970) Discrete Distributions, John Wiley.

23. (1970) Continuous Univariate Distributions, 1, 2, John Wiley.

24. (1970) Continuous Multivariate Distributions, John Wiley.

25. *Johnson, N. L.* and *Kotz, S. M.* (1977) Urn Models and their Applications : An Approach to Modern Discrete Probability Theory, John Wiley.

26. *Lamperti, J.* (1966) Probability, W. A. Benjamin, Inc.

27. *Larson, H. J.* (1982) Introduction to Probability Theory and Statistical Inference, 3rd edition, John Wiley.

28. *Lo'eve, M.* Probability Theory, D Van Nostrand, 1963 and East-West Press 1968.

29. *Lukacs, E.* (1960) Characteristic Functions, Charles Griffin and Co. Ltd.

30. (1968) Stochastic Convergence, Health Mathematical Monograph.

31. *McCord J. R.* (III), *Moroney R. M.* (Jr.) (1964) Probability Theory, The Macmillan Co. N. Y.

32. *McShane, E. J.* (1944) Integration, Princeton University Press.

33. *Meyer, P. L.* (1965) Introductory Probability and Statistical Application, Addison-Wesley.

34. *Mood, A. M., Graybill, F. A.* and *Boes, D. C.* (1974) An Introduction to Theory of Statistics, McGraw-Hill and Tata McGraw-Hill.

35. *Mosteller, F., Rourke, R. E. K.* and *Thomas, G. B.* (Jr.) (1961) Probability and Statistics, Addison-Wesley.

36. *Munroe, M. E.* (1953) Introduction to Measure and Integration, Addison-Wesley.

37. *Papoulis, A.* (1965) Probability, Random Variables and Stochastic Process, McGraw-Hill.

38. *Pergen, F.* (1960) Modern Probability Theory and its Applications, John Wiley and Sons, Inc.

39. *Pfeiffer, P. E.* (1965) Concepts of Probability Theory, McGraw-Hill.

40. *Pitt, H. R.* (1963) Integration, Measure and Probability Theory, Oliver and Boyd.

41. *Rahaman, N. A.* (1967) Exercise in Probability and Statistics, Griffin

42. *Rao, C. R.* (1965) Linear Statistical Inference and its Applications, John Wiley and Sons.

43. *Rohatgi, V. K.* (1976) An Introduction to Probability, Theory and Mathematical Statistics, Wiley Eastern.

44. *Ross, S. M.* (1972) Introduction to Probability Models, Academic Press.

45. *Rozanov, Y. A.* (1969) Introductory Probability Theory, (Translated and edited by R. A. Silverman) Prentice Hall.

46. *Sankaranarayanan, G.* (1986) Elementary Theory of Mathemetical Statistics and its application. Oxford IBH Publication.

47. *Savage, L. J.* (1954) The Foundations of Statistics, John Wiley.

48. *Sevastyanov, B. A., Chistyakov, V. P. and Zubkov, A. M.* (1985) Problems in the Theory of Probability, Mir Publishing.

49. *Uspensky, J. V.* (1937) Introduction to Mathematical Probability McGraw-Hill and Tata McGraw-Hill.

50. *Whittle, P.* (1976) Probability, John Wiley and Sons.

51. *Wilks, S. S.* (1976) Mathematical Statistics, John Wiley and Toppan.

APPENDIX

TABLE 1 ORDINATES OF THE NORMAL DENSITY FUNCTION

$$\phi(x) = \frac{1}{\sqrt{2\pi}} \; e^{-x^2/2}$$

X	.00	.01	.02	.03	.04	.05	.06	.07	.08	.09
.0	.3989	.3989	.3989	.3988	.3986	.3984	.3982	.3980	.3977	.3973
.1	.3970	.3965	.3961	.3956	.3951	.3945	.3939	.3932	.3925	.3918
.2	.3910	.3902	.3894	.3885	.3876	.3867	.3857	.3847	.3836	.3825
.3	.3814	.3802	.3790	.3778	.3765	.3752	.3739	.3725	.3712	.3697
.4	.3683	.3668	.3653	.3637	.3621	.3605	.3589	.3572	.3555	.3538
.5	.3521	.3503	.3485	.3467	.3448	.3429	.3410	.3391	.3372	.3352
.6	.3332	.3312	.3292	.3271	.3251	.3230	.3209	.3187	.3166	.3144
.7	.3123	.3101	.3079	.3056	.3034	.3011	.2989	.2966	.2943	.2920
.8	.2897	.2874	.2850	.2827	.2803	.2780	.2756	.2732	.2709	.2685
.9	.2661	.2637	.2613	.2589	.2565	.2541	.2516	.2492	.2468	.2444
1.0	.2420	.2396	.2371	.2347	.2323	.2299	.2275	.2251	.2227	.2203
1.1	.2179	.2155	.2131	.2107	.2083	.2059	.2036	.2012	.1989	.1965
1.2	.1942	.1919	.1895	.1872	.1849	.1826	.1804	.1781	.1758	.1736
1.3	.1714	.1691	.1669	.1647	.1626	.1604	.1582	.1561	.1539	.1518
1.4	.1497	.1476	.1456	.1435	.1415	.1394	.1374	.1354	.1334	.1315
1.5	.1295	.1276	.1257	.1238	.1219	.1200	.1182	.1163	.1145	.1127
1.6	.1109	.1092	.1074	.1057	.1040	.1023	.1006	.0989	.0973	.0957
1.7	.0940	.0925	.0909	.0893	.0878	.0863	.0848	.0833	.0818	.0804
1.8	.0790	.0775	.0761	.0748	.0734	.0721	.0707	.0694	.0681	.0669
1.9	.0656	.0644	.0632	.0620	.0608	.0596	.0584	.0573	.0562	.0551
2.0	.0540	.0529	.0519	.0508	.0498	.0488	.0478	.0468	.0459	.0449
2.1	.0440	.0431	.0422	.0413	.0404	.0396	.0387	.0379	.0371	.0363
2.2	.0355	.0347	.0339	.0332	.0325	.0317	.0310	.0303	.0297	.0290
2.3	.0283	.0277	.0270	.0264	.0258	.0252	.0246	.0241	.0235	.0229
2.4	.0224	.0219	.0213	.0208	.0203	.0198	.0194	.0189	.0184	.0180
2.5	.0175	.0171	.0167	.0163	.0158	.0154	.0151	.0147	.0143	.0139
2.6	.0136	.0132	.0129	.0126	.0122	.0119	.0116	.0113	.0110	.0107
2.7	.0104	.0101	.0099	.0096	.0093	.0091	.0088	.0086	.0084	.0081
2.8	.0079	.0077	.0075	.0073	.0071	.0069	.0067	.0065	.0063	.0061
2.9	.0060	.0058	.0056	.0055	.0053	.0051	.0050	.0048	.0047	.0046
3.0	.0044	.0043	.0042	.0040	.0039	.0038	.0037	.0036	.0035	.0034
3.1	.0033	.0032	.0031	.0030	.0029	.0028	.0027	.0026	.0025	.0025
3.2	.0024	.0023	.0022	.0022	.0021	.0020	.0020	.0019	.0018	.0018
3.3	.0017	.0017	.0016	.0016	.0015	.0015	.0014	.0014	.0013	.0013
3.4	.0012	.0012	.0012	.0011	.0011	.0010	.0010	.0010	.0009	.0009
3.5	.0009	.0008	.0008	.0008	.0008	.0007	.0007	.0007	.0007	.0006
3.6	.0006	.0006	.0006	.0005	.0005	.0005	.0005	.0005	.0005	.0004
3.7	.0004	.0004	.0004	.0004	.0004	.0004	.0003	.0003	.0003	.0003
3.8	.0003	.0003	.0003	.0003	.0003	.0002	.0002	.0002	.0002	.0002
3.9	.0002	.0002	.0002	.0002	.0002	.0002	.0002	.0002	.0001	.0001

TABLE 2 CUMULATIVE NORMAL DISTRIBUTION

$$\phi(x) = \int_{-\infty}^{x} \frac{1}{\sqrt{2\pi}} \; e^{-t^2/2} \, dt$$

x	.00	.01	.02	.03	.04	.05	.06	.07	.08	.09
.0	.5000	.5040	.5080	.5120	.5160	.5199	.5239	.5279	.5319	.5359
.1	.5398	.5438	.5478	.5517	.5557	.5596	.5636	.5675	.5714	.5753
.2	.5793	.5832	.5871	.5910	.5948	.5987	.6026	.6064	.6103	.6141
.3	.6179	.6217	.6255	.6293	.6331	.6368	.6406	.6443	.6480	.6517
.4	.6554	.6591	.6628	.6664	.6700	.6736	.6772	.6808	.6844	.6879
.5	.6915	.6950	.6985	.7019	.7054	.7088	.7123	.7157	.7190	.7224
.6	.7257	.7291	.7324	.7357	.7389	.7422	.7454	.7486	.7517	.7549
.7	.7580	.7611	.7642	.7673	.7704	.7734	.7764	.7794	.7823	.7852
.8	.7881	.7910	.7939	.7967	.7995	.8023	.8051	.8078	.8106	.8133
.9	.8159	.8186	.8212	.8238	.8264	.8289	.8315	.8340	.8365	.8389
1.0	.8413	.8438	.8461	.8485	.8508	.8531	.8554	.8577	.8599	.8621
1.1	.8643	.8665	.8686	.8708	.8729	.8749	.8770	.8790	.8810	.8830
1.2	.8849	.8869	.8888	.8907	.8925	.8944	.8962	.8980	.8997	.9015
1.3	.9032	.9049	.9066	.9082	.9099	.9115	.9131	.9147	.9162	.9177
1.4	.9192	.9207	.9222	.9236	.9251	.9265	.9279	.9292	.9306	.9319
1.5	.9332	.9345	.9357	.9370	.9382	.9394	.9406	.9418	.9429	.9441
1.6	.9452	.9463	.9474	.9484	.9495	.9505	.9515	.9525	.9535	.9545
1.7	.9554	.9564	.9573	.9582	.9591	.9599	.9608	.9616	.9625	.9633
1.8	.9641	.9649	.9656	.9664	.9671	.9678	.9686	.9693	.9699	.9706
1.9	.9713	.9719	.9726	.9732	.9738	.9744	.9750	.9756	.9761	.9767
2.0	.9772	.9778	.9783	.9788	.9793	.9798	.9803	.9808	.9812	.9817
2.1	.9821	.9826	.9830	.9834	.9838	.9842	.9846	.9850	.9854	.9857
2.2	.9861	.9864	.9868	.9871	.9875	.9878	.9881	.9884	.9887	.9890
2.3	.9893	.9896	.9898	.9901	.9904	.9906	.9909	.9911	.9913	.9916
2.4	.9918	.9920	.9922	.9925	.9927	.9929	.9931	.9932	.9934	.9936
2.5	.9938	.9940	.9941	.9943	.9945	.9946	.9948	.9949	.9951	.9952
2.6	.9953	.9955	.9956	.9957	.9959	.9960	.9961	.9962	.9963	.9964
2.7	.9965	.9966	.9967	.9968	.9969	.9970	.9971	.9972	.9973	.9974
2.8	.9974	.9975	.9976	.9977	.9977	.9978	.9979	.9979	.9980	.9981
2.9	.9981	.9982	.9982	.9983	.9984	.9984	.9985	.9985	.9986	.9986
3.0	.9987	.9987	.9987	.9988	.9988	.9989	.9989	.9989	.9990	.9990
3.1	.9990	.9991	.9991	.9991	.9992	.9992	.9992	.9992	.9993	.9993
3.2	.9993	.9993	.9994	.9994	.9994	.9994	.9994	.9995	.9995	.9995
3.3	.9995	.9995	.9995	.9996	.9996	.9996	.9996	.9996	.9996	.9997
3.4	.9997	.9997	.9997	.9997	.9997	.9997	.9997	.9997	.9997	.9998

x	1.282	1.645	1.960	2.326	2.576	3.090	3.291	3.891	4.417
$\phi(x)$	.90	.95	.975	.99	.995	.999	.9995	.99995	.999995
$2[1-\phi(x)]$	.20	.10	.05	.02	.01	.002	.001	.0001	.000001

INDEX

22

N

O

P

Q

R

S

T

U

V

W